TRIPLE HELIX FORMING OLIGONUCLEOTIDES

PERSPECTIVES IN ANTISENSE SCIENCE

Why a series of new volumes on antisense oligonucleotides? Because of the enduring fascination with the antisense biotechnology, which in theory gives the scientific and therapeutically-oriented communities the ability to sequence-specifically inhibit protein translation and hence the expression of genes.

At any rate, that's the theory. In practice, as is well known, the application of that theory to solve real biological problems presents a series of tortuous new problems, many of which are just now beginning to be understood, and hopefully resolved. Nevertheless, the progress in antisense biotechnology in the past few years alone has been impressive, and indeed the first antisense drug, fomivirsen (for cytomegalovirus retinitis: Isis, Carlsbad, CA) has recently been approved by the FDA.

The reader will find no dearth of well designed experiments in these volumes that demonstrate, to the best of current technology, sequence specific inhibition of gene expression. Much effort has also been expended by many authors in critical analysis of their results, a process always necessary for proper interpretation of data derived from antisense experiments.

There is little doubt that the coming years will witness further improvements and refinements in this dynamic technology, driven not only by the power of the idea, but also by the necessity generated by the sequencing of the human genome. These volumes therefore represent only the beginning of the harnessing of this impressive potential.

It was an honor for me to be asked to be series editor for these volumes, none the least because it gave me a chance to extensively interact with an excellent series of individual volume editors, who, at the time of this writing, included Stefan Endres, Peg McCarthy, Claude Malvy and LeRoy Rabbani. The results are mostly a product of their efforts, and of course even more so those of large number of authors. Finally, on behalf of all contributors, I would like to thank Charles Schmieg of Kluwer, who conceived of and drove this project, and without whom this collection would not exist.

C. A. Stein, Series Editor

Recently Published Book in the Series

Margaret M. McCarthy:
Modulating Gene Expression By Antisense Oligonucleotides
To Understand Neural Functioning

TRIPLE HELIX FORMING OLIGONUCLEOTIDES

edited by

Claude Malvy
Annick Harel-Bellan
Linda L. Pritchard

CNRS France

PERSPECTIVES IN ANTISENSE SCIENCE
Series Editor: C. A. Stein
The College of Physicians & Surgeons, Columbia University

KLUWER ACADEMIC PUBLISHERS
Boston / Dordrecht / London

Distributors for North, Central and South America:
Kluwer Academic Publishers
101 Philip Drive
Assinippi Park
Norwell, Massachusetts 02061 USA
Telephone (781) 871-6600
Fax (781) 871-6528
E-Mail <kluwer@wkap.com>

Distributors for all other countries:
Kluwer Academic Publishers Group
Distribution Centre
Post Office Box 322
3300 AH Dordrecht, THE NETHERLANDS
Telephone 31 78 6392 392
Fax 31 78 6546 474
E-Mail <services@wkap.nl>

Electronic Services <http://www.wkap.nl>

Library of Congress Cataloging-in-Publication Data

Triple helix forming oligonucleotides / edited by Claude Malvy, Annick
   Harel-Bellan, Linda L. Pritchard.
        p. cm. -- (Perspectives in antisense science)
     Includes index.
     ISBN 0-7923-8418-0 (alk. paper)
     1. Triple-helix-forming oligonucleotides. I. Malvy, Claude,
1947-   . II. Harel-Bellan, Annick, 1951-   . III. Pritchard, Linda L.
IV. Series.
QP625.T75T75 1999
572.8'5--dc21
                                                      98-50990
                                                         CIP

Contents

Section I: BACKGROUND AND STRUCTURAL ASPECTS

Section II: THE BIOLOGY OF TRIPLE HELICES

List of Contributors

Ulysse Asseline
Centre de Biophysique Moléculaire, CNRS UPR 4301, rue Charles-Sadron, 45071 Orléans Cedex 2, France

François-Xavier Barre
CNRS UPR 9079 - IFR Y1221, Institut de Recherches sur le Cancer, 7 rue Guy Moquet, 94801 Villejuif Cedex, France

Thomas Bentin
Center for Biomolecular Recognition, Department of Medical Biochemistry and Genetics, Biochemistry Laboratory B, The Panum Institute, Blegdamsvej 3, 2200 Copenhagen N, Denmark

K. J. Breslauer
Rutgers, The State University of New Jersey, Department of Chemistry, Wright-Rieman Laboratories, 610 Taylor Road, Piscataway, NJ 08854-8087, USA

Dmitry Cherny
Institute of Molecular Genetics, Russian Academy of Sciences, Kurchatov's Square, 123182 Moscow, Russia

David T. Curiel
Department of Medicine, Division of Hematology / Oncology, The University of Alabama at Birmingham, 520 Wallace Tumor Institute, 1824 Sixth Avenue South, Birmingham, AL 35294-3300, USA

Scot W. Ebbinghaus
Department of Medicine, Division of Hematology / Oncology, The University of Alabama at Birmingham, 520 Wallace Tumor Institute, 1824 Sixth Avenue South, Birmingham, AL 35294-3300, USA

Christophe Escudé
Laboratoire de Biophysique, Muséum National d'Histoire Naturelle, INSERM U.201 - CNRS URA 481, 43 rue Cuvier, 75005 Paris, France

Maxim D. Frank-Kamenetskii
Center for Advanced Biotechnology, Department of Biomedical Engineering, Boston University, 36 Cummington St., Boston, MA 02215, USA

Thérèse Garestier
Laboratoire de Biophysique, Muséum National d'Histoire Naturelle, INSERM U.201 - CNRS URA 481, 43 rue Cuvier, 75005 Paris, France

Carine Giovannangeli
*Laboratoire de Biophysique, Muséum National d'Histoire Naturelle, INSERM U.201
- CNRS URA 481, 43 rue Cuvier, 75005 Paris, France*

Peter M. Glazer
*Departments of Therapeutic Radiology and Genetics, Yale University School of
Medicine, P. O. Box 208040, New Haven, CT 06520-8040, USA*

Sergei M. Gryaznov
*Director, Nucleic Acids Chemistry, Geron Corporation, 230 Constitution Drive,
Menlo Park, CA 94025, USA.*

Anne-Laure Guieysse
*Laboratoire de Biophysique, Muséum National d'Histoire Naturelle, INSERM U.201
- CNRS URA 481, 43 rue Cuvier, 75005 Paris, France*

Annick Harel-Bellan
*CNRS UPR 9079 - IFR Y1221, Institut de Recherches sur le Cancer, 7 rue Guy
Moquet, 94801 Villejuif Cedex, France*

Claude Hélène
*Laboratoire de Biophysique, Muséum National d'Histoire Naturelle, INSERM U.201
- CNRS URA 481, 43 rue Cuvier, 75005 Paris, France*

Jörg Jendis
*Institute of Medical Virology, University of Zurich, Gloriastrasse 30, CH-8028
Zurich, Switzerland*

Marina Kochetkova
*Department for Cytogenetics and Molecular Genetics, Women's and Children's
Hospital, 72 King William Road, North Adelaide, SA 5006, Australia*

Asya Levina
*Laboratory of Nucleic Acids Chemistry, Novosibirsk Institute of Bioorganic
Chemistry, pr. Lavrentjev 8, Novosibirsk 630090, Russia*

L. James Maher, III
*Department of Biochemistry and Molecular Biology, Gugg. 16, Mayo Foundation,
200 First St. SW, Rochester, MN 55905, USA*

Claude Malvy
*CNRS URA 147, Institut Gustave Roussy, 39 rue Camille Desmoulins, 94805
Villejuif Cedex, France*

Charles M. Mayfield
*Department of Medicine, Division of Hematology / Oncology, The University of
Alabama at Birmingham, 520 Wallace Tumor Institute, 1824 Sixth Avenue South,
Birmingham, AL 35294-3300, USA*

Donald M. Miller
Department of Medicine, Division of Hematology / Oncology, The University of Alabama at Birmingham, 520 Wallace Tumor Institute, 1824 Sixth Avenue South, Birmingham, AL 35294-3300, USA

Sergei M. Mirkin
University of Illinois at Chicago, Department of Genetics (MC 669), College of Medicine, 900 South Ashland Avenue, Chicago, IL 60607-7170, USA

Karin Moelling
Director, Institute of Medical Virology, University of Zurich, Gloriastrasse 30, CH-8028 Zurich, Switzerland

Peter E. Nielsen
Center for Biomolecular Recognition, Department of Medical Biochemistry and Genetics, Biochemistry Laboratory B, The Panum Institute, Blegdamsvej 3, 2200 Copenhagen N, Denmark

G. Eric Plum
Rutgers, The State University of New Jersey, Department of Chemistry, Wright-Rieman Laboratories, 610 Taylor Road, Piscataway, NJ 08854-8087, USA

Horea Porumb
Laboratoire de Chimie Structurale et Spectroscopie Biomoléculaire - CSSB, UPRESA CNRS 7031, UFR Médecine - Santé - Biologie Humaine, Université Paris XIII, 93017 Bobigny Cedex, France; Unité de Biochimie Enzymologie - Physicochimie et Pharmacologie des Macromolécules Biologiques, UMR CNRS 1772, Institut Gustave Roussy, 94805 Villejuif Cedex, France

Danièle Praseuth
Laboratoire de Biophysique, Muséum National d'Histoire Naturelle, INSERM U.201 - CNRS URA 481, 43 rue Cuvier, 75005 Paris, France

Linda L. Pritchard
CNRS UPR 9079 - IFR Y1221, Institut de Recherches sur le Cancer, 7 rue Guy Moquet, 94801 Villejuif Cedex, France

Mary Frances Shannon
Division of Biochemistry and Molecular Biology, JCMR ANU, Canberra, ACT 2601, Australia

Jian-sheng Sun
Laboratoire de Biophysique, Muséum National d'Histoire Naturelle, INSERM U.201 - CNRS URA 481, 43 rue Cuvier, 75005 Paris, France

Fedor Svinarchuk
CNRS URA 147, Institut Gustave Roussy, 39 rue Camille Desmoulins, 94805 Villejuif Cedex, France

xii

Karen M. Vasquez
Departments of Therapeutic Radiology and Genetics, Yale University School of Medicine, P. O. Box 208040, New Haven, CT 06520-8040, USA

Nadarajah Vigneswaran
Department of Medicine, Division of Hematology / Oncology, The University of Alabama at Birmingham, 520 Wallace Tumor Institute, 1824 Sixth Avenue South, Birmingham, AL 35294-3300, USA

Kyonggeun Yoon
Director, Cutaneous Gene Therapy, Department of Dermatology and Cutaneous Biology, 233 South 10th Street, Jefferson Medical College, Philadelphia, PA 19107, USA

Valentina Zarytova
Director, Laboratory of Nucleic Acids Chemistry, Novosibirsk Institute of Bioorganic Chemistry, pr. Lavrentjev 8, Novosibirsk 630090, Russia

Foreword

The interest of forming artificial double helices between messenger RNA and synthetic oligonucleotides in order to control gene expression, the so-called antisense approach, is now well established. A similar and more recent concept is the use of synthetic oligonucleotides to form triple helices with DNA or RNA; this approach should ultimately permit modulation of gene expression at the level of transcription (the "anti-gene strategy"), as well as *via* antisense mechanisms. Today we are in the beginnings of this fascinating field.

Short synthetic oligonucleotides can be designed to form local triple helices on long double-stranded DNA target sequences. These triple helix forming oligonucleotides (TFOs) are highly sequence-specific DNA-binding ligands, and as such they have many potential applications, whether in biotechnology or as therapeutic tools. In biotechnology, TFOs can be used *in vitro* to enrich specific DNA sequences in a highly selective manner, and also have a number of other possible applications, for example in diagnostics.

As for eventual therapeutic applications, it is already clear that natural triple helices (H DNA) exist in living organisms. This has been more particularly shown in bacteria; and in eukaryotes, potential H-DNA-forming regions have been found within regulatory (promoter) regions of the genome, suggesting a potential role in modulating gene expression. In order to artificially regulate a gene, the possibility of specifically targeting a DNA sequence *via* triple helix formation is very attractive and, as therapeutic agents, TFOs could be used to specifically inhibit the expression of deleterious genes. Indeed, in experimental models, they can chase regulatory DNA-binding proteins from their specific sequence, thus locally and specifically influencing gene activity. In addition, TFOs can be coupled to molecules such as psoralen, for example, which introduce irreversible lesions into the target sequence. If one succeeded in irreversibly canceling the message contained by the targeted sequence, the consequence would logically be total gene inactivation. This immediately raises the questions of induced cellular toxicity and of the efficiency of the cell system in processing these lesions. Different strategies could be applied depending on whether toxicity was desired or not. For instance, TFOs could be used to selectively eliminate cells in the genome of which a retroviral sequence (or any other extracellular and specific sequence) had been inserted, a tool which is conspicuously missing at the moment. Since no enzyme equivalent to RNase H for antisense oligonucleotides has been described with TFOs targeting DNA, one might expect that much research will be directed toward developing effective artificial scissors. However, RNA might also be a target; and here the concept of triple-helix technology joins that of the antisense field. The clamp-forming oligonucleotides should be more specific than antisense oligonucleotides, and might allow a block of translation without the need for RNase H.

Triple helices of DNA, whether intramolecular or intermolecular, have been extensively studied from a biophysical point of view. They form *in vitro* on specific sequences and under defined physico-chemical conditions, but several of their characteristics have been found to impair their efficiency *in vivo*. First, there are still limitations on the type of DNA sequences able to form triple helices: at present TFOs can act only on homopurine/homopyrimidine target sequences, which greatly reduces the number of potential targets. Second, the affinity of TFOs for their target sequences is strongly influenced by physico-chemical parameters and, in fact, intracellular conditions are not likely to be favorable to this interaction.

For all these reasons, reports of biological effects of TFOs on cellular genes are scarce. Proof of a triple-helix-mediated mechanism is even harder to establish. Indeed, due to the numerous side effects generated in cells by synthetic oligonucleotides, only the use of a mutated target sequence as negative control under physiological conditions, correlating triple helix formation (or lack thereof) with biological response, can provide such proof. Thus one might be tempted to point out that, from a rigorously Cartesian point of view, an irrefutable demonstration of a triple-helix-mediated effect on gene expression *in situ* in living organisms has yet to be provided.

However, several cellular results give hope for the future of triple helices with oligonucleotides: 1) triple helices can form in eukaryotic cells; 2) preformed triple helices are stable in living cells; and 3) once in the cytoplasm, oligonucleotides go directly to the nucleus – although the means of vectorization chosen to introduce them into the cell may influence this. Occasionally conflicting results are reported for similar systems, or different target genes respond differently under seemingly identical conditions, highlighting the fact that parameters important for determining the cellular outcome may be escaping notice. Consequently it is apparent that a better understanding of how the cell deals with these molecules (intracellular trafficking, degradation, kinetics of target binding, DNA repair, mutagenesis and its consequences...) is now necessary in order to reach intracellular rates of triple helix formation compatible with the concept of artificial gene regulation. The projected endpoint is certainly worth the effort: TFOs are among the rare tools which could be specific and efficient anti-viral agents, and other *in vivo* applications may also be envisaged.

Attainment of this goal implies overcoming obstacles related to the aforementioned physico-chemical characteristics of triple helices that limit their efficacy *in vivo*. Development of TFOs and TFO analogs whose chemistries allow them to be more resistant to degradation and the resultant triple helix to be more stable once formed is a priority. These chemistries include backbone and base modifications as well as addition of stabilizers. A number of studies are devoted to the progression of this active field of research, and the conjoined efforts of physical chemists and biologists will hopefully result in a broadening of the panel of potential targets and in an increased TFO efficiency in live cells.

Finally, many hurdles have still to be passed before gene regulation at the DNA level by triple helices becomes effective, but the current state of research in the field allows one to be cautiously optimistic that it will indeed become a reality.

Our intent in choosing the subjects included in this volume was to provide the reader with broad-based coverage of the field of research on triple helix forming oligonucleotides (and oligonucleotide analogs), including both physico-chemical and cellular, conceptual and practical/experimental concerns, and relating the promises but also the problems encountered. We wish to take this opportunity to thank all the contributors whose efforts made this possible, and our publisher for taking into account our specific needs and concerns. By holding to a tight publication schedule, we have tried to present a volume that will be timely and useful to the widest possible audience. *A vous de juger* – you be the judge.

Villejuif, 21 September 1998

Annick Harel-Bellan
Claude Malvy
Linda L. Pritchard

BACKGROUND AND STRUCTURAL ASPECTS

1 THE ANTIGENE STRATEGY: PROGRESS AND PERSPECTIVES IN SELECTIVE GENE SILENCING

Claude Hélène

INTRODUCTION

It was only four years after the discovery of the double helix (1953) that a publication by Felsenfeld, Davis and Rich described the formation of a triple helix involving one poly A and two poly U chains (*1*). This was the starting point for a number of studies devoted to multi-stranded structures of polynucleotides. These studies were limited to homopolynucleotides and alternating polynucleotides which were the only available molecules at that time (for a review see references *2* and *3*). The interest in triple-helical structures then decreased due to the lack of diversity in the sequences that could be investigated and the absence of a demonstration of biological function for such structures.

In 1987, the interest in triple-helical structures was rekindled by three independent publications. Two of them showed that short oligonucleotides could bind in a sequence-specific manner to the major groove of the DNA double helix where they formed a triple-helical complex (*4, 5*). At the same time a triple-helical structure was shown to be involved in H-DNA formation, a structure observed, *e.g.*, under torsional stress, at polypurine•polypyrimidine sequences of DNA with mirror symmetry (*6*). The mechanism of recognition involved the formation of Hoogsteen hydrogen bonds between the bases of the third strand and the purine bases of the double helix, when the third strand contained an oligopyrimidine sequence. This model was then extended to triple helices where the third strand engaged reverse Hoogsteen hydrogen bonds when it contained purine bases. In the Hoogsteen mode the third strand runs parallel to the oligopurine sequence of the double helix, whereas in the reverse Hoogsteen configuration it adopts an antiparallel orientation. Oligonucleotides containing C and T form Hoogsteen hydrogen bonds, whereas oligonucleotides containing G and A form reverse Hoogsteen hydrogen bonds. The situation is more complex when the third-strand oligonucleotide contains G and T, in which case a parallel or an antiparallel orientation may be observed depending on the length of the G and T tracts and on the number of GpT + TpG steps (see reference *7* for a review).

4

THE ANTIGENE STRATEGY FOR SELECTIVE CONTROL OF GENE EXPRESSION

The two publications on triple-helix-directed recognition of the major groove of DNA pointed out the potential use of oligonucleotides to control gene expression in a sequence-specific manner:

> *"The results presented in the (above) study open the way for the design of sequence-specific DNA-binding drugs which could selectively recognize polypurine•polypyrimidine sequences in double-stranded DNA... Such sequences constitute good targets for site-specific reagents which could interfere with gene expression" (4).*

> *"The work reported here demonstrates that homopurine•homopyrimidine double helical tracts can be recognized within large DNA by triple helix formation under physiological conditions... Moreover, as molecular biology defines specific disease states at the DNA level, a chemotherapeutic strategy of 'artificial repressors' based on triple helix-forming DNA analogs becomes a possibility" (5).*

This formed the basis of what we later called *the antigene strategy* for sequence-specific control of gene expression at the transcriptional level (8). The first *in vitro* experiment was reported in 1988 and used a (G, T)-containing oligonucleotide to inhibit the *in vitro* transcription of the *myc* gene (9).

Ten years later, where are we? The development of the antigene strategy faces several problems, some of which are similar to those raised by the development of antisense oligonucleotides (*e.g.*, cell uptake, pharmaco-kinetics, bioavailability) while others are specific to antigene oligonucleotides. The following discussion focuses on the latter aspects.

Cellular localization

Obviously the targets of antigene oligonucleotides are located in the cell nucleus. Therefore these oligonucleotides must be able to find their way to the nucleus. Microinjection experiments have shown earlier that oligonucleotides spontaneously concentrate from the cytoplasm into the nucleus. Cell permeabilization (with digitonin or streptolysin-O) or electroporation lead to high concentrations of oligonucleotide in the nucleus, as does microinjection. Some cationic lipids used as vectors also lead to nuclear localization of oligonucleotides; the result depends not only on the cationic lipid but also on the cell type. There is much work yet to be done to improve the nuclear delivery of oligonucleotides.

It should be noted that nuclear localization of some *antisense* oligonucleotides has been demonstrated from an analysis of the mechanism of action. Oligonucleotides which have been shown to inhibit splicing reactions must bind to the pre-mRNA in the cell nucleus (*10, 11*).

Accessibility of DNA targets within the chromatin structure

In order to exert the expected biological effect, the antigene oligonucleotide must not only reach the nucleus but also bind to its target sequence within the structure of nuclear DNA.

When the target sequence is located in the control region of the gene (promoter or enhancer sequences) the oligonucleotide could bind its target sequence in much the same way as transcription factors do (*12*). However kinetic parameters might become rate-limiting. For example it is known that (C, T)-rich oligonucleotides bind DNA with a slow on-rate and this is, at least in part, due to the requirement for cytosine protonation in the third strand oligonucleotide. In contrast, (G A)-containing oligonucleotides bind much faster under physiological conditions (*13*).

When the target sequence is located within the transcribed region of the gene, then the question arises as to whether this sequence is accessible before transcription starts or becomes accessible only during transcription. A triple-helical complex of sufficient length is not expected to form on a nucleosomal structure (at least on a continuous sequence), because one side of the DNA double helix is facing inward toward the histone core. Sequence-specific control of gene transcription and site-directed mutagenesis on *plasmids* have shown unambiguously that antigene oligonucleotides can access plasmidic targets both in gene control regions and in transcribed regions. For endogenous genes there are still few data available. In our laboratory we have used oligonucleotide-psoralen conjugates to demonstrate that the polypurine tract (PPT) of the HIV provirus is accessible when its DNA is integrated in the genome of chronically infected cells (*14*). Similarly, the gene for the chemokine receptor CCR5 was shown to be accessible to an oligonucleotide attached to a nitrogen mustard (*15*). In both cases the photochemical or chemical reaction with DNA was obtained after permeabilized cells had been incubated with the oligonucleotide. However, the yield of the reaction remains far from 100%. The highest reported yield is 30% for HIV PPT DNA cross-linked to an oligonucleotide-psoralen conjugate (*14*). This might of course be due to limitations in the photochemical reaction itself but also to differences in accessibility in a heterogeneous cell population. We do not yet know whether 100% of the cells exhibit an accessible target at any given time. In any event this question also remains to be answered for the accessibility of mRNA targets to antisense oligonucleotides.

Recent experiments using site-directed mutagenesis on an integrated λ-phage in mouse cells have also provided evidence for accessibility to a PNA (*16*) but these data do not allow us to infer the fraction of DNA that is accessible to the oligomer.

Proof-of-concept that antigene oligonucleotides can inhibit gene transcription

There is good evidence that genes borne by a plasmid vector can be inhibited at the transcriptional level by antigene oligonucleotides (*17*). However, there is (still) limited information on endogenous genes. Even in cases where the expected biological response has been observed, it remains to be established whether this can be unambiguously ascribed to triplex formation. The same biological response is expected if, *e.g.*, the oligonucleotide binds to DNA and inhibits transcription factor binding or binds to the transcription factor and prevents its binding to DNA, as has been suggested for the *myc* gene in human cells (*18*). The design of control experiments to prove that the observed biological response is due to triplex formation is as difficult - if not more so - with antigene as with antisense oligonucleotides. When an antisense oligonucleotide works by inducing RNase H cleavage of a target mRNA, the identification of the cleavage products provides a direct proof of its binding to the target sequence. Only a few reports have demonstrated this effect. For antigene oligonucleotides there is no reported reaction that would allow us to detect the effect of oligonucleotide binding to DNA (unless the oligonucleotide is equipped with a reactive substituent). The best available control is to use a mutant of the target sequence, provided the mutation does not alter gene expression and its control. This can be easily achieved with plasmidic constructs or with the HIV provirus in chronically infected cells, but is of course more difficult with an endogenous gene. For example, we have shown that a target sequence in the gene coding for the α-subunit of the interleukin-2 receptor (IL-2Rα) can be mutated between the binding sites of two transcription factors, NFκB and SRF, without affecting the control of IL-2Rα expression. The mutations were introduced in a 15 bp oligopyrimidine•oligopurine sequence that was shown to be a target for triplex formation by a 15 nt oligonucleotide. No triplex was detected with the mutant sequence. This oligonucleotide conjugated either to an acridine derivative (to stabilize the triple helix) (*19*) or to a psoralen (to induce a photo-cross-linking reaction) (*20*) was shown to be active at inhibiting the wild-type plasmid but not the mutated one. This experiment clearly proves that binding to the wild-type and not to the mutated DNA was responsible for the biological response.

In accessibility studies where the PPT sequence of HIV was targeted with an oligonucleotide-psoralen conjugate in chronically infected cells, it was shown that mutations in the PPT sequence abolished the photo-cross-linking reaction, as expected for a triplex-mediated reaction (*14*).

Selectivity of antigene oligonucleotides

We cannot (yet) measure the expression of all genes in a living cell when the cell is submitted to a perturbation. Hopefully this will soon be available from genome sequencing and the development of DNA chip technology. Then we will be able to measure in a quantitative way the selectivity of action of antisense and antigene oligonucleotides. However it should be kept in mind that in many if not most cases, even the sequence-specific effect of an oligonucleotide on a single gene will have consequences on the expression of other genes either directly (*e.g.*, through transduction cascades) or indirectly (*e.g.*, by modulating the expression of a growth factor or a growth factor receptor).

What selectivity might we expect for antigene oligonucleotides? Some information is available from *in vitro* studies on oligonucleotide binding to DNA sequences containing mismatches (*21, 22*). The differences in binding energy are in the range 1-4 kcal·mol^{-1}, which means that binding constants for the (fully matched) target sequence and a sequence differing by one base pair differ by a factor of 10 to 10^3. These numbers are of the same order of magnitude as those for antisense oligonucleotides targeted to mRNA sequences. However, when an irreversible reaction is induced, *e.g.*, RNase H cleavage of the mRNA-antisense hybrid, then a kinetic amplification of discrimination is expected between matched and mismatched sequences, provided the binding of the oligonucleotide is not too strong (the lifetime of the complex with the mismatched sequence should not be long enough to allow for RNase H binding and/or cleavage as compared to its complex with the matched sequence). For antigene oligonucleotides no such irreversible reaction has yet been demonstrated. Therefore, with our present knowledge, the selectivity of antigene oligonucleotides is expected to follow the strength of their binding to matched and mismatched sequences. However, when the target sequence is located in the transcribed region, the transcription machinery stalls when arrested by the triple-helical complex and may resume transcription if the oligonucleotide dissociates from the DNA before the RNA polymerase with its truncated transcript is released. Again, a kinetic amplification of discrimination between complexes with different binding strengths can be achieved notwithstanding the fact that the structural perturbation due to triple helix formation might also change the properties of RNA polymerase stalled at the triple helix site (*23*).

If the oligonucleotide is equipped with a reactive substituent, the kinetic parameters of the irreversible reaction may play a dominant role in the discrimination between the target and related sequences containing mismatches. This is true also when the oligonucleotide is conjugated to a substituent that may recruit a nuclease or any other DNA-modifying enzyme, *e.g.*, oligonucleotide-camptothecin conjugates which may induce topoisomerase I cleavage of the target DNA (*24*).

Another important aspect of efficacy and selectivity in the mechanism of action of antigene oligonucleotides is related to the recognition and stabiliz tion of triple-helical complexes by cellular proteins. Proteins which bind triple-helical DNA have been described (*25*). The answer to the question of whether these proteins have a physiological role must await further characterization. Even if the recognition of a triplex is not part of their function under normal physiological conditions, the simple fact that they bind to the triple-helical complexes formed when the cells are treated with an antigene oligonucleotide may stabilize these complexes and amplify their ability to arrest transcription. At the same time they may lead to unwanted effects if they bind and stabilize mismatched complexes sufficiently to arrest transcription at mismatched sites.

If triple-helical complexes have a sufficiently long lifetime, they might be recognized as permanent lesions and recruit proteins/enzymes involved in repair processes. There is some evidence in the literature for induction of mutations by transcription-coupled repair of non-covalent triplexes (*16*). If they bind to triple-helical structures, repair proteins may play an important role in oligonucleotide-directed arrest of transcription even if the whole repair cascade is not induced.

How to improve binding of antigene oligonucleotides to their DNA target sequence?

In many cases the stability of triple-helical complexes is not strong enough to compete with transcription factor binding or, even more so, to arrest the transcription machinery once it is launched on its DNA template. With natural

8

oligodeoxynucleotides, only a few (G, A)-containing oligonucleotides form very stable complexes (*26*). The requirement for cytosine protonation makes (C, T)-containing oligonucleotides unstable under physiological conditions, except if they contain only a few cytosines with thymine neighbors. In order to increase the stability of the complexes, the third strand oligonucleotide has to be modified.

a) Covalent attachment of intercalators at the end of oligonucleotides has been shown to strongly stabilize triple-helical complexes due to intercalation at the junction between triplex and duplex regions along the DNA template (*27, 28*). The design of triple-helix-specific intercalating agents has opened the possibility of stabilizing triplexes by attachment of such a triplex ligand at any site along the third-strand oligonucleotide (*29*).

b) Chemical modifications of the backbone of the oligonucleotide may also stabilize triplex formation . Oligonucleotides with ribose or 2'-O-methyl derivatives instead of deoxyribose give more stable complexes than oligodeoxynucleotides when they bind in the Hoogsteen configuration (*30*). No stable complex was detected when reverse Hoogsteen hydrogen bonds were expected to form, *e.g.*, with (G, A)-oligoribonucleotides. A similar behavior was observed with N3'→P5' phosporamidate (pn) linkages (*31*). In our hands, oligophosphoramidates gave the most stable triplexes ever observed in the Hoogsteen configuration for (C, T)-or (T, C, G)-containing oligonucleotides. For example the 15-mer (pn) $T_4 CT_4 G_6$ which binds to the PPT sequence of HIV-1 in the Hoogsteen mode is at least 10 times more efficient at arresting transcription of the *nef* gene than was a phosphodiester oligodeoxynucleotide-intercalator conjugate (*31*). Covalent attachment of an intercalator to the oligophosphoramidate improves binding and transcription inhibition. There is no evidence presently available showing that a (G, A)-containing oligophosphoramidate binds to its DNA target. Of course one cannot exclude the possibility that by changing the experimental conditions, in particular polycations, one might be able to detect oligophosphoramidate binding in the reverse Hoogsteen configuration.

Peptide Nucleic Acids (PNAs) represent another backbone modification which opens new possibilities for triple helix formation. Pyrimidine PNAs can displace the oligopyrimidine strand of DNA to form a 2:1 (PNA:DNA) complex with the oligopurine sequence. This strand invasion reaction is different in its mechanism from that of the triple-helical complexes described above. Strand invasion is strongly salt dependent in its kinetics of formation. Nevertheless other potential applications of PNAs to control gene expression are presently being investigated (*32*).

c) Chemical modifications of the bases may also improve the stability of triple-helical complexes. The first studies were aimed at replacing cytosine by an analog that would not require protonation to recognize G in a C.G base pair. Subsequently, base modifications were introduced in third-strand oligonucleotides to extend the range of recognition sequences beyond oligopyrimidine•oligopurine sequences. However, we are still far from a general solution to major groove recognition of all four base pairs in DNA sequences (*33*). This remains a challenge for the future development of the antigene strategy, even though oligopyrimidine•oligopurine sequences are overrepresented in the human genome and all genes are expected to contain potential target sequence(s). It should be kept in mind that, within a gene, control regions (promoter, enhancer) and all the transcribed region, *including introns*, must be considered when choosing a target site for a triple helix-forming oligonucleotide.

d) Anomeric effects in triplex formation. Oligonucleotides synthesized with the α-anomers of nucleotides form triple helices with double-stranded DNA which are as stable as those formed with the natural β-anomers only in the Hoogsteen configuration [(T, C)- or (T, C, G)-containing oligonucleotides] *(34)*. The α oligomers bind – as expected – in the opposite orientation as compared to the β oligomers, except in the case of oligo T, where a parallel orientation with respect to the oligopurine (oligo A) target sequence was observed for both the α and β oligomers *(4)*.

In the reverse Hoogsteen configuration, only (G, T)-containing but not (G, A)-containing α oligonucleotides were shown to form triple helices with double-helical DNA *(35)*. The stability of α (G, T) triple helices is less than that of the natural β analog, and binding occurs in the opposite orientation, *i.e.*, parallel to the oligopurine target sequence. Self-association of α (G, A) oligonucleotides is less significant than that of β (G, A) oligonucleotides. Therefore, the absence of a detectable triple-helical complex with α (G, A) oligomers under conditions where the β (G, A) triple helix is observed shows that other (conformational) parameters play an important role in preventing triplex formation. The data available deal only with a few sequences, and more studies are required before a general conclusion can be reached on the ability of α oligonucleotides to form triple helices in the reverse Hoogsteen configuration.

α oligonucleotides are much more stable with respect to nuclease digestion than their natural β counterparts. Therefore they present an advantage over natural oligonucleotides for *in vivo* studies.

Competing structures in triple helix-forming oligonucleotides

The ability of an oligonucleotide to bind to the major groove of DNA depends on its availability as a single-stranded structure. Due to their particular base composition, oligonucleotides designed to form triple helices with DNA often form intra- or inter-molecular self-associated structures. Oligonucleotides containing runs of G form tetra-stranded structures due to the formation of G quartets (four G's associated by hydrogen bonds in the same plane). (G, A)-containing oligonucleotides may also form stable self-associated duplexes where two oligonucleotides bind parallel to each other. C-rich oligonucleotides may form a tetra-stranded structure called i-DNA where two double-helices are associated with their base pairs intercalated. These base pairs involve one neutral and one protonated cytosine (triple helix formation with (C, T)-oligonucleotides also requires cytosine protonation). Self-associated structures may also be observed with (G, T)-oligonucleotides. All these structures compete with triple helix formation and may lead to unexpected effects. For example, some (G, A)-oligonucleotides have been shown to form stable triple-helical complexes at 37° C, whereas no triplex was detected at 4° C *(36)*. This is due to the difference in thermodynamic parameters (enthalpy) for self-association of the (G, A)-oligonucleotide *versus* triplex formation.

10

Therefore intra- and intermolecular interactions of the third-strand oligonucleotides should always be investigated if one wants to obtain meaningful thermodynamic and kinetic parameters for triplex formation.

CLAMP AND CIRCULAR OLIGONUCLEOTIDES: ANOTHER APPLICATION OF TRIPLE-HELICAL COMPLEXES

A triple-helical complex can be formed on a single-stranded oligopurine sequence by an oligonucleotide made of two portions: one portion forms Watson-Crick base pairs with the oligopurine target; the other one binds to the double helix to form Hoogsteen hydrogen bonds with the oligopurine sequence (*37*). It was proposed that:

> *"formation of a triple-stranded structure on a single-stranded nucleic acid such as messenger RNA or viral RNA or DNA might prove more efficient to arrest translation, reverse transcription or replication than double helix formation"* (*37*).

Replication arrest by a clamp oligonucleotide was demonstrated on a single-stranded DNA template under conditions where an antisense oligonucleotide had no effect (*38*). The polypurine tract of HIV-1 was chosen as a target in these original studies (*37, 38*). A similar strategy was developed by Karin Möelling and co-workers using longer oligonucleotides (54 nt) with a hairpin-loop structure and single-stranded 5'- and 3'-ends (*39, 40*). It is not yet clear whether the observed inhibition of HIV-1 in cell cultures was due to clamp formation on the viral RNA (which could arrest reverse transcription), to RNase H-induced cleavage of the target RNA by the antisense portion of the oligonucleotide, or to oligonucleotide binding to other cellular or viral proteins, *e.g.*, HIV reverse transcriptase (*40*).

A high binding affinity and high selectivity is expected for clamp oligonucleotides. The bases A and G in the target sequence are recognized by 4 and 5 hydrogen bonds, respectively. The two oligonucleotide portions can be linked by an oligonucleotide sequence or by a non-nucleotidic linker such as an oligoethyleneglycol moiety. An intercalating agent can be attached to the Hoogsteen portion to stabilize the complex (*37*). A psoralen-directed photo-cross-linking reaction can be used to lock the complex in place (*38*).

At the time when clamp oligonucleotides were proposed, Eric Kool and co-workers developed circular oligonucleotides that are able to recognize an oligopurine sequence by forming both Watson-Crick and Hoogsteen hydrogen bonds as do the clamp oligonucleotides (*41*). Circular oligonucleotides form more stable complexes than clamp oligonucleotides, due to entropic factors linked to preorganization of the interacting parts. The complexes can even be made more robust by adding a disulfide bridge across the center of the macrocyclic ring (*42*). Clamp and circular oligonucleotides can also be constructed to recognize an oligopyrimidine sequence. It turns out that a closed duplex binds to an oligopyrimidine sequence much more tightly than a simple Watson-Crick complement. In these triple-helical complexes, the oligopyrimidine target sequence is engaged in Hoogsteen hydrogen bonding interactions with the circular duplex (*43*).

FROM GENE TARGETING TO GENE THERAPY

The review presented in the previous sections deals with synthetic oligonucleotides designed to bind the major groove of the DNA double helix. The question arises as to whether the third strand component could be generated *in situ*. Obviously using a DNA vector one can obtain an RNA transcript that could bind to genomic DNA to form a local triple helix, provided the DNA vector contains the appropriate sequences to synthesize the RNA with the proper base sequence and orientation to obey the rules described above for intermolecular triplexes. There are several ways to engineer the DNA vector in such a way that a stable triple-helix-forming RNA is transcribed. Two examples have been recently described where the triple-helix-forming RNA sequence was included in a longer RNA transcribed from an RNA Polymerase II promoter on an episomal vector. The genes for both the growth factor IGF1 (*44*) and its receptor IGF1-R (*45*) were chosen as targets. The cassette corresponding to the antigene sequence was inserted in both orientations so that transcription expressed either the (C, U)-containing sequence or the (G, A)-containing sequence within the RNA transcript. Only the (G, A)-containing RNA exhibited an inhibition of either IGF1 or IGF1-R. The targeted sequences were different for the two genes. Therefore, crossed experiments could be carried out to confirm the specificity of the biological response. The RNA targeted to IGF1 induced a complete disappearance of the IGF1 mRNA, as expected if transcription was inhibited by triple helix formation. It had no effect on IGF1-R gene expression (nor on control genes such as actin). In contrast, the RNA transcript targeted to IGF1-R inhibited both IGF1-R and IGF1 mRNAs. This was not unexpected inasmuch as IGF1-R blockage by an antisense RNA was also shown to induce a down-regulation of IGF1. Although these experiments do not prove that a triple helix is involved in the mechanism of action of the RNA transcripts, they do provide a basis to further explore this "gene therapy" approach, analogous in its concept to gene therapy protocols using antisense RNA transcripts to control translation of specific mRNAs. Expression vectors other than the episomal vectors used in the IGF1 and IGF1-R studies can be considered. For short target sequences (such as those involved in triplex formation), Pol III promoters might be more appropriate than Pol II promoters. For example, the Pol II promoter for U7 snRNA was recently shown to be efficient at producing short antisense RNA sequences targeted to aberrant splice sites in thalassemic β-globin pre-mRNA (*46*).

A more basic problem is raised by the experiments with RNA transcripts in cell cultures. *In vitro* studies using synthetic RNAs as third strands failed to show any binding of (G, A)-containing RNA or their 2'-O-methyl derivatives to double-helical DNA in the reverse Hoogsteen configuration, whereas (C, U)-containing RNAs or their 2'-O-methyl derivatives formed much stronger (Hoogsteen) complexes than an isosequential DNA (*30*). It is not too surprising that (C, U)-containing RNA transcripts do not form a triplex *in vivo*, because the requirement for cytosine protonation makes such triplexes too unstable at physiological pH. However, it is more surprising that (G, A)-containing RNAs form a triplex stable enough to arrest transcription in cell cultures. There are several possibilities that can be investigated to help understand this observation: (i) the proper conditions for triplex formation have yet to be found *in vitro*; (ii) the triple-helical complex is stabilized *in vivo* because either the triplex-forming RNA sequence is presented by the rest of the RNA transcript in a structure that fits the major groove of DNA or the triple-helical

12

complex is stabilized by proteins; (iii) the mechanism of mRNA down-regulation does not involve triplex formation but another mechanism, for example the sequestration of proteins by the RNA transcript or the DNA vector. Further studies are clearly required to clarify these points.

CONCLUSION

This review has presented briefly some of the questions raised by the development of the antigene strategy to control gene expression at the transcriptional level. This chapter was meant to be a brief introduction to this book, and the reader is strongly encouraged to consult the chapters which follow for more in-depth coverage of the literature on each of the topics. The development of antigene oligonucleotides as drugs is facing some of the same problems that have faced antisense oligonucleotides. There is no major theoretical obstacle to targeting specific sequences in DNA. The field is much younger, however, and we still need more basic studies to clearly appreciate the breadth of potential applications of the antigene strategy.

Meanwhile, other applications of triplex-forming oligonucleotides are being developed. They can be used as tools in molecular and cellular biology, *e.g.*, to probe the accessibility of DNA sequences in the cell nucleus under different experimental conditions (*15*), to induce site-directed mutations in cell cultures (*47*), to recruit proteins to specific sites on DNA (*48*), or to design artificial sequence-specific bending ligands (*49, 50*).

For diagnostic purposes, the formation of a triple-helical complex may allow the detection of DNA sequences without any opening of the double helix, thus avoiding one of the rate-limiting steps in the hybridization process. A triplex strategy can be used to purify plasmids used in gene therapy protocols (*51*). This technology utilizes to its advantage the requirement for cytosine protonation to obtain stable triple helices in the pyrimidine motif. The plasmid containing the appropriate target (oligopyrimidine•oligopurine sequence) is selectively retained on an affinity column by a (C, T)-containing oligonucleotide at low pH and then eluted at higher pH.

The scope of the potential applications of triplex-forming oligonucleotides would undoubtedly be broadened if the molecular mechanisms for recognition of the major groove side of the DNA double helix could be extended to sequences containing all four base pairs. This has recently been achieved for the recognition of the minor groove by hairpin polyamides (*52*). Clearly more work is needed on both the chemical and biological sides to fully exploit the potentials of triplex-forming oligonucleotides and their sequence-specific recognition of DNA.

ACKNOWLEDGMENTS

I thank all my co-workers who have made major contributions to the work summarized in the present chapter. Their names can be found in the references and some of them have authored other chapters in this book. Our investigations have been supported by the Institut National de la Santé et de la Recherche Médicale

(INSERM), the Centre National de la Recherche Scientifique (CNRS), the Muséum National d'Histoire Naturelle (MNHN), the Agence Nationale de Recherche sur le SIDA (ANRS), the Ligue Nationale Contre le Cancer, the Association pour la Recherche sur le Cancer and Rhône-Poulenc.

REFERENCES

1. Felsenfeld, G., Davies, D. R. and Rich, A. (1957). Formation of a three-stranded poly-nucleotide molecule. *J. Am. Chem. Soc.* **79**, 2023-2024.
2. Michelson, A. M., Massoulié, J. and Guschlbauer, W. (1967). Synthetic polynucleotides. *Prog. Nucl. Acids Res. Mol. Biol.* **6**, 83-141.
3. Maher, L. J., (1996). Prospects for the therapeutic use of antigene oligonucleotides. *Cancer Investigation.* **14**, 66-82.
4. Le Doan, T., Perrouault, L., Praseuth, D., Habhoub, N., Decout, J.-L., Thuong, N. T., Lhomme, J. and Hélène, C. (1987). Sequence-specific recognition, photocrosslinking and cleavage of the DNA double helix by an oligo-[α]-thymidylate covalently linked to an azidoproflavine derivative. *Nucleic Acids Res.* **15**, 7749-7760.
5. Moser, H. E. and Dervan, P. B. (1987). Sequence-specific cleavage of double helical DNA by triple helix formation. *Science* **238**, 645-650.
6. Mirkin, S. M., Lyamichev, V. I., Drushlyak, K. N., Dobrynin, V. N., Filippov, S. A. and Frank-Kamenetskii, M. D. (1987). DNA H form requires a homopurine-homopyrimidine mirror repeat. *Nature* **330**, 495-497.
7. Sun, J. S., Garestier, T. and Hélène, C. (1996). Oligonucleotide-directed triple helix formation. *Curr. Opin. Struct. Biol.* **6**, 327-333.
8. Hélène, C. (1991). The anti-gene strategy: control of gene expression by triplex forming-oligonucleotides. *Anticancer Drug Des.* **6**, 569-584.
9. Cooney, M., Czernuszewicz, G., Postel, E. H., Flint, S. J., Hogan, M. E. (1988). Site specific oligonucleotide binding represses transcription of the hyman c-*myc* gene *in vitro. Science* 241, 456-459.
10. Condon, T. P. and Bennett, C. F. (1996). Altered mRNA splicing and inhibition of human E-selectin expression by an antisense oligonucleotide in human umbilical vein endothelial cells. *J. Biol. Chem.* **48**, 30398-30403.
11. Sierakowska, H., Sambade, M. J., Agrawal, S. and Kole, R. (1996). Repair of thalassemic human α-globin mRNA in mammalian cells by antisense oligonucleotides. *Proc. Natl. Acad. Sci. USA* **93**, 12840-12844.
12. Beato, M. and Eisfeld, K. (1997). Transcription factor access to chromatin. *Nucleic Acids Res.* **25**, 3559-3563.
13. Faucon, B., Mergny, J.-L. and Hélène, C. (1996). Effect of third strand composition on triple helix formation: purine *versus* pyrimidine oligodeoxynucleotides. *Nucleic Acids Res.* **24**, 3181-3188.
14. Giovannangeli, C., Diviacco, S., Labrousse, V., Gryaznov, S., Charneau, P. and Hélène, C. (1997). Accessibility of nuclear DNA to triplex-forming oligonucleotides: the integrated HIV-1 provirus as a target. *Proc. Natl. Acad. Sci. USA* **94**, 79-84.
15. Belousov, E. S., Afonina, I. A., Kutyavin, I. V., Gall, A. A., Reed, M. W., Gamper, H. B., Wydro, R. M. and Meyer, R. B. (1998). Triplex targeting of a native gene in permeabilized intact cell: covalent modification of the gene for the chemokine receptor CCR5. *Nucleic Acids Res.* **26**, 1324-1328.
16. Faruqi, A. F., Egholm, M. and Glazer, P. M. (1998). Peptide nucleic acid-targeted mutagenesis of a chromosomal gene in mouse cells. *Proc. Natl. Acad. Sci. USA* **95**, 1398-1403.

17. Giovannangeli, C. and Hélène C. (1997). Progress in developments of triplex-based strategies. *Antisense & Nucleic Acid Drug Devel.* **7**, 413-421.

18. Michelotti, E. F., Tomonaga, T., Krautzsch, H. and Levens, D. (1995). Cellular nucleic acid binding protein regulates the CT element of the human c-*myc* protooncogene. *J. Biol. Chem.* **271**, 9494-9499.

19. Grigoriev, M., Praseuth, D., Guieysse, A.-L., Robin, P., Thuong, N. T. and Hélène, C. (1993). Inhibition of interleukin-2 receptor α-subunit gene expression by oligonucleotide-directed triple helix formation. *C. R. Acad. Sci. Paris Life Sci.* **316**, 492-495.

20. Grigoriev, M., Praseuth, D., Guieysse, A.-L., Robin, P., Thuong, N. T., Hélène, C. and Harel-Bellan, A. (1993). Inhibition of gene expression by triple helix-directed site-specific cross-linking. *Proc. Natl. Acad. Sci. USA* **90**, 3501-3505.

21. Mergy, J.-L., Sun, J. S., Rougée, M., Montenay-Garestier, T., Barcelo, F., Chomilier, J. and Hélène, C. (1991). Sequence specificity in triple helix formation: theoretical and experimental studies of the effect of mismatches on triplex stability. *Biochemistry* **30**, 9791-9798.

22. Greenberg, W. A. and Dervan, P. B. (1995). Energetics of formation of sixteen triple helical complexes which vary at a single position within a purine motif. *J. Am. Chem. Soc.* **117**, 5016-5022.

23. Wang, Z., Rana, T. M. (1997). DNA damage-dependent transcriptional arrest and termination of RNA polymerase II elongation complexes in DNA template containing HIV-1 promoter. *Proc. Natl. Acad. Sci. USA* **94**, 6688-6693.

24. Matteucci, M., Lin H. Y., Huang, T., Wagner, R., Sternbach, D. D., Mehrotra, M. and Besterman, J. M. (1997). Sequence-specific targeting of duplex DNA using a camptothecin-triple helix forming oligonucleotide conjugate and topoisomerase I. *J. Am. Chem. Soc.* **119**, 6939-6940.

25. Guieysse, A. L. and Praseuth, D. "Triplex-binding proteins." In *Triple Helix Forming Oligonucleotides*, Claude Malvy and Annick Harel-Bellan, ed. Norwell, MA: Kluwer Academic Publishers (Chapter 16, this volume).

26. Svinarchuk, F. and Malvy, C. "Gene targeting triple helix forming purine oligonucleotides." In *Triple Helix Forming Oligonucleotides*, Claude Malvy and Annick Harel-Bellan, ed. Norwell, MA: Kluwer Academic Publishers (Chapter 11, this volume).

27. Sun, J. S., François, J. C., Montenay-Garestier, T., Saison-Behmoaras, T., Roig, V., Thuong, N. T. and Hélène, C. (1989). Sequence-specific intercalating agents: intercalation at specific sequences on duplex DNA *via* major groove recognition by oligonucleotide-intercalator conjugates. *Proc. Natl. Acad. Sci. USA* **86**, 9198-9202.

28. Thuong, N. T. and Hélène, C. (1993). Sequence-specific recognition and modification of double-helical DNA. *Angew. Chem. Int. Ed. Engl.* **32**, 666-690.

29. Silver, G., Nguyen, C. H., Boutorine, S., Bisagni, E., Garestier, T. and Hélène, C. (1997). Conjugates of oligonucleotides with triplex-specific intercalating agents. Stabilization of triple-helical DNA in the promoter region of the gene for the (subunit of interleukin 2 (IL-2Rα). *Bioconjugate Chem.* **8**, 15-22.

30. Escudé, C., François, J. C., Sun, J. S., Ott, G., Sprinzl, M., Garestier, T. and Hélène, C. (1993). Stability of triple helices containing RNA and DNA strands: experimental and molecular modeling studies. *Nucleic Acids Res.* **21**, 5547-5553.

31. Escudé, C., Giovannangeli, C., Sun, J. S., Lloyd, D. H., Chen, J. K., Gryaznov, K., Garestier, T. and Hélène, C. (1996). Stable triple-helices formed by oligonucleotide N3'→P5' phosporamidates inhibit transcription elongation. *Proc. Natl. Acad. Sci. USA* **93**, 4365-4369.

32. Good, L. and Nielsen, P. E. (1997). Progress in developing PNA as a gene-targeted drug. *Antisense & Nucleic Acid Drug Devel.* **7**, 431-437.

33. Doronina, S. O. and Behr, J.-P. (1997). Towards a general triple helix mediated DNA recognition scheme. *Chemical Society Reviews* 63-71.

34. Sun, J. S., Giovannangeli, C., François, J. C., Kurfurst, R., Montenay-Garestier, T., Saison-Behmoaras, T., Thuong, N. T. and Hélène, C. (1991) Triple-helix formation by α oligodeoxynucleotide-intercalator conjugates. *Proc. Natl. Acad. Sci. USA* **88**, 6023-6027.

35. Noonberg, S. B., François, J. C., Praseuth, D., Guieysse, A. L., Lacoste, J., Garestier, T. and Hélène, C. (1995). Triplex formation with α anomers of purine-rich and pyrimidine-rich oligodeoxynucleotides. *Nucleic Acids Res.* **23**, 4042-4049.

36. Noonberg, S. B., François, J. C., Garestier, T. and Hélène, C. (1995). Effect of competing self-structure on triplex formation with purine-rich oligodeoxynucleotides containing GA repeats. *Nucleic Acids Res.* **23**, 1956-1956.

37. Giovannangeli, C., Montenay-Garestier, T., Rougée, M., Chassignol, M., Thuong, N. T. and Hélène, C. (1991). Single-stranded DNA as a target for triple helix formation. *J. Am. Chem. Soc.* **113**, 7775-7777.

38. Giovannangeli, C., Thuong, N. T. and Hélène, C. (1993). Oligonucleotide clamps arrest DNA synthesis on a single-stranded DNA target. *Proc. Natl. Acad. Sci. USA* **90**, 10013-10017.

39. Volkman, S., Jendis, J., Frauendorf, A. and Moelling, K. (1995). Inhibition of HIV-1 reverse transcription by triple-helix forming oligonucleotides with viral RNA. *Nucleic Acids Res.* **23**, 1204-1212.

40. Jendis, J., Strack, B., Volkmann, S., Böni, J. and Mölling, K. (1996). Inhibition of Replication of Fresh HIV Type 1 Patient Isolates by a Polypurine Tract-Specific Self-Complementary Oligodeoxynucleotide. *AIDS Res. and Hum. Retrovirus.* **12**, 1161-1168.

41. Kool, E. T. (1991) Molecular recognition by circular oligonucleotides: Increasing the selectivity of DNA binding. *J. Am. Chem. Soc.* **113**, 6265-6266.

42. Chaudhuri, N. C. and Kool, E. T. (1995). Very high affinity DNA recognition by bicyclic and cross-linked oligonucleotides. *J. Am. Chem. Soc.* **117**, 10434-10442.

43. Wang, S. H., Kool, E. T. (1994). Recognition of single-stranded nucleic acids by triplex formation. The binding of pyrimidine-rich sequences. *J. Am. Chem. Soc.* **116**, 8857-8858.

44. Sheveley, A., Burfeind, P., Schulze, E., Rininsland, F., Johnson, T., Trojan, J., Chernicky, C., Hélène, C., Ilan, J. and Ilan, J. (1997). Potential triple helix-mediated inhibition of IGF-I gene expression significantly reduces tumorigenicity of glioblastoma in an animal model. *Cancer Gene Ther.* **4**, 105-112.

45. Rininsland, F., Johnson, T. R., Chernicky, C. L., Schulze, E., Burfeind, P., Ilan, J. and Ilan, J. (1997). Suppression of insulin-like growth factor type I receptor by a triple-helix strategy inhibits IGF-I transcription and tumorigenic potential of rat C6 glioblastoma cells. *Proc. Natl. Acad. Sci. USA* **94**, 5854-5859.

46. Gorman, L., Suter, D., Emerick, V., Schümperli, D. and Kole, R. (1998). Stable alteration of pre-mRNA splicing patterns by modified U7 small nuclear RNAs. *Proc. Natl. Acad. Sci. USA* **95**, 4929-4934.

47. Raha, M., Wang, G., Seidman, M. M. and Glazer, P. M. (1996). Mutagenesis by third-strand-directed psoralen adducts in repair-deficient human cells: high frequency and altered spectrum in a xeroderma pigmentosum variant. *Proc. Natl. Acad. Sci. USA* **93**, 2941-2946.

48. Svinarchuk, F., Nagibneva, I., Cherny, D., Ait-Si-Ali, S., Pritchard, L. L., Robin, P., Malvy, C. and Harel-Bellan, A. (1997). Recruitment of transcription factors to the target site by triplex-forming oligonucleotides. *Nucleic Acids Res.* **25**, 3459-3464.

49. Liberles, D. A. and Dervan, P. B. (1996). Design of artificial sequence-specific DNA bending ligands. *Proc. Natl. Acad. Sci. USA* **93**, 9510-9514.

50. Akiyama, T. and Hogan, M. E. (1996). The design of an agent to bend DNA. *Proc. Natl. Acad. Sci. USA* **93**, 12122-12127.
51. Wils, P., Escriou, V., Warnery, A., Lacroix, F., Lagneaux, D., Olliver, M., Crouzet, J., Mayaux, J. F. and Scherman, D. (1997). Efficient purification of plasmid DNA for gene transfer using triple-helix affinity chromatography. *Gene Ther.* **4**, 323-330.
52. White, S., Szewczyk, J. W., Turner, J. M., Baird, E. E. and Dervan, P. B. (1998). Recognition of the four Watson-Crick base pairs in the DNA minor groove by synthetic ligands. *Nature* **391**, 468-471.

2 TRIPLE-HELIX STRUCTURE. THE TRIPLE-HELIX-FORMING OLIGONUCLEOTIDE

Horea Porumb

SUMMARY

While waiting for more detailed information on the triple-helix structure to be provided by techniques such as fiber X-ray diffraction and nuclear magnetic resonance, this chapter summarizes some basic principles regarding the construction of triple helices and draws attention to parasite interactions of the triple-helix-forming oligonucleotide that might divert it from its intended target. It does not purport to be exhaustive and, as will be obvious at each step, every generalization should be made with caution.

INTRODUCTION

The basic triplets

It is currently accepted that triple helices require a purine-rich strand within the target duplex. The third strand of the triple helix is attached via hydrogen bonds mostly to this strand, usually termed *strand-2*. *Strand-1* is the pyrimidine-rich strand of the duplex. Theoretical models of the basic triplets have been put forward and discussed elsewhere and will only briefly be recalled (*1-3*).

For practical purposes, in this review we distinguish among three classes of triple helices:

i. The "acid" triple helix, containing T and protonated C in the third strand, is based on the triplets $C^+(GC)$ and T(AT), where the parentheses contain the Watson-Crick base pair. The orientation of the third base in the triplet distinguishes between direct (cis)- Hoogsteen and reverse (trans)- Hoogsteen pairing of this nucleotide to the Watson-Crick base pair. It is generally accepted that triplets based on *direct-Hoogsteen* hydrogen bonding, as is the case here, require *parallel* orientation of the third strand with respect to the purine-rich strand-2 of the duplex, whereas the opposite is true for the reverse-Hoogsteen-based triplets (*4*).

ii. Triple helices with G,A-containing third strands, the pairing scheme of which involves triplets of the G(GC) and A(AT) types, are derived from the classical Watson-Crick base pairs plus "Hoogsteen-type" hydrogen bonds, where G pairs to G and A to A. It seems that the G,A-containing third strands realize stronger interactions with the DNA double helix than the pyrimidine-rich ones (*5*).

iii.Triple helices with G,T-containing third strands, sometimes rather misleadingly referred to as "purine-rich" (R) strands, are based on G(GC) and T(AT) triplets but, occasionally and less rationally, also on G(CG) and T(TA) triplets - see for example (*6*).

Strand orientation

It should be kept in mind that the stability of the planar triplet does not necessarily imply stability of the helix. With purine-rich oligonucleotides, the most favorable orientation of the third strand (and implicitly the choice between Hoogsteen and reverse-Hoogsteen bonding) depends on the particular sequence employed. Model building showed that, on energetic grounds, the parallel orientation is preferred when the third strand contains identical repeats of the same nucleotide; on the contrary, because of the better isomorphism of the triplets based on reverse-Hoogsteen interactions, a mixed sequence should probably favor the antiparallel orientation of the strands (*7*); the parallel orientation should nevertheless be remarkably stable (*8, 9*). When targeting the d(G3A4G3).d(C3T4C3) duplex, the parallel orientation of either G,A-containing or G,T-containing third strands was more favorable. The stability of the former triple helix was greater than that of the latter (*10*). When targeting the Friend leukemia virus sequence, close to its LTR, AGAAAAAGGGGGG by antiparallel triple-helix-forming oligonucleotides, the G,A-containing oligonucleotide leads to a more stable triple helix than the G,T-containing one (*11*).

STRUCTURAL FEATURES - WHAT WE KNOW FOR SURE

As will be seen below, detailed knowledge on triple-helix structure has by now become classic and comes primarily from fiber X-ray diffraction and nuclear magnetic resonance. The basic principles and the potentials of these techniques have been described elsewhere (*12, 13*). The incorporation of information provided by other techniques was and still remains indispensable - e.g., infrared spectroscopy data is valuable for the conformation of sugars. Energy minimization and/or molecular dynamics calculations represented in every case the final stage of the analysis. The combination of several techniques will eventually allow one to extract "signatures", criteria of triplex stability, basic rules relative to the conception of a triple-helix-forming oligonucleotide. Here is an account of the current consensus as regards the building blocks of the main types of triple helices.

Parallel and antiparallel structures based on the G(GC) triplet

The packing of the decamer d(GGCCAATTGG) happens to provide, within the same crystal, the basic repeat units of both parallel and antiparallel triple helices based on the G(GC) triplet (*14*). The Watson-Crick components of the triplets appear undisturbed by the presence of the third base. In the *parallel* triplets, there are three direct-Hoogsteen hydrogen bonds between N2, N1 and O6 of the guanine in the third strand and, respectively, the N7 and O6 of guanine and N4 of cytosine in the Watson-Crick duplex, which means that the hydrogen bonding involves both bases of the Watson-Crick pair. In the *antiparallel* triplets, two reverse-Hoogsteen hydrogen bonds form between the N1 and N2 of the guanine in the third strand and, respectively, the N7 and O6 of the guanine in the Watson-Crick duplex. The glycosidic angles are all *anti*. The crystal structure provided the helical parameters that allowed the modelization of full, right-handed triple helices based on G(GC) triplets with either parallel or antiparallel orientation of the third strand. In both cases the Watson-Crick portions resemble B-DNA. The parallel triplex has a twist of 26.8° and a rise of 3.4 Å, with the widths of the Crick-Hoogsteen and Watson-Hoogsteen grooves equal to 6 Å and 7 Å, respectively. The antiparallel one has a twist of 29.5°, a rise of 3.45 Å and unequal grooves, equal to 7 Å and 3 Å, respectively.

Parallel structure based on the T(AT) triplet

Fourier Transform Infrared Spectroscopy (FTIR) has predicted that the parallel structure based on the T(AT) triplet has South-type sugars, like B-DNA (*15*). A structure for the parallel triple helix $d(T)_n.d(A)_n.d(T)_n$, consistent with available infrared spectroscopy and crystal data was calculated by a linked atom least squares procedure, in which T in the third strand is direct-Hoogsteen hydrogen-bonded to A and related to it by a pseudo dyad axis, which, in conjunction with the existent symmetry of the Watson-Crick duplex, determines that the three strands of the helix adopt identical backbone conformations, similar to that of B-DNA, with C2'-endo sugar puckers (*16*). The helix has a rise of 3.26 Å per residue and 12 residues/turn.

Antiparallel triple helices containing G and T in the third strand

Nuclear magnetic resonance (N.M.R.) characterization of an intramolecular triple helix containing G and T in the third strand confirmed the existence of the G(GC) and T(AT) triplets, with the glycosidic torsion angles of all bases in *anti* conformation. The triple helix is based on reverse-Hoogsteen hydrogen bonding of the third strand, which runs antiparallel to the purine strand of the Watson- Crick duplex. (*17*). When combined with a molecular dynamics approach, the study provided the solution structure of this helix at high resolution. A third groove is formed between the G,T-containing strand and the pyrimidine strand of the double helix, wider and deeper than the two grooves present in B-DNA(*18*)

It was possible to replace one T(AT) triplet with an A(AT) triplet in the above molecule, which then provided the link to the modeling of a triple helix with a G,A-containing third strand (*19*).

Homonuclear two-dimensional N.M.R. spectroscopy identified distinct patterns of hydration in the grooves of both G,T-containing (R.RY) and C^+,T-containing (Y.RY) triple helices in aqueous solution (*20*).

Subsequent N.M.R. or X-ray diffraction studies mainly concentrated on the occurrence of accidental (*21*) or exceptional triplets, usually sandwiched within homonucleotidic sequences, or modified bases, such as the N7-glycosylated guanine which, in the third strand, mimics the protonated cytosine (*22*). Triple helices containing complex sequences give fiber-type X-ray diffraction patterns (*23*).

Generalization of the above data for triple helices of mixed sequences is difficult to make at this stage of knowledge. The amount of structural data that has been obtained from X-ray diffraction and N.M.R. may appear modest. This is basically because of the difficulty of obtaining crystals and respectively unique species under the high concentrations required by these techniques. The use of self-folding oligonucleotide is hardly sufficient to prevent self associations or partial folding of the molecule. The obtention of specimens suitable for N.M.R. analysis is the exception rather than the rule. The example that will be provided at the end of this review illustrates the possible interactions into which a particular triple-helix-forming oligonucleotide is found to indulge itself. On the one hand, by demonstrating the impossibility of avoiding such interactions, one will justify the scarcity of structural data on triple helices available to date. On the other hand, this demonstration draws attention to the fact that the side interactions are inseparable from the conceived role of a triple-helix-forming oligonucleotide. For that matter, the side interaction of the triple-helix-forming oligonucleotide might sometimes represent the real cause of its biological activity or, under different circumstances, the cause of its failure to show biological activity.

OTHER TECHNIQUES, ADDITIONAL INFORMATION

FTIR

The power of FTIR spectroscopy may be illustrated by the work on parallel and antiparallel A(AT) intramolecular triple helices (*24*). The stability of this antiparallel triplex, as measured by the melting temperature, is 6° C higher than that of the parallel one. The former, based on reverse-Hoogsteen pairing of the third strand, only contains South-type sugars, as in B-DNA. The latter has mixed sugar geometry, the adenosine strand within the Watson-Crick duplex possessing North-type sugars, as in A-DNA. The organization of water within the parallel T(AT) triple helix was analyzed by infrared spectroscopy, gravimetric measurements, and molecular dynamics simulations (*25*). In aqueous medium the triple helix is less solvated than the corresponding Watson-Crick duplex (17 vs. 21 water molecules per nucleotide). Below 81% relative humidity, which coincides with the emergence of North-type sugars in the adenosine strand of the triplex, the triple helix remains more hydrated than the duplex. In particular, a spine of water molecules is seen by molecular dynamics simulations in the groove created between the thymidine strands. Details of a similar nature were obtained from films of a pyrimidine-purine-pyrimidine hairpin triplex (*26*).

An FTIR study of the triple helix $(dG)_{20}.(dG)_{20}.(dC)_{20}$ has shown dramatic hydration-dependent vibrational changes taking place in the spectral region associated with the glycosidic linkages, reflecting the effect of water in the Watson-Hoogsteen groove, drawing the attention once more to the interdependence between hydration and conformation (*27*).

The role of water

The importance of the explicit inclusion of water molecules in molecular modeling, as opposed to the treatment in vacuum with a distance-dependent dielectric constant to mimic the shielding effect of the solvent, was proven when the sequence $(dC)_{10}.(dG)_{10}.(dG)_{10}$ was shown to correctly prefer the *anti* conformation of the glycosidic angles in the absence of any external constraints, for either Hoogsteen or reverse-Hoogsteen base pairing schemes (*28*). Taking into account the solvent effects on model $d(CG.G)_7$ and $d(TA.T)_7$ DNA triple helices, free energy minimization predicted that in order to accommodate a third strand in triplex formation, the backbone of the otherwise B-type DNA duplex has to adjust to create a deep, A-like major groove (*29*). Inclusion of explicit water molecules is necessary in order to predict the interaction between the third strand and duplex DNA as well as the stability of mismatch triplets (*30, 31*).

Kinetics of triplex formation

Whereas the relatively modest specificity in target recognition by triple-helix-forming oligonucleotides requires sequences that exceed 20 nucleotides in length, from the point of view of the kinetics of triplex formation the optimal length of a G,T-containing oligonucleotide was found by DNase I footprinting to be 12 bases. Longer oligonucleotides were slightly weaker in binding affinity, while shorter ones were considerably weaker. The remarkable information is that point mutations near the 3' end of the oligonucleotide had a more detrimental effect on triplex formation than did changes to the 5' end, suggesting that the zipping-up of the triple helix begins by the interaction of the 3' end of the oligonucleotide (*32*).

The kinetics of triple-helix formation was studied by electrophoresis in the presence of various salt concentrations. The second-order association constants of the duplex with (A,G)- and (G,T)-containing triple-helix-forming oligonucleotides were found to be lower by an order of magnitude than those obtained with (C^+,T)-containing oligonucleotides, which in turn were three orders of magnitude lower than those obtained for the formation of a Watson-Crick duplex (*33*). According to a different study, G,T-containing oligonucleotides bind much faster than G,A- or C^+,T-containing third strands (*34*). By N.M.R., the lifetimes of the Hoogsteen base pairs in a cross-linked, intramolecular pyrimidine-purine-pyrimidine triple helix were in the range of 3-370 ms, whereas the exchange times for the Watson-Crick imino protons were of the order of 1 hr (*35*).

The role of salt

The effect of salt on triplex formation is rather complex, as NaCl concentrations above 50 mM decreased, rather than enhanced the rate of triplex formation (*33*). It may be envisaged that, whereas charge neutralization is important, too high a concentration of counterion could mask certain groups essential for strand recognition and, in addition, the presence of several counterions in the same solution cannot avoid the competition of these ions among themselves. It is very likely that monovalent ions tend to compete with divalent ones, whence the interest in not mixing them up when wishing to study individual effects.

Polyamines, such as spermine, tend to increase the rate of triplex formation (*36, 37*).

Monovalent ions other than lithium (particularly K^+ and Rb^+, less so NH_4^+ and Na^+) are inhibitory to the association rate without significantly affecting the rate of dissociation (*38*). Contrary to general belief, the inhibitory effect of specific monovalent ions in this particular study was not accompanied by the induction of tetraplex association reactions, showing that certain salts down-modulate the rate of triplex formation otherwise than by simply providing competitive association pathways. It is known in this respect that K^+ and Rb^+ favor G-tetrad formation, whereas Na^+ does not, simply because the energetic cost of the dehydration of the latter is too high (*39*). In general, triple-helix formation is supported by magnesium. In addition, the induction/stabilization of quadruplex structure by K^+ is counteracted by low concentrations of Mn^{2+}, Co^{2+} or Ni^{2+} (*40*).

While theoretical and computational studies of the ion atmosphere of DNA are still in progress, the role of ions in promoting or preventing triple-helix formation is still an unsolved problem (*41*). Many variations on the theme of the salt effects are found in the literature and will not be developed here. The legend goes that divalent ions are somehow essential to strand recognition and/or triple-helix stabilization. In the example that will be treated below it will be seen that, at least in dilute solution, the formation of a triple helix did not require divalent ions.

BEHAVIOR OF A TRIPLE-HELIX-FORMING OLIGONUCLEOTIDE

Designing an oligonucleotide targeted to the HER2 oncogene.

The problems involved in designing a triple-helix-forming oligonucleotide (TFO) is illustrated by our results on targeting a natural, G,A-rich, 28-mer oligodeoxyribonucleotide to the promoter of the human HER2/neu oncogene (Schema 1). Oligonucleotide 28(C) and the appropriate target duplex have been shown to form a triple helix *in vitro*, where the third-strand oligonucleotide is antiparallel with respect to the purine-rich strand of the duplex, to which it is joined by reverse-Hoogsteen hydrogen bonds (*42*). When administered with the aid of the lipofectin vector, the third-strand oligonucleotide penetrated across the cell membrane and accumulated in the nuclei of MCF7 breast cancer cells, where it inhibited in a concentration-dependent and time-dependent manner the expression of the HER2 gene – as measured by Northern analysis of the HER2 mRNA level and ELISA analysis of the corresponding protein. When designing the TFO sequence, one had to overcome the pur-pyr inversions within the target duplex. This demanded making the choice among the four nucleotides for the one to be to be put in the third strand against the site of inversion. In the case of the HER2 oligonucleotide, molecular modeling suggested the C(CG) triplet. The gain in energy over the other postulants depended somewhat on the manner of approaching the calculation. The attachment of C to the duplex relies on one hydrogen-bond. All other nucleotides, while still able to form one hydrogen bond, would have distorted the uniformity of the helix. The result was confirmed by a systematic DNase I footprinting approach (*43*).

Side interactions of the TFO

Formation of the triple helix, T_{123}, in 0.1 M LiCl (with no divalent ion present during hybridization) is confirmed by electrophoretic titration, as shown in Figure 1. Because of the repetitive nature of the sequence, when titrating the labeled TFO with increasing concentrations of target duplex, a duplex is first generated at intermediate concentrations of the target. It was demonstrated that this duplex, D_{13}, is based on an imperfect Watson-Crick pairing of the TFO to the pyrimidine-rich strand of the target (Schema 1). Quantitative reasons were given elsewhere for this duplex being visible only in a certain domain of the titration range, before the triple helix becoming predominant (*44*). The interaction leading to the imperfect duplex could

24

Oligonucleotide 28(C), 28 nt TFO, S_3:
5'-ggg Agg Agg Agg _C_gg Agg Agg AAg Agg A-3'

Parallel duplex, *direct-Hoogsteen* base pairing, D_{33}:
5'-ggg Agg Agg Agg _C_gg Agg Agg AAg Agg A-3'
5'-ggg Agg Agg Agg _C_gg Agg Agg AAg Agg A-3'

Antiparallel duplex, *reverse-Hoogsteen* base pairing, D_{23}:
5'-ggg Agg Agg Agg _C_gg Agg Agg AAg Agg A-3'
3'-ggg Agg Agg Agg _T_gg Agg Agg AAg Agg A-5'

Hairpin, antiparallel strand orientation, *reverse-Hoogsteen* base pairing, P_3. There are 3 unconfirmed possibilities, each with 10 links per molecule, of which only one is shown:

```
5'-gggAggAggAgg Cg
   xIIIxIIIIII   )
3'-AggAgAAggAgg Ag
```

Imperfect duplex, D_{13}, Watson-Crick base pairing. There are 28 base pairs with 6 wobble-type mismatches:

```
5'-GGGAGGAGGAGGCGGAGGAGGAAGAGGA  -strand-3
   X****X********X**X********X****X
3'-TCCTCTTCCTCCTCCACCTCCTCCTCCC  -strand-1
```

Triple-helix, T_{123}. The purine-pyrimidine inversion in the target duplex is in heavy characters. The strand numbers are indicated in brackets:

```
3'-TCC TCT TCC TCC TCC ACC TCC TCC TCC C-5'.(1)
5'-Agg AgA Agg Agg Agg Tgg Agg Agg Agg g-3'.(2)
3'-Agg AgA Agg Agg Agg Cgg Agg Agg Agg g-5' (3)
```

Schema 1. Oligonucleotide sequences and possible pairing schemes conceivably take place in the cellular DNA unwound during replication, thus augmenting by a second function the biological efficacy of this oligonucleotide.

One will already retain from the gel pattern shown in Figure 1 that, in the presence of just 0.1 M LiCl in the incubation medium, when present alone at low concentration (right hand lane, marked 0 nM dimer concentration), oligonucleotide 28(C) exists under two forms of related mobility. We tentatively assign them to the *unwound* form, S_3, and the reverse-Hoogsteen *hairpin*, P_3. Note that other pairing schemes for an alternating G,A sequence are available (*45, 46*), but in this review we do not wish to debate on the nature of this and of the other species identified in gels. At higher concentrations of the TFO, a *duplex* is obtained, D_{33}, supposedly *parallel* and Hoogsteen bonded. No higher order oligomer is seen (left hand lane in Figure 1). The dissociation constant of duplex D_{33} has already been determined and is equal to 80 nM; the dissociation constant of the triple helix and the TFO is of the same order of magnitude (*42*).

In the presence of magnesium in the incubation medium, when alone (right hand lane in Figure 2), oligonucleotide 28(C) adopts additional structures: it presumably exists as a non-associated, self-folded hairpin, P_3, as a Hoogsteen-bonded, *parallel duplex, D_{33}*, as a *parallel tetraplex, Q3* and as a higher order *aggregate, A_3*.

With the exception of one single nucleotide, oligonucleotide 28(C) has the same sequence, but of opposite polarity, as the purine-rich *strand-2* of the target duplex (Schema 1). An *antiparallel duplex, D_{12}*, presumably reverse-Hoogsteen bonded, forms in the presence of magnesium ions in the incubation medium, when *strand-2*, of opposite polarity to 28(C) is provided (see the other lanes in Figure 2). Interestingly, the antiparallel duplex forms at the expense of the parallel associations of the radioactive strand, but not of the hairpin. Indeed, in the presence of magnesium, the non-associated oligo 28(C) is present only as reverse-Hoogsteen hairpin, P_3, (compare the band of high mobility in Figure 2 to the two bands of highest mobility in Figure 1). This hairpin is not driven into duplex association with the strand of opposite polarity, since both the hairpin and the antiparallel duplex are based on the same type of interactions (reverse-Hoogsteen); the duplex has more base pairs but the hairpin takes advantage of the entropy term.

Note that the electrophoresis gels in Figures 1 and 2 contain magnesium in order to stabilize the structures under the non-equilibrium conditions of migration. The presence of magnesium in the gel but not in the incubation medium is not sufficient to induce formation of an antiparallel duplex between oligonucleotide 28(C) and 28(+) (not shown). Refraining from generalizations, at least as far as the species generated by oligonucleotide 28(C) are concerned, magnesium in the gel did not induce formation of associations other than those already present in the incubation medium.

Some generalizations

One obtains several pieces of general information by carefully examining Figures 1 and 2:

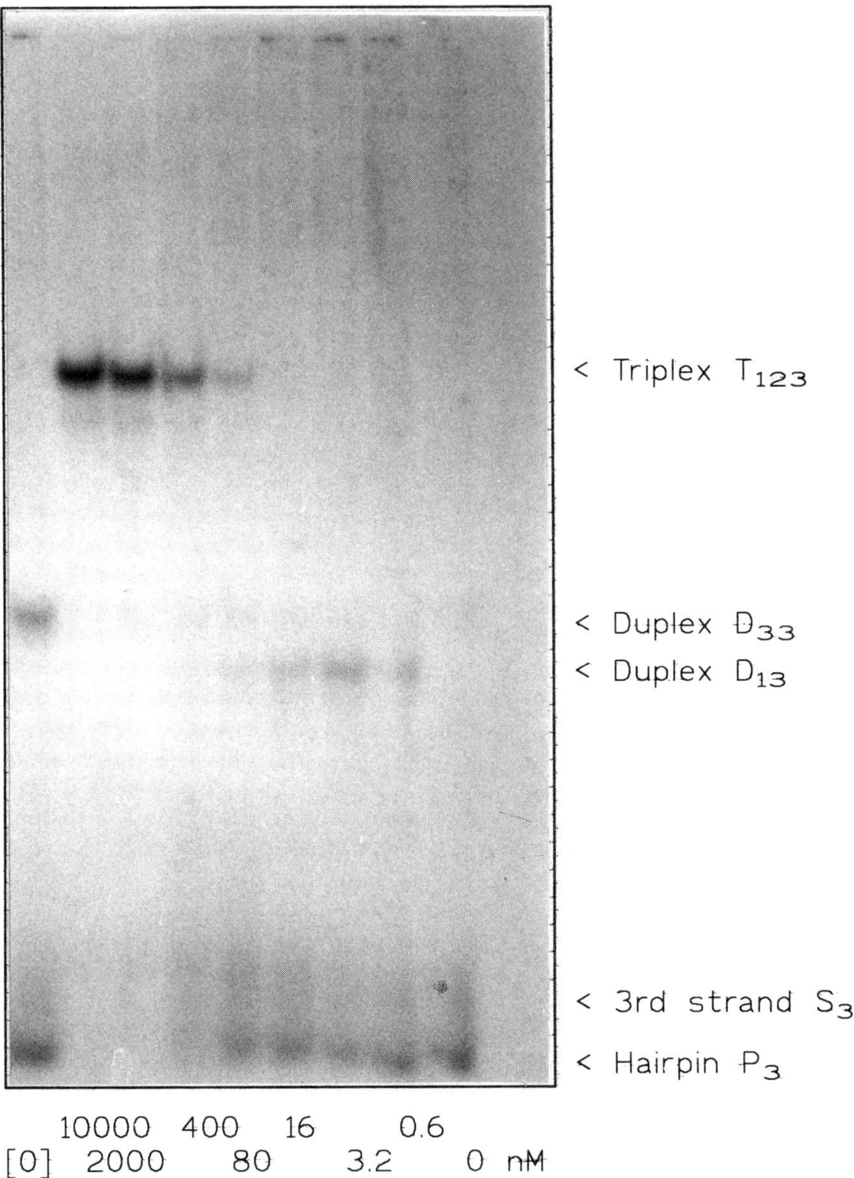

Figure 1. Titration of 5×10^{-10} M, ^{32}P-labelled oligonucleotide 28(C) with unlabelled target duplex, D_{12}, of increasing concentrations, as indicated beneath each lane. Lane [O] contains free oligonucleotide 28(C) at 30 nM strand concentration. Hybridization was performed in 0.1 M LiCl and the 20% polyacrylamide electrophoresis gel was run in TBE buffer containing 10 mM $MgCl_2$. The temperature was 4° C throughout the experiment. The bands corresponding to single-stranded unwound, S_3, hairpin, P_3, imperfect duplex, D_{13} and triple-helix species, T_{123} are indicated. Technical details as in (42). For symbols, see Schema 1.

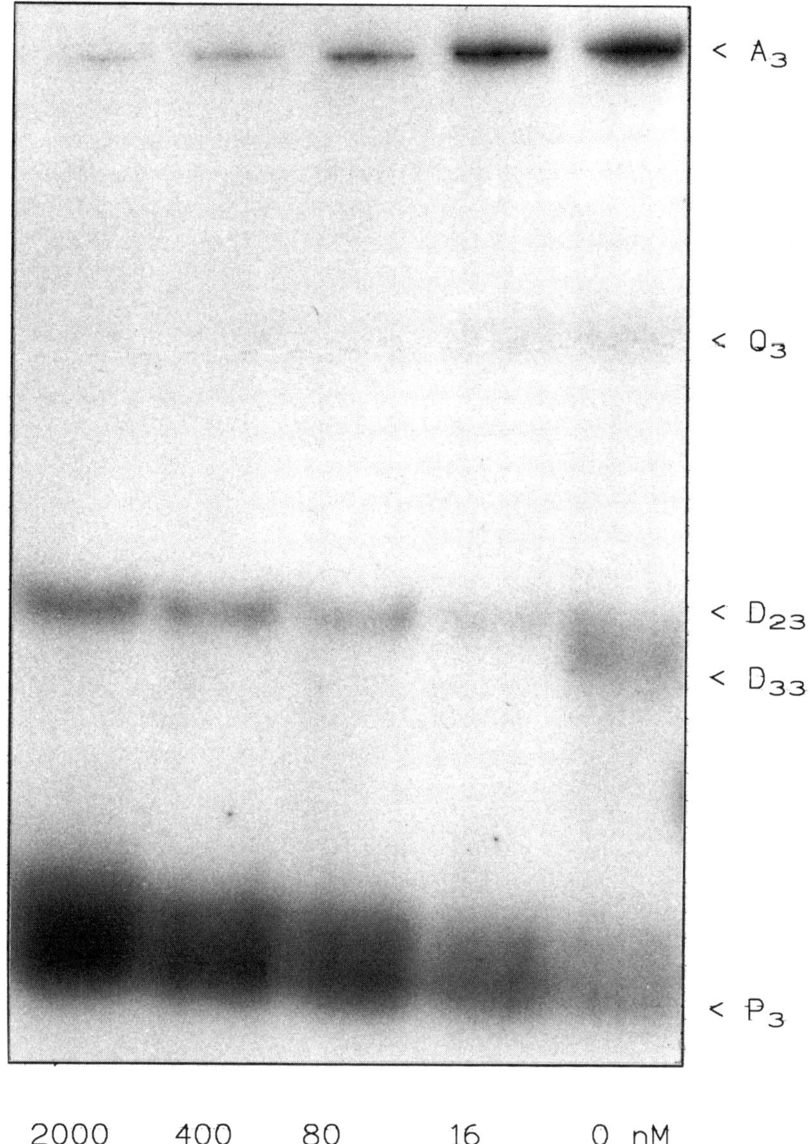

Figure 2. Titration of 5×10^{-10} M, ^{32}P-labelled oligonucleotide 28(C) (*strand-3*) with unlabelled *strand-2*, of increasing concentrations, as indicated. Hybridization was performed in 10 mM $MgCl_2$, and the 20% polyacrylamide electrophoresis gel was run in TBE buffer containing 10 mM $MgCl_2$. The temperature was 4° C throughout the experiment. The bands corresponding to hairpin, P_3, parallel duplex, D_{33}, antiparallel duplex, D_{23}, parallel tetraplex, $Q3$ and highly aggregated species, A_3 are indicated. Technical details as in (*42*). For symbols, see Schema 1.

- Formation of a triple helix by a G,A-rich TFO does take place in the absence of divalent ions in the incubation medium. Titration of the TFO with excess target duplex leads, within a well-defined concentration range, to a disproportionation reaction that generates an imperfect Watson-Crick duplex, D_{13}.

- When dilute, in the presence of lithium ions, the single-stranded G,A-rich TFO is present as an unstructured chain, S_3 and as a hairpin, P_3 (hairpin formation is a unimolecular folding reaction). In the presence of magnesium, only the hairpin form, P_3, is present. Note that structures D_{13} and P_3 are consequences of the particular sequence of the TFO.

- A parallel Hoogsteen homoduplex, D_{33}, *does* form in the *absence of divalent ions* (or monovalent ions other than lithium) by a bimolecular reaction involving identical strands, when the concentration is above the dissociation constant of the duplex.

- The parallel homoduplex, D_{33}, forms at the expense both of the hairpin, P_3, and of the higher order associations, A_3. It thus emerges that these higher order associations are *reversible*. It also follows that the hairpin, P_3, based on a reverse-Hoogsteen stem, with several inherent mismatches, is less stable than the 28 bp parallel duplex, D_{33}, stabilized by direct-Hoogsteen bonds.

- In the presence of lithium ions and above the K_d of homodimerization no parallel associations of the oligonucleotide of an order higher than the duplex D_{33} are formed.

- The *antiparallel* duplex, D_{23}, composed of strands of identical sequence but of opposite polarity, *does not* form in the absence of divalent (or other specific) ions in the incubation medium. Note that formation of both D_{23}, and D_{13}, by disproportionation-type reactions with cellular DNA *in vivo,* might be relevant to the augmentation of the biological effect of the TFO.

- In the presence of *magnesium* ions in the incubation medium, there are traces of *parallel tetraplex, Q_3.*, but not of antiparallel tetraplex, even though the two strands of appropriate polarity are available. On the contrary, in the presence of only *lithium* ions in the incubation medium, we demonstrated the potential formation of *antiparallel* tetraplex by our sequence (not shown). The lithium ion appears therefore to oppose formation of the *parallel* but not of the *antiparallel* tetraplex.

Altogether, the behavior of the TFO is complex, and there are many possibilities of self-associations and non-triple-helix associations with the target. All these compete with the formation of the triple helix itself, so that it is difficult to assess which of the species is the biologically active one!

ACKNOWLEDGMENTS

The author is grateful to E. Taillandier, C. Malvy and S. Fermandjian for provision of excellent facilities. Thanks are due to Dr M. Ouali for stimulating discussions.

REFERENCES

1. Cheng, Y. K. and Pettitt B. M. (1992). Stabilities of double- and triple- stranded helical nucleic acids. *Prog. Biophys. Mol. Biol.* **58**, 225-257.
2. Piriou, J. M., Ketterle, C., Gabarro-Arpa, J., Cognet, J. A. and Le Bret, M. (1994). A database of 32 DNA triplets to study triple helices by molecular mechanics and dynamics. *Biophys. Chem.* **50**, 323-343.
3. Plum, G. E., Pilch, D. S., Singleton, S. F. and Breslauer, K. J. (1995). Nucleic acid hybridization: triplex stability and energetics. *Ann. Rev. Biophys. Biomolec. Struc.* **24**, 319-350.
4. Lavery, R., Zakrzewska, K., Sun, J.-S. and Harvey, S. C. (1992). A comprehensive classification of nucleic acid structural families based on strand direction and base pairing. *Nucleic Acids Res.* **20**, 5011-5016.
5. Pilch, D. S., Levenson, C. and Shafer, R. H. (1991). Structure, stability, and thermodynamics of a short intermolecular purine-purine-pyrimidine triple helix. *Biochemistry* **30**, 6081-6087.
6. Orson, F. M., Thomas, D. W., McShan, W. M, Kessler, D. and Hogan, M. E. (1991). Oligonucleotide inhibition of IL2Rα mRNA transcription by promoter region collinear triplex formation in lymphocytes. *Nucleic Acids Res.* **19**, 3435-3441.
7. Sun, J.-S., De Bizemont, T., Duval-Valentin, G., Montenay-Garestier, T. and Hélène, C. (1991). Extension of the range of recognition sequences for triple helix formation by oligonucleotides containing guanines and thymines. *C.R. Acad. Sci. Paris* ser. III **313**, 585-590.
8. Ouali, M., Bouziane, M., Ketterle, C., Gabarro-Arpa, J., Auclair, C. and Le Bret, M. (1996) A molecular mechanics and dynamics study of alternate triple-helices involving the integrase-binding site of HIV-1 virus and oligonucleotides having a 3'-3' internucleotide junction. *J. Biomol. Struct. Dyn.* **13**, 835-853.
9. Ouali, M., Letellier, R., Sun, J.-S., Akhebat, A., Adnet, F., Liquier, J. and Taillandier, E. (1993) Determination of G*G.C triple-helix structure by molecular modeling and vibrational spectroscopy. *J. Am. Chem. Soc.* **115**, 4264-4270.
10. Scaria, P. V., Will, S., Levenson, C. and Shafer, R. H. (1995). Physicochemical studies of the d(G3T4G3).d(G3A4G3).d(C3T4C3) triple helix. *J. Biol. Chem.* **270**, 7295-7303.
11. Porumb, H., Dagneaux, C., Letellier, R., Malvy, C. and Taillandier, E. (1994). Triple-helices targeted to the polypurine tract of a murine retrovirus. *Gene* **149**, 101-107.
12. Kennard, O. and Salisbury, S. A. (1993). Oligonucleotide X-ray structures in the study of conformation and interactions of nucleic acids. *J. Biol. Chem.* **268**, 10701-10704.
13. Feigon, J., Koshlap, K. M. and Smith, F. W. (1995). 1H NMR spectroscopy of DNA triplexes and quadruplexes. *Meth. Enzymol.* **261**, 225-255.
14. Vlieghe, D., Van Meervelt, L., Dautant, A., Gallois, B., Precigoux, G. and Kennard, O. (1996). Parallel and antiparallel $(G.GC)_2$ triple helix fragments in a crystal structure. *Science* **273**, 1702-1705.
15. Liquier, J., Coffinier, P., Firon, M. and Taillandier, E. (1991). Triple helical polynucleotidic structures: sugar conformations determined by FTIR spectroscopy. *J. Biomol. Struct. Dyn.* **9**, 437-445.
16. Raghunathan, G., Miles, H. T. and Sasisekharan, V. (1993). Symmetry and molecular structure of a DNA triple helix: $d(T)_n.d(A)_n.d(T)_n$. *Biochemistry* **32**, 455-462.

30

17. Radhakrishnan, I. and Patel, D. J. (1993). Solution structure of an intramolecular purine.purine.pyrimidine DNA triplex. *J Amer. Chem. Soc.* **115**, 1615-1617.
18. Radhakrishnan, I. and Patel, D. J. (1993). Solution structure of a purine:purine:pyrimidine DNA triplex containing G.GC and T.AT triples. *Structure* **1**, 135-152.
19. Radhakrishnan I, de los Santos, C, and Patel D. J. (1993). Nuclear magnetic resonance structural studies of A.AT base triple alignments in intramolecular purine.purine.pyrimidine DNA triplexes in solution. *J. Mol. Biol.* **234** 188-97.
20. Radhakrishnan, I. and Patel, D. J. (1994). Hydration sites in purine.purine.pyrimidine and pyrimidine.purine.pyrimidine DNA triplexes in aqueous solution. *Structure* **2**, 395-405.
21. Nunn, C. M., Trent, J. O. and Neidle, S. (1997). A model for the [C$^+$-GxC]n triple helix derived from observation of the C$^+$-GxC base triplet in a crystal structure. *FEBS Lett.* **416**, 86-89.
22. Koshlap, K. M., Schultze, P., Brunar, H., Dervan, P. B. and Feigon, J. (1997). Solution structure of an intramolecular DNA triplex containing an N7-glycosylated guanine which mimics a protonated cytosine. *Biochemistry* **36**, 2659-2668.
23. Liu, K., Sasisekharan, V., Miles, H. T. and Raghunathan, G. (1996). Structure of Py.Pu.Py DNA triple helices. Fourier transforms of fiber-type x-ray diffraction of single crystals. *Biopolymers* **39**, 573-589.
24. Dagneaux, C., Gousset, H., Schyolkina, A. K., Ouali, M., Letellier, R., Liquier, J., Florentiev, V. L. and Taillandier, E. (1996). Parallel and antiparallel A*A-T intramolecular triple helices. *Nucleic Acids Res.* **24**, 4506-4512.
25. Ouali, M., Gousset, H., Geiguenaud, F., Liquier, J., Gabarro-Arpa, J., Le Bret, M. and Taillandier, E. (1997). Hydration of the dTn.dAn*dTn parallel triple helix: a Fourier transform infrared and gravimetric study correlated with molecular dynamics simulations. *Nucleic Acids Res.* **25**, 4816-4824.
26. Fang, Y., Wei, Y., Bai, C., Tang, Y, Lin, S. B. and Kan, L. S. (1997). Hydrated water molecules of pyrimidine/purine/pyrimidine DNA triple helices as revealed by FT-IR spectroscopy: a role of cytosine methylation. *J. Biomol. Struct. Dyn.* **14**, 485-493.
27. White, A. P. and Powell, J. W. (1995). Observation of the hydration-dependent conformation of the (dG)20.(dG)20.(dC)20 oligonucleotide triplex using FTIR spectroscopy. *Biochemistry* **34**, 1137-1142.
28. Laughton, C. A., and Neidle, S. (1992). Prediction of the structure of the Y$^+$.R$^-$.R$^+$ -type DNA triple helix by molecular modelling. *Nucleic Acids Res.* **20**, 6565-6541.
29. Cheng, Y. K. and Pettitt, B. M. (1995). Solvent effects on model d(CG.G)7 and d(TA.T)7 DNA triple helices. *Biopolymers* **35**, 457-473.
30. Weerasinghe, S., Smith, P. E. and Pettitt, B. M. (1995). Structure and stability of a model pyrimidine-purine-purine DNA triple helix with a GC.T mismatch by simulation. *Biochemistry* **34**, 16269-16278.
31. van Gunsteren, W. F., Luque, F. J., Timms, D. and Torda, A. E. (1994). Molecular mechanics in biology: from structure to function, taking account of solvation. *Annu. Rev. Biophys Biomol. Struct.* **23**, 847-863.
32. Cheng, A. J. and Van Dyke, M. W. (1994). Oligodeoxyribonucleotide length and sequence effects on intermolecular purine-purine-pyrimidine triple-helix formation. *Nucleic Acids Res.* **22**, 4742-4747.
33. Xodo, L. E., Pirulli, D. and Quadrifoglio, F. (1997). A kinetic study of triple-helix formation at a critical R*Y sequence of the murine c-Ki-ras promoter by (A,G)- and (G,T) oligonucleotides. *Eur. J. Biochem.* **248**, 424-432.

34. Paes, H. M. and Fox, K. R. (1997). Kinetic studies on the formation of intermolecular triple helices. *Nucleic Acids Res.*, **25**, 3269-3274.
35. Cain, R. J. and Glick, G. D. (1998). Use of cross-links to study the conformational dynamics of triplex DNA. *Biochemistry*, **37**, 1456-1464.
36. Singleton, S. F. and Dervan, P. B. (1993). Equilibrium association constant for oligonucleotide-directed triple helix formation at single DNA sites: linkage to cation valence and concentration. *Biochemistry* **32**, 13171-13179.
37. Musso, M. and Van Dyke, M. W. (1995). Polyamine effects on purine-purine-pyrimidine triple helix formation by phosphodiester and phosphorothioate oligodeoxyribonucleotides. *Nucleic Acids Res.* **23**, 2320-2327.
38. Cheng, A. J. and Van Dyke, M. W. (1993). Monovalent cation effects on intermolecular purine-purine-pyrimidine triple-helix formation. *Nucleic Acids Res.* **21**, 5630-5635.
39. Gresh, N. and Pullman, B. (1986). A theoretical study of the selective entrapment of alkali and ammonium cations between guanine tetramers. *Int. J. Quantum. Chem.* **12**, 49-56.
40. Blume, S. W., Guarcello, V., Zacharias, W. and Miller, D. M. (1997). Divalent transition metal cations counteract potassium-induced quadruplex assembly of oligo(dG) sequences. *Nucleic Acids Res.*, **25**, 617-625.
41. Jarayam, B. and Beyeridge, D. L. (1996). Modeling DNA in aqueous solutions: theoretical and computer simulation studies on the ion atmosphere of DNA. *Annu. Rev. Biomol. Struct.* **25**, 367-394.
42. Porumb, H., Gousset, H., Letellier, R., Salle, V., Briane, D., Vassy, J., Amor-Gueret, M., Israël, L. and Taillandier, E. (1996). Temporary ex vivo inhibition of the expression of the human oncogene HER2 (NEU) by a triple helix-forming oligonucleotide. *Cancer Res.*, **56**, 515-522.
43. Chandler, S. P. and Fox, K. R. (1996). Specificity of antiparallel triple helix formation. *Biochemistry*, **35**, 15038-15048.
44. Porumb, H., Gousset, H. and Taillandier, E. (1998). Triple helix-forming oligonucleotides which make imperfect Watson-Crick duplexes that compete with the creation of the triplex. *Electrophoresis*, in press.
45. Huertas, D., Bellsolell, L., Casasnovas, J. M., Coll, M. and Azorin, F. (1993). Alternating d(GA)n DNA sequences form antiparallel stranded homoduplexes stabilized by the formation of G.A base pairs. *EMBO J.*, **12**, 4029-4038.
46. Shiber, M. C., Braswell, E. H., Klump, H. and Fresco, J. R. (1996). Duplex-tetraplex equilibrium between a hairpin and two interacting hairpins of d(A-G)10 at neutral pH. *Nucleic Acids Res.*, **24**, 5004-5012.

3 THERMODYNAMIC STATE DIAGRAMS OF OLIGONUCLEOTIDE TRIPLE HELICES

G. Eric Plum and K. J. Breslauer

SUMMARY

Understanding of the cellular role of nucleic acid triple helices and utilization of triple-helix forming oligonucleotides in biotechnology, diagnostics, and therapeutics depend on development of an understanding of triple-helix formation as a function of the nucleic acid components and solution conditions. This article reviews developments in nucleic acid triple-helix thermodynamics with emphasis on the construction and interpretation of state diagrams as a means of characterizing the complex behavior of triple-helix forming oligonucleotides. [1]

INTRODUCTION

Triple-helical nucleic acid structures were first identified in the 1950s and 1960s. After a flurry of research characterizing polymer triple helices, activity declined due to the apparent lack of biological function. The resurgence of interest in triple helices stems from the identification of a possible biological role and potential biotechnology applications. Sufficiently long stretches of purine•pyrimidine duplex with mirror symmetry can disengage and then refold as a triple helix with an excluded pyrimidine single strand. This so-called H-DNA may be involved in cellular function including control of gene expression (2-5). The sequence-specific association of an oligonucleotide probe to form a triple helix with a DNA duplex or a messenger RNA has a variety of potential applications (5-8), including site-specific delivery of chemical functionality, blocking of regulatory or endonuclease cleavage sites, blocking of transcription at a DNA duplex target (antigene), and the blocking of translation on a mRNA target (antisense).

---

[1] This paper is a truncated version of a more extensive review of triple helix thermodynamics (1).

34

Desirable characteristics for triple-helix forming oligonucleotides include high binding affinity, sequence specificity, resistance to degradation by cellular nucleases, bioavailability, and efficient cellular uptake. Affinity and specificity are essentially thermodynamic characteristics. Modifications designed to avoid degradation or provide functionality can affect affinity and specificity. Bioavailability and cellular uptake are critical for therapeutic applications but will not be considered here.

Understanding of the cellular role of triplexes, as well as the rational design of third-strand oligonucleotides and analogues for site-specific recognition and modulation of cellular activity at the level of DNA duplex or messenger RNA, will require an understanding of the stabilities of triple helices in the contexts of their components and of solution conditions. The thermodynamics of triple-helix formation has been reviewed extensively (*1, 4, 9, 10*).

USE OF STATE DIAGRAMS TO EVALUATE THE BEHAVIOR OF NUCLEIC ACID COMPLEXES

The stability of a (pur•pyr)•pyr triple helix is a complex function of temperature, salt concentration, and pH. Under some solution conditions the three-stranded complex is intact. Under different solution conditions, only the duplex form is stable, while under still other conditions only single strands are present. A useful way of assessing this complex behavior is by construction of a state diagram (*11-13*). Because there are more than two parameters defining the triplex and duplex stability, a complete characterization of the system requires a multidimensional state diagram.

Most thermodynamic studies of nucleic acids have relied on uv absorbance spectroscopy and calorimetry (*14-16*). Temperature-dependent uv absorbance monitored melting curves provide a convenient means of assessing the thermal stability T_m of nucleic acid complexes, whether duplex or triplex. Differential scanning calorimetry (DSC) measures the heat capacity as a function of temperature. In addition to direct measurement of the thermodynamic quantities ΔH and ΔCp associated with the transition, DSC provides a measurement of T_m.

Transition curves are created by plotting the T_m's of the various structural transitions of the nucleic acid complex as a function of pH or salt concentration. This exercise allows one to construct a series of two-dimensional subdiagrams of the complete multidimensional state diagram. The curves plotted on the state diagram represent the midpoints of each melting transition, therefore delineating regions of the pH *vs.* salt *vs.* temperature space where different states dominate. When the curves are well separated on the state diagram, they represent the conditions under which an equimolar mixture of the two states is present. Under conditions where the transition curves are not well separated, a more complex mixture will be present. The breadth of the transition region is determined by the enthalpy change associated with the transition. If the $\Delta H°$ for disruption of the cooperative unit is large, as in a polymer, the transitions are sharp. However, for oligonucleotides, the transition region can be quite broad.

The positions of the transition curves on the diagrams will be independent of oligonucleotide concentration only for intramolecular triple helices and polymers. For constructs in which two or three oligonucleotides combine to form the triplex, the transition curves will shift to higher temperature as nucleic acid concentration increases. However, the shapes of the curves should not change significantly in the absence of large heat capacity differences between the various states. Triple-helix constructs containing hairpins may display more complex concentration-dependent behavior, particularly at high salt concentration where self-complementary duplex formation may be favored.

SALT DEPENDENCE OF TRIPLE-HELIX FORMATION

The salt-dependent behavior of triple-helix stability can be quite complex (*10*). This complexity stems in part from the interactions of monovalent cations with the nucleic acid and in part from the monovalent cation concentration dependence of oligovalent cation binding. In the (pur•pyr)•pyr case, additional complexity stems from the influence of pH on the salt dependence of third strand association.

Because the triple helix is highly negatively charged, as is the duplex, cations associate preferentially with the complex relative to the constituent single strands. When the complex is disrupted, the net negative linear charge density on the oligonucleotides decreases. This leads to a release of cations into the solution. An increase in bulk salt concentration results in a stabilization of the complex, due to the unfavorable entropy of mixing associated with the release of counterions. This behavior is manifest in a positive dependence of T_m on salt concentration. The dependence of T_m on salt concentration may be positive, zero, or even negative depending on the number of positively charged bases relative to uncharged bases in the third strand. This behavior has been observed with C^+ residues in the (pur•pyr)•pyr family of duplexes (*10*) and it can be anticipated for any modification that imparts positive charge to the triple helix.

The extent and nature of the association of cations with the triple helix depends on the cation valence. For simple monovalent cations (i.e., Na^+, K^+, etc.) the association with nucleic acids typically is nonspecific. The monovalent cation concentration in the vicinity of the nucleic acid is elevated relative to bulk concentration; there are no apparent differences in binding of the various simple monovalent cations. For divalent and higher valence cations, specific interactions often become important (*17*). The influence of monovalent cation concentration on the affinity of multivalent cations can be significant, with increasing monovalent cation concentration reducing the affinity of oligovalent cations (*18*).

pH AND SALT CONCENTRATION DEPENDENCE OF (PUR•PYR)•PYR TRIPLEX FORMATION

In the (pur•pyr)•pyr family, association of the third strand is accompanied by protonation of the third-strand cytosine residues. When cytosine residues are present

36

in the Hoogsteen pairing strand, this protonation event results in a dependence of the three-stranded complex's stability on pH. The pH-dependent behavior of the (pur•pyr)•pyr family of triple helices may be quite complex (*10*).

The sensitivity of (pur•pyr)•pyr triplex stability to pH is dependent on the relative number of third strand cytosines, as well as their positions relative to each other and relative to the ends of the third strand. Because the pK_a of cytosine in single-stranded DNA is about 4.3, third strand dissociation at pH above that value is accompanied by release of protons from the protonated dC residues. The apparent pK_a of third-strand association is typically much higher than the intrinsic pK_a of the cytosine (*11, 19*), indicating that the protonation event and folding of the third strand are coupled in some way and/or the local environment of the cytosine residue shifts the intrinsic pK_a to a value higher than the apparent pK_a of the third strand. The intrinsic pK_a of the cytosine proton in the triple helix may depend on its neighbors. While this issue is critical in evaluation of triple-helix energetics, thermodynamic measurements alone cannot address it. The pH-dependent studies of isotope-labeled nucleic acids by nmr (*20*) will be required to understand this issue.

There are three domains of the pH scale in which (pur•pyr)•pyr triple helices display distinct behavior (*11*). Temperature vs. salt concentration (on a logarithmic scale) state diagrams that correspond to these pH regimes are presented in Figure 1. The specific pH values that define these regions depend on the composition of the triple helix.

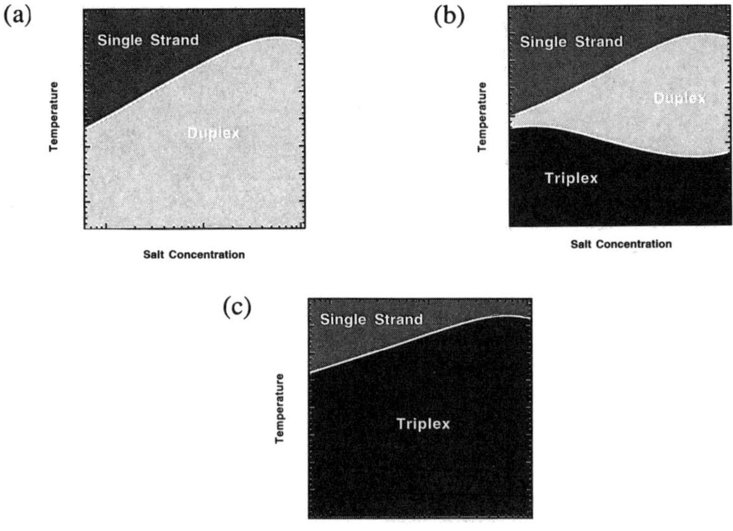

Figure 1. Typical state diagrams for (pur•pyr)•pyr triple helices as a function of pH. (A) High pH (~ 8), (B) intermediate pH, and (C) low pH (~ 5). The variously shaded regions are labeled to indicate which structures predominate under those conditions.

At high pH (Figure 1A), typically in the vicinity of pH 8, one finds no triple helix formed and a single transition curve defines a state diagram consisting of two regions in which the duplex and coil states dominate. One sees the salt dependence typical of the duplex melting transition. The change in slope at high salt concentration is due to significant nonideality effects (lowered water activity) at high salt concentration and is also typical of duplex melting.

At intermediate pH (Figure 1B), the state diagram is defined by two transition curves that divide the diagram into three sections. The lower temperature curve describes the triplex-to-duplex transition and the higher temperature curve the duplex-to-coil transition. Because the duplex-to-coil transition does not depend on pH in this range, this curve is superimposable with the transition curve observed in the high pH state diagram. The melting curves that correspond to the solution conditions on this state diagram are biphasic.

This diagram can be used to evaluate the apparent pK_a of the triple helix. Because the lower temperature transition curve corresponds to an equimolar mixture of triplex and duplex, the apparent pK_a is equal to the pH of the state diagram at the salt concentration and temperature values that fall on the transition curve. A composite state diagram (isotherms plotted on pH *vs.* salt concentration coordinates) from which apparent pK_a values are more readily determined has been presented (*11*).

At low pH (Figure 1C), typically in the vicinity of pH 5, the state diagram is again defined by a single transition curve. This curve describes the transition of triplex directly to single strand without a significant population of duplex intermediate. The transition curve is shifted to higher temperature relative to the duplex-to-coil transition curves observed in the other two diagrams. Because under these conditions the triplex is more thermally stable than the duplex, the pH-independent duplex to coil transition is no longer observed.

The pH dependence of the triple-helix melting temperature (dT_m/dpH) is negative when cytosines are present in the third strand and zero in their absence (*21-23*). The specific value of dT_m/dpH for cytosine-containing triplexes is dependent on the relative number of third-strand cytosines and on their positions relative to each other (*23*). Völker and Klump (*23*) have further shown that two intervening thymidines are required to make the cytosines act independently. This observation is important because it suggests that interactions (presumably, electrostatic in nature) extending beyond the nearest neighbors are important when developing predictive thermodynamic models for third-strand cytosine-containing triple helices. This contrasts with the modeling of the base sequence dependence of DNA duplex thermodynamic properties that has been described very effectively by consideration of only nearest-neighbor base pair interactions (*24, 25*).

Complete thermodynamic characterization of a triple helix as a function of solution conditions has been reported only infrequently. Typically, only T_m's are measured. Such studies make the implicit assumption when comparing triple helices that the enthalpy change ($\Delta H°$) associated with melting is identical. Unfortunately, this is rarely true. Furthermore, the complex dependence of T_m on salt species and concentrations, on the concentration of oligonucleotide, and on pH makes even comparison of T_m's between different triple helices problematic, unless the data were collected under identical solution conditions. The complex salt- and pH-dependent behavior observed for triple-helix formation emphasizes the necessity of construction of multidimensional state diagrams. Only when the state behavior of triple helices is determined over a wide range of solution conditions can one hope to develop a useful predictive understanding of triple-helix thermodynamics.

MODELS OF TRIPLE-HELIX STATE BEHAVIOR

As described above, the salt and pH dependence (other variables also may be included) of T_m can be used to construct a state diagram for triple-helical nucleic acid systems. Several formalisms have been devised to connect the observed state diagram for triple-helix systems to microscopic phenomena. These phenomena, which include strand association/dissociation, release/uptake of counterions, protons, and water molecules, are not necessarily independent of one another. Each model represents a fundamentally different approach to the problem of describing the dependence of triple-helix thermal stability on solution conditions. Each model relies on a number of assumptions and is somewhat limited in scope. Ultimately, when sufficient experimental data are available, some features of the models will require reevaluation.

Hüsler and Klump (26) used matrix methods to derive a partition function for oligonucleotide triple-helix formation. The advantage of such an approach is that explicit sequence information can be exploited and interactions between neighboring protonation sites can be explored. In addition, their partition function can include end effects explicitly. The advantage of the ability to treat each nucleotide triple explicitly is tempered by the large number of parameters required to achieve such detail. It is possible to reduce the number of parameters simply by treating the base triples of one type as identical without regard to sequence. The model as currently formulated does not address counterion uptake/release that can have a significant impact on pH-dependent behavior. Treatment of the salt concentration dependence of triplex stability is necessary for construction of state diagrams.

The model presented by Lavelle and Fresco (27), which describes the salt and pH dependence of T_m, models nucleic acids of variable charge density and triple helices using a modification of the formalism of Record and co-workers (28). The counterion condensation model for nucleic acids, developed by Manning (29), models DNA as an infinite line of uniform charge density. The effective linear charge density (parameterized in the model as the average charge spacing) is one of the primary variables that determine the extent of counterion association, the other being the counterion charge. The effective charge density of the polymer is increased by association of the third strand. Protonation of third-strand cytosine reduces the net charge density. These effects are averaged over the length of the polymer, which eliminates any sequence-dependent information. In long polymers, the loss of sequence-specific effects is not a significant problem because sequence effects will average out. However, in oligonucleotides interactions between neighboring cytosines are likely to be important. A further problem with application of the condensation model to oligonucleotides is the influence of end effects. Olmsted and co-workers (*30, 31*) have demonstrated that the association of counterions to oligonucleotides is overestimated significantly by the condensation model.

A model for describing the state behavior of an intramolecular DNA triple-helix as a function of pH, electrolyte activity, and temperature has been presented by Plum and Breslauer (*11*). The model, based on the Wyman and Gill model for independent structural transitions in proteins (*32, 33*), treats the triplex-to-duplex and duplex-to-coil transitions of the oligonucleotide as conformational transitions associated with uptake or release of some numbers of cations, hydrogen ions, and water molecules. With independently measured enthalpy values, the model successfully fits the entire pH, salt, temperature state diagram. It does not require knowledge of the number of binding sites *a priori* and makes testable predictions of the number of cations, hydrogen ions, and water molecules taken up or released upon change of state. The model is quite flexible and can be extended easily for additional effects such as drug or protein binding. On the other hand, no provision is made for inclusion of sequence information. The model is currently formulated for intramolecular triple helices only. Extension to molecularities greater than 1, however, is not difficult.

FREE ENERGY OF TRIPLEX FORMATION

For the potential applications of sequence-specific recognition of duplex DNA by triple-helix formation, single mismatches must significantly destabilize the three-stranded complex. Sufficient data are not available to assess quantitatively the third-strand binding affinity (free energy) penalty for all possible natural base mismatched as a function of nearest neighbors and solution conditions. However, it is clear that the impact of an interruption in the required base triple motif depends on the triplex family. A single base triple deviation in the center of a (pur•pyr)•pyr triplex formed by binding of a 15-mer reduces the association constant by a factor ranging from 15 to 1500, which corresponds to a $\Delta\Delta G°$ of 1.6-4.3 kcal/mole, depending on the substitution (*34*). A similarly constructed triplex of the (pur•pyr)•pur family was destabilized by a factor ranging from 4 to 200 in the association constant, which corresponds to a $\Delta\Delta G°$ of 0.8–3.2 kcal/mol (*35*). While it is premature to draw a definitive conclusion, these data suggest that the (pur•pyr)•pyr family of triple helices may provide superior discrimination at the single base level.

The nature of the oligonucleotide backbone is an important determinant of triple-helix stability. There are eight possible combinations when DNA or RNA is substituted for each of the three strands in a (pur•pyr)•pyr triple helix while maintaining the nucleotide base sequence. Several reports have appeared of relative stabilities [$\Delta\Delta G°$ (*36-38*) or ΔT_m (*39*)] of (pur•pyr)•pyr triple helices as a function of backbone identity either DNA or RNA and 2'-O-Me-RNA (*38*). In addition to different base sequences and solution conditions, various nucleic acid constructs were used to evaluate backbone effects. These include three oligonucleotides (*37-39*), a hairpin duplex and a linear oligonucleotide third strand (*36*), and a chimeric circular oligonucleotide that provided both pyrimidine strands and an purine-rich oligonucleotide (*38*). Nonetheless, there are significant similarities among the several sets of data. In the aggregate, triple helices with RNA third strands were the most stable, although the all-DNA triplexes were quite stable. With a DNA third strand and RNA purine strand, the triple helix did not form regardless of the pyrimidine strand. Interestingly, the 2'-O-methylation of any one of the strands suppressed the differences in stability among the triplexes (*38*).

ENTHALPY OF TRIPLEX FORMATION

The free energy change ($\Delta G°$) comes from association/dissociation constants (K) that typically are model dependent. These models generally are for simplified reaction schemes and usually ignore numerous important contributions to the equilibrium. These include, but are not limited to, counterion/coion effects, nonideality effects, reorganization of hydrating waters, and water activity effects, etc. Free energy values are sensitive to solution conditions.

Because nonspecific counterion effects are largely due to mixing entropy effects, they are not expected to influence $\Delta H°$ strongly. $\Delta H°$ can be measured directly by calorimetry in a model-independent manner. While $\Delta H°$ is obtainable in a model independent manner, great care must nonetheless be exercised in comparing $\Delta H°$ values. Standard states in biological applications of thermodynamics typically are defined operationally. This leads to difficulties in comparison. The traditional way of overcoming this problem is to define a reference state at sufficiently high temperature for single-stranded structures to be obliterated resulting in a state with thermodynamically negligible differences due to sequence, etc. Comparisons are made in terms of differences in $\Delta H°$ or $\Delta G°$ with respect to this reference state – that is, $\Delta\Delta H°$ or $\Delta\Delta G°$ values are compared. Thus, the high temperature disruption of single-strand structure is advantageous because it results in a relatively well-defined reference state, the influence of which can be subtracted out in the $\Delta\Delta$ comparisons. In low temperature isothermal experiments both initial and final states may differ between oligonucleotides; therefore, even $\Delta\Delta$ values cannot be compared unambiguously.

The enthalpy change associated with disruption of a triple helix into a single strand and a Watson–Crick duplex depends on the base composition, base sequence, length, and to an unknown extent, on solution conditions. Roberts and Crothers (*40*) applied a van't Hoff analysis to evaluate the stability ($\Delta G°$) and its temperature dependence ($\Delta H°$) for 23 (pur•pyr)•pyr triple helices. A set of nearest-neighbor values that allow one to estimate $\Delta H°$ and $\Delta G°$ in 100 mM Na^+ as a function of pH and third-strand nucleotide sequence were derived. This set of rules for predicting triple-helix thermodynamic parameters should provide useful estimates of $\Delta G°$, $\Delta H°$, K, and T_m.

Because the heat of cytosine protonation is significant, $\Delta H°$ measured for cytosine-containing (pur•pyr)•pyr triple-helix formation is expected to be sensitive to protonation equilibria. The correction for this effect is not trivial because $\Delta H_{protonation}$ for dC incorporated into the triple helix is not known and it is not known whether all possible sites are protonated at moderate pH (*11, 41*).

Table 1 is a compilation of measured enthalpy values for DNA triple-helix third strand dissociation [primarily from the (pur•pyr)•pyr family].

While the various constructs and solution conditions make detailed comparisons difficult, one sees that the enthalpy change associated with third-strand dissociation (normalized per base triple) is smaller than that observed for DNA duplex disruption (*24*). In addition, one sees that the model-independent calorimetrically determined enthalpy values are systematically smaller than the van't Hoff enthalpy values, whether determined optically or from the shapes of the differential scanning calorimetry profiles. There are several possible sources of such discrepancies. These include the failure of the two-state assumption upon which the van't Hoff model is founded, buffer effects due to uptake of hydrogen ions released upon triple-helix disruption (*21*), and/or neglect of small $\Delta C_p°$ effects (*50*). As discussed in detail elsewhere (*10*), none of these provides a satisfactory explanation for the discrepancy between the different enthalpy values. Discrepancies between van't Hoff and calorimetric enthalpy values have been observed in a variety of biological systems (*51-53*).

There is further evidence that simple two-state models may not in all cases be adequate to describe triple-helix formation. Kamiya and co-workers (*42*) have studied triple-helix formation at low temperature by isothermal titration calorimetry. A 15-mer homopyrimidine oligonucleotide was mixed with a 23-mer duplex to form a triple helix over a range of temperature between 15 and 35° C. The resulting titration curves (heat per mole of injectant *vs.* molar ratio of titrant to titrate) were fit to a two-state model to obtain values for $\Delta H°$ and K_a ($\Delta G°$). A strong temperature dependence was observed for $\Delta H°$, indicating a large negative heat capacity change, about -1 kcal/mole•K. Interestingly, there was essentially no temperature dependence observed for $\Delta G°$. This observation is inconsistent with the apparently large $\Delta C_p°$ value. The authors reasonably interpret these observations as indicating that the triple-helical complex formation is not simply a two-state process. They further interpret their data in terms of a model based on the work of Eftink and coworkers (*54*), who demonstrated that coupling of equilibria could lead to an apparent $\Delta C_p°$ even when there is no intrinsic $\Delta C_p°$. The Eftink model has recently been employed by Ferrari and Lohman (*55*) to describe the association of single-stranded DNA-binding protein with single-stranded $(dA)_n$ oligomers. The observations of Kamiya and co-workers (*42*) emphasize that single-stranded structures, which may be present at low temperatures, can affect the association of triple helices. Such effects, which may be obliterated by methods that employ high temperatures (i.e., uv melting and DSC), also have been observed to contribute to duplex formation (*56*).

Table 1. Enthalpy change associated with oligonucleotide triple helix disruption

Motif	3rd strand sequence	ΔH^1 (kcal/mol)	Solution Conditions	Method[2]	ref
(UY)Y	dCTCTTCTTTTCTTTC	-5.6*	10 mM acetate/10 mM cacodylate, 0.2 M NaCl, 20 mM MgCl$_2$, pH 4.8, 25° C, *ΔH of association	ITC	42
(UY)Y	dCTCTCTCTCTCT	9.7	10 mM cacodylate, 0.15 M NaCl, 0.005 M MgCl$_2$, pH 7.0	UV$_c$	27
(UY)Y	dCCCTTTTCCC	3.8 5.1 4.2* 3.4**	10 mM cacodylate, 50 mM MgCl$_2$, pH 5.3 *pH 5.5 **2.0 M NaCl, 0 mM MgCl$_2$, pH 5.5	DSC DSC$_s$ UV$_c$ UV$_c$	43 43 44 44
(UY)Y	dTTTTTTTTTT	2.3	10 mM cacodylate, 50 mM MgCl$_2$, 0.1 mM EDTA, pH 5.5	UV$_c$	44
(UY)Y	dCTTCCTCCTCT	7.5 6.6*	0.1 M Na acetate, 50 mM NaCl, 10 mM MgCl$_2$, pH 5.0 * after protonation "correction"	UV$_s$	45
(UY)Y	dCTCTCTCTTT	4.0 5.9 5.2* 4.8*	20 mM cacodylate, 0.1 M NaCl, pH 6.7 20 mM cacodylate, 0.1 M NaCl, pH 4.5 *after subtraction of duplex contribution	DSC DSC$_s$ DSC DSC$_s$	12
(UY)Y	dCTCTTCTTTC	5.5* 6.6* 6.9* 5.7* 5.7 6.5	0.1 M Na acetate, 50 mM NaCl, 10 mM MgCl$_2$, pH 5.0 *after subtraction of 74 kcal/mol contribution of duplex 0.1 M Na acetate, 50 mM NaCl, 10 mM MgCl$_2$, pH 6.0 C^m = 5-methylcytosine	DSC DSC$_s$ UV$_c$ UV$_s$ DSC UV$_c$	46 19
(UY)Y	dT$_4$CT$_4$CT$_4$CT$_4$	-2.9 to -4.2*	1.0 M NaCl, various buffers, pH 7.0 *ΔH of association	ITC	21
(UY)Y	dCCTCTTC	2.9	1.0 M NaCl, 10 mM pyrophosphate 0.1 mM EDTA, pH 6.34	DSC	11
(UY)Y	dTTTTTCTCTCTCTCT	2.0 2.0	10 mM phosphate, 200 mM NaCl, pH 6.5 varied solution conditions	DSC quantitative affinity cleavage	22 25
(UY)Y	dT$_{19}$	2.4	0.01 M PIPES, 0.001 M EDTA, pH 7.0	DSC	47
(UY)Y	avg. of 12 triple helices	6.3	0.1 M Na$^+$, 1 mM EDTA (acetate buffer, pH 5.0)	UV$_s$	48
(UY)Y	23 triple helices	4.9(CC) +8.9(TC+CT) +7.4(TT)	0.1 M Na$^+$, 1 mM EDTA (acetate buffer, pH 4.75-6.0) (cacodylate buffer, pH 6.25-7.0)	UV$_s$	40
(UY)U	dGGGAAAAGGG	4.5	10 mM cacodylate, 50 mM MgCl$_2$, pH 7.5	DSC	43
(UY)U/Y	dGGGTTTTGGG	2.4 3.3 1.9	10 mM tris, 50 mM MgCl$_2$, pH 7.5	DSC DSC$_s$ UV$_c$	43 49

1   per third strand base

2   ΔH from calorimetry: DSC, differential scanning calorimetry; DSC$_s$, van't Hoff value from shape of DSC curve; ITC, isothermal titration calorimetry. ΔH from uv-monitored melting: UV$_c$ van't Hoff value from concentration dependence of T$_m$, UV$_s$ van't Hoff value from shape of melting curve.

CHALLENGES ASSOCIATED WITH USE OF TRIPLE-HELIX PROBES

The desire to engineer high affinity binding while retaining exact sequence specificity is to some extent contradictory. A balance must be struck between increasing the binding affinity and toleration of mismatches. Binding affinity can be modulated in the laboratory quite easily by adjustment of the temperature and solution conditions (pH, nature, and concentration of salts). The cellular milieu, however, is not under the experimenter's control. Therefore, application of third-strand association in the cell will require adaptation of the oligonucleotide or analogue probe to hybridize with the target under the conditions imposed by the biological system. In addition, transport into the cell and nuclease resistance are of critical importance to successful application of antisense and antigene strategies.

The strategies that have developed to address the challenges to realization of the promise of sequence-directed triple-helix formation have centered around modification of the oligonucleotide. Modifications to the bases, to the backbone, and topological constraints have been employed to enhance binding affinity. Another strategy to stabilize triple helices is coupling third-strand association to drug binding equilibria or covalent attachment to various intercalators, groove binders, or cationic peptides and polyamines.

Another issue, the importance of which has only recently been appreciated, is the avoidance of self-structure in the oligonucleotide that competes with the desired binding event. At acidic or neutral pH, cytosine-rich oligodeoxyribonuclotides can form a hemiprotonated four-stranded complex referred to as i-DNA, which can compete effectively with triple-helix formation (*57*). In RNA, the equilibrium is shifted away from the i-motif and toward the triple helix (*58*). G-rich oligonucleotides are very prone to self-structure and -aggregation, which can inhibit triple-helix formation (*59, 60*). Oligonucleotides containing GA repeats also can participate in competing self-structures (*61-63*). Taken together, these problems have severely limited the use and thermodynamic characterization of purine-rich oligonucleotides and (pur•pyr)•pur triplexes. There is some hope for remedy (*64-66*), but a general solution is not yet apparent.

REFERENCES

1. Plum, G. E. (1997). Thermodynamics of oligonucleotide triple helices. *Biopolymers (Nucleic Acid Sciences)* **44**, 241-256.
2. Wells, R. D., Collier, D. A., Hanvey, J. C., Shimizu, M. and Wohlrab, F. (1988). The chemistry and biology of unusual DNA structures adopted by oligopurine•oligopyrimidine sequences. *FASEB J.* **2**, 2939-2949.
3. Htun, H. and Dahlberg, J. E. (1989). Topology and formation of triple-stranded H-DNA. *Science* **243**, 1571-1576.

4. Frank-Kamenetskii, M. D. and Mirkin, S. M. (1995). Triplex DNA structures. *Ann. Rev. Biochem.* **64**, 65-95.

5. Soyfer, V. N. and Potaman, V. N. *Triple-Helical Nucleic Acids.* New York: Springer-Verlag, 1996.

6. Sun, J.-S. and Hélène, C. (1993). Oligonucleotide-directed triple helix formation. *Curr. Opin. Struct. Biol.* **3**, 345-356.

7. Lönnberg, H. and Vuorio, E. (1996). Towards genomic drug therapy with antisense oligonucleotides. *Ann. Med.* **28**, 511-522.

8. Giovannangeli, C. and Hélène, C. (1997). Progress in developments of triplex-based strategies. *Antisense Nucleic Acid Drug Dev.* **7**, 413-421.

9. Cheng, Y.-K. and Pettitt, B. M. (1992). Stabilities of double- and triple-strand helical nucleic acids *Prog. Biophys. Mol. Biol.* **58**, 225-257.

10. Plum, G. E., Pilch, D. S., Singleton. S. F. and Breslauer, K. J. (1995). Nucleic acid hybridization: triplex stability and energetics. *Ann. Rev. Biophys. Biomol. Struct.* **24**, 319-350.

11. Plum, G. E. and Breslauer, K. J. (1995). Thermodynamics of an intramolecular DNA triple helix: a calorimetric and spectroscopic study of the pH and salt dependence of thermally induced structural transitions. *J. Mol. Biol.* **248**, 679-695.

12. Völker, J., Botes, D. P., Lindsey, G. C. and Klump, H. H. (1993). Energetics of a stable intramolecular DNA triple helix formation. *J. Mol. Biol.* **230**, 1278-1290.

13. Mills, M., Völker, J. and Klump, H. H. (1996). Triple helical structures involving inosine: there is a penalty for promiscuity. *Biochemistry* **35**, 13338-13344.

14. Puglisi, J. D. and Tinoco, I., Jr. (1989). Absorbance melting curves of RNA. *Methods Enzymol.* **180**, 304-325.

15. Marky, L. A. and Breslauer, K. J. (1987). Calculating thermodynamic data for transitions of any molecularity from equilibrium melting curves. *Biopolymers* **26**, 1601-1620.

16. Plum, G. E., Breslauer, K. J. and Roberts, R. W. "Thermodynamics and kinetics of nucleic acid association/dissociation and folding processes." In *Comprehensive Natural Products Chemistry.* Oxford, England: Elsevier Science, 1998, in press.

17. Thomas, T. and Thomas, T. J. (1993). Selectivity of polyamines in triplex DNA stabilization. *Biochemistry* **32**, 14068-14074.

18. Singleton, S. F. and Dervan, P. B. (1993). Equilibrium association constants for oligonucleotide-directed triple helix formation at single DNA sites: linkage to cation valence and concentration. *Biochemistry* **32**, 13171-13179.

19. Xodo, L. E., Manzini, G., Quadrifoglio, F., van der Marel, G. A. and van Boom, J. H. (1991). Effect of 5-methylcytosine on the stability of triple-stranded DNA–a thermodynamic study. *Nucleic Acids Res.* **19**, 5625-5631.

20. Gaffney, B. L., Kung, P. P., Wang, C. and Jones. R. A. (1995). Nitrogen-15-labeled oligodeoxynucleotides. 8. Use of ^{15}N NMR to probe Hoogsteen hydrogen bonding at guanine and adenine N7 atoms of a DNA triplex. *J. Am. Chem. Soc.* **117**, 12281-12283.

21. Wilson, W. D., Hopkins, H. P., Mizan, S., Hamilton, D. D. and Zon, G. (1994). Thermodynamics of DNA triplex formation in oligomers with and without cytosine bases: influence of buffer species, pH, and sequence. *J. Am. Chem. Soc.* **116**, 3607-3608.

22. Plum, G. E., Park. Y.-W., Singleton, S. F., Dervan, P. B. and Breslauer, K. J. (1990). Thermodynamic characterization of the stability and the melting behavior of

a DNA triplex: a spectroscopic and calorimetric study. *Proc. Natl. Acad. Sci. USA* **87**, 9436-9440.

23. Völker, J. and Klump. H. H. (1994). Electrostatic effects in DNA triple helices. *Biochemistry* **33**, 13502-13508.

24. Owczarzy, R., Vallone, P. M., Gallo, F. J., Paner, T. M., Lane, M. and Benight, A. S. (1997). Predicting sequence-dependent melting stability of short duplex DNA oligomers. *Biopolymers (Nucleic Acid Sciences)* **44**, 217-239.

25. Singleton, S. F. and Dervan, P. B. (1994). Temperature dependence of the energetics of oligonucleotide-directed triple-helix formation at a single DNA site. *J. Am. Chem. Soc.* **116**, 10376-10382.

26. Hüsler, P. L. and Klump, H. H. (1995). Prediction of pH-dependent properties of DNA triple helices. *Arch. Biochem. Biophys.* **317**, 46-56.

27. Lavelle, L. R. and Fresco, J. R. (1995). UV spectroscopic identification and thermodynamic analysis of protonated third strand deoxycytidine residues at neutrality in the triplex $d(C^+\text{-}T)_6{:}[d(A\text{-}G)_6{\bullet}d(C\text{-}T)_6]$; evidence for a proton switch. *Nucleic Acids Res.* **23**, 2692-2705.

28. Record, M. T., Jr., Woodbury, C. P. and Lohman, T. M. (1976). Na^+ effects on transition of DNA and polynucleotides of variable linear charge density. *Biopolymers* **15**, 893-915.

29. Manning, G. S. (1978). The molecular theory of polyelectrolyte solutions with applications to the electrostatic properties of polynucleotides. *Quart. Rev. Biophys.* **11**, 179-246.

30. Olmsted, M. E., Anderson, C. F. and Record, M. T., Jr. (1989). Monte Carlo description of oligoelectrolyte properties of DNA oligomers: range of the end effect and the approach of molecular and thermodynamic properties to the polyelectrolyte limits. *Proc. Natl. Acad. Sci. USA* **86**, 7766-7770.

31. Olmsted, M. E., Anderson, C. F. and Record, Jr., M. T. (1991). Importance of oligoelectrolyte end effects for the thermodynamics of conformational transitions of nucleic acid oligomers: a grand canonical Monte Carlo analysis. *Biopolymers* **31**, 1593-1604.

32. Gill, S. J., Richey, B., Bishop, G. and Wyman, J. (1985). Generalized binding phenomena in an allosteric macromolecule. *Biophys. Chem.* **21**, 1-14.

33. Wyman, J. and Gill, S. J. *Binding and Linkage: Functional Chemistry of Biological Macromolecules*, Mill Valley, CA: University Science Books, pp. 168-178, 1990.

34. Best, C. C. and Dervan, P. B. (1995). Energetics of formation of sixteen triple helical complexes which vary at a single position within a pyrimidine motif. *J. Am. Chem. Soc.* **117**, 1187-1193.

35. Greenberg, W. A. and Dervan, P. B. (1995). Energetics of formation of sixteen triple helical complexes which vary at a single position within a purine motif. *J. Am. Chem. Soc.* **117**, 5016-5022.

36. Roberts, R. W. and Crothers, D. M. (1992). Stability and properties of double and triple helices: dramatic effects of RNA or DNA backbone composition. *Science* **258**, 1463-1466.

37. Han, H. and Dervan, P. B. (1993). Sequence-specific recognition of double helical RNA and RNA•DNA by triple helix formation. *Proc. Natl. Acad. Sci. USA* **90**, 3806-3810.

38. Wang, S. and Kool, E. T. (1995). Relative stabilities of triple helices composed of combinations of DNA, RNA and 2'-O-methyl-RNA backbones: chimeric circular oligonucleotides as probes. *Nucleic Acids Res.* **23**, 1157-1164.

39. Escudé, C., Francois, J.-C., Sun, J.-S., Ott, G., Sprinzl, M., Garestier, T. and Hélène, C. (1993). Stability of triple helices containing RNA and DNA strands: experimental and molecular modeling studies. *Nucleic Acids Res.* **21**, 5547-5553.

40. Roberts, R. W. and Crothers, D. M. (1996). Prediction of the stability of DNA triplexes. *Proc. Natl. Acad. Sci. USA* **93**, 4320-4325.

41. Bartley, J. P., Brown, T. and Lane, A. N. (1997). Solution conformation of an intramolecular DNA triplex containing a nonnucleotide linker: comparison with the DNA duplex. *Biochemistry* **36**, 14502-14511.

42. Kamiya, M., Torigoe, H., Shindo, H. and Sarai, A. (1996). Temperature dependence and sequence specificity of DNA triplex formation: An analysis using isothermal titration calorimetry. *J. Am. Chem. Soc.* **118**, 4532-4538.

43. Scaria, P. V. and Shafer, R. H. (1996). Calorimetric analysis of triple helices targeted to the $d(G_3A_4G_3) \cdot d(C_3T_4C_3)$ duplex. *Biochemistry* **35**, 10985-10994.

44. Pilch, D. S., Brousseau, R. R. and Shafer, R. H. (1990). Thermodynamics of triple helix formation: spectrophotometric studies on the $d(A)_{10} \cdot 2d(T)_{10}$ and $d(C^+_3T_4C^+_3) \cdot d(G_3A_4G_3) \cdot d(C_3T_4C_3)$ triple helices. *Nucleic Acids Res.* **18**, 5743-5750.

45. Manzini, G., Xodo, L. E., Gasparotto, D., Quadrifoglio, F., van der Marel, G. A. and van Boom, I. H. (1990). Triple helix formation by oligopurine-oligopyrimidine DNA fragments. Electrophoretic and thermodynamic behavior. *J. Mol. Biol.* **213**, 833-843.

46. Xodo, L. E., Manzini, G. and Quadrifoglio, F. (1990). Spectroscopic and calorimetric investigation on the DNA triplex formed by d(CTCTTCTTTCTTTTCTTTCTTCTC) and d(GAGAAGAAAGA) at acidic pH. *Nucleic Acids Res.* **18**, 3557-3564.

47. Hopkins, H. P., Hamilton, D. D., Wilson, W. D. and Zon, G. (1993). Duplex and triplex formation with dA_{19} and dT_{19}. Thermodynamic parameters from calorimetry, NMR, and circular dichroism studies. *J. Phys. Chem.* **97**, 6555-6563.

48. Roberts, R. W. and Crothers, D. M. (1991). Specificity and stringency in DNA triplex formation. *Proc. Natl. Acad. Sci. USA* **88**, 9397-9401.

49. Scaria, P. V., Will, S., Levenson, C. and Shafer, R. H. (1995). Physicochemical studies of the $d(G_3T_4G_3) \cdot d(G_3A_4G_3) \cdot d(C_3T_4C_3)$ triple helix. *J. Biol. Chem.* **270**, 7295-7303.

50. Chaires, J. B. (1997). Possible origin of differences between van't Hoff and calorimetric enthalpy estimates. *Biophys. Chem.* **64**, 15-23.

51. Liu, Y. and Sturtevant, J. M. (1995). Significant discrepancies between van't Hoff and calorimetric enthalpies. III. *Biophys. Chem.* **64**, 121-126.

52. Liu, Y. and Sturtevant, J. M. (1995). Significant discrepancies between van't Hoff and calorimetric enthalpies. II. *Protein Sci.* **4**, 2559-2561.

53. Naghibi, H., Tampa, A. and Sturtevant, J. M. (1995). Significant discrepancies between van't Hoff and calorimetric enthalpies. *Proc. Natl. Acad. Sci. USA* **92**, 5597-5599.

54. Eftink, M. R., Anusiem, A. and Biltonen. R. L. (1983). Enthalpy-entropy compensation and heat capacity changes for protein-ligand interactions: general thermodynamic models and data for the binding of nucleotides to ribonuclease A. *Biochemistry* **22**, 3884-3896.

55. Ferrari, M. E. and Lohman, T. M. (1994). Apparent heat capacity change accompanying a nonspecific protein-DNA interaction. Escherichia coli SSB tetramer binding to oligodeoxyadenylates. *Biochemistry* **33**, 12896-12910.

56. Vesnaver, G. and Breslauer, K. J. (1991). The contribution of DNA single-stranded order to the thermodynamics of duplex formation. *Proc. Natl. Acad. Sci. USA* **88**, 3569-3573.

57. Mergny, J. L., Lacroix, L., Han, X., Leroy, I.-L. and Hélène, C. (1995). Intramolecular folding of pyrimidine oligodeoxynucleotides into an i-DNA motif. *J. Am. Chem. Soc.* **117**, 8887-8898.

58. Lacroix, L., Mergny, J. L., Leroy, J. L. and Hélène, C. (1996). Inability of RNA to form the i-motif: implications for triplex formation. *Biochemistry* **35**, 8715-8722.

59. Olivas, W. M. and Maher, L. J. (1995). Competitive triplex/quadruplex equilibria involving guanine-rich oligonucleotides. *Biochemistry* **34**, 278-284.

60. Alunni-Fabbroni, M., Manzini, G., Quadrifoglio, F. and Xodo, L. E. (1996). Guanine-rich oligonucleotides targeted to a critical R . Y site located in the Ki-ras promoter. The effect of competing self-structures on triplex formation. *Eur. J. Biochem.* **238**, 143-151.

61. Noonberg, S. B., Francois, J. C., Garestier, T. and Hélène, C. (1995). Effect of competing self-structure on triplex formation with purine-rich oligodeoxynucleotides containing GA repeats. *Nucleic Acids Res.* **23**, 1956-1963.

62. Shiber, M. C., Braswell, E. H., Klump, H. and Fresco, J. R. (1996). Duplex-tetraplex equilibrium between a hairpin and two interacting hairpins of d(A-G)$_{10}$ at neutral pH. *Nucleic Acids Res.* **24**, 5004-5012.

63. Mukerji, L., Shiber, M. C., Fresco, J. and Spiro, T. G. (1996). A UV resonance Raman study of hairpin dimer helices of d(A-G)$_{10}$ at neutral pH containing intercalated dA residues and alternating dG tetrads. *Nucleic Acids Res.* **24**, 5013-5020.

64. Kandimalla, E. R. and Agrawal, S. (1995). Single strand targeted triplex-formation. Destabilization of guanine quadruplex structures by foldback triplex-forming oligonucleotides. *Nucleic Acids Res.* **23**, 1068-1074.

65. Faruqi, A. F., Krawczyk, S. H., Matteucci, M. D. and Glazer, P. M. (1997). Potassium-resistant triple helix formation and improved intracellular gene targeting by oligodeoxyribonucleotides containing 7-deazaxanthine. *Nucleic Acids Res.* **25**, 633-640.

66. Svinarchuk, F., Cherny, D., Debin, A., Delain, E. and Malvy, C. (1996). A new approach to overcome potassium-mediated inhibition of triplex formation. *Nucleic Acids Res.* **24**, 3858-3865.

4 IMAGING OF TRIPLEXES BY ELECTRON AND SCANNING FORCE MICROSCOPY

Dmitry Cherny

INTRODUCTION

During the last decade numerous attempts have been made to use synthetic oligonucleotides, primarily triple-helix-forming oligonucleotides (TFO; *1-3*), and more recently peptide nucleic acids (PNA; *4, 5*) as tools for exploring DNA structure and for creating various methods for regulation of gene expression. TFOs and PNAs form triplexes with DNA, albeit of differing natures (*6* and *herein*). It was assumed that an unusual parallel triplex, or R-DNA, is assembled as an intermediate structure during specific recombination mediated by RecA protein (*7, 8*). Genome analysis and labeling using various types of triplexes is an important microscopy project with great potential.

Conventional transmission electron microscopy (EM) has proved to be an efficient method in studying recognition of dsDNA by numerous sequence-specific ligands and oligonucleotides in particular (*9*). Scanning force microscopy (SFM) is now becoming widely used for analysis of sequence-specific recognition of dsDNA by several proteins and oligonucleotides (*10*).

Microscopy techniques allow: (a) the visualization of particular features of the specific complexes such as size, shape of the bound ligand, violations of DNA path (e.g., bending and flexibility induced by specific complex formation); (b) the localization of specific complexes over DNA fragments up to several tens of thousands of bp; and (c) determination of rate and affinity constants for the specific complexes and their selectivity. The latter features are essential for the creation of microscopy methods of genome labeling.

There exist a variety of EM techniques for mounting DNA molecules under different salt and pH conditions, which are often close to those needed for the specific complex formation (*11, 12*). Routine viewing of DNA molecules requires enhancement of DNA contrast, usually achieved either by interaction of DNA with uranyl acetate in solution (staining) or covering with heavy metals, e.g., Pt, W, in a vacuum (shadowing). Stained molecules are usually viewed in a dark-field mode and

50

have very high contrast. However, some proteins (streptavidin in particular) are not clearly seen with this procedure. Shadowing significantly enhances the apparent DNA and bound ligand geometrical sizes, though the resolution on the images is less than that of stained samples. Small violations of the DNA double helix, spanning over 10-20 bp, are usually difficult to detect in EM, with minor exceptions – for instance, for bending events which occur even at shorter distances. The minimal size of proteins in a complex with dsDNA, detected with certainty by EM, is about 40-50 kDa. Note that determination of site-specifically bound ligand positions needs analysis of the ensemble of the DNA molecules with their proper alignment, i.e., mapping (13). Taking into account variations in DNA lengths and errors of measurements, routine accuracy of site positioning is about 10-20 bp or less for short (several hundreds or a few thousand bp long) DNA fragments. Specific sites over long pieces of DNA fragments (up to several tens of thousands of bp or more) can be located with an accuracy of a few hundred bp.

Likewise, DNA and triplex formation can be detected by scanning force microscopy. Most SFM techniques use an old EM procedure for imaging DNA molecules, i.e., adsorption to the surface of freshly cleaved mica in the presence of magnesium (11), with slight modifications. Despite the fact that the resolution of biological objects by SFM has not achieved its full potential and is still inferior to that of EM (10, 14-17), SFM observations have particular advantages. First, the captured picture is inherently digital. Second, SFM reveals the topography of the object (or in a strict sense the topography of the object subjected to its interaction with the scanning tip). Third, complexes of DNA with sequence-specific bound ligands can be directly detected by SFM techniques *via* their coupling with small proteins, a fact that does not apply routinely for EM, especially for DNA molecules with highly interwound chains. Lastly and probably most importantly, biological objects can be imaged under liquid (18-20). It is clear that EM and SFM are complementary methods; therefore, studying the specific complexes and triplexes using both approaches in parallel can be very useful.

Here the results of viewing various triplexes both by electron and scanning force microscopies are presented.

IMAGING OF REGULAR TRIPLEXES

The intermolecular triplexes first observed under EM involved circular plasmids containing a (TC)45•(GA)45 insert and poly[d(Tm⁵C)]. In the EM the complexes appeared as a rosette of petals (21). Intermolecular triplexes formed by short 10-20 nt-long TFOs with purine-pyrimidine tracts of dsDNA are not seen directly by EM due to their small size. Biotinylated derivatives of TFOs form site-specific complexes with the plasmid DNAs which can be detected using streptavidin as a marker. Within experimental errors of measurements, the position of the streptavidin beads coincides well with the position of the purine-pyrimidine tract, thus confirming the triplex formation. This was first demonstrated both for the triplexes formed by biotin-5'-dAGGAGGAGGAGA-3' (PyPuPu triplex) and biotin-5'-dTCTCCTCCTCCTCCTTTT-3' (PyPuPy triplex) oligonucleotides with the same 17 bp-long purine·pyrimidine tract (22). Note that complexes that appeared virtually identical were formed and mounted for EM imaging (Figure 1) under different conditions: PyPuPu triplex was assembled under neutral pH and stabilized by Mn^{2+} ions, wherease PyPuPy triplex was formed under acidic pH and stabilized by Mg^{2+} ions. For the case where the purine·pyrimidine tract was three times longer than the length of the TFO used, some DNA molecules had two streptavidin beads within the tract (23). Efficiency of target site detection *via* streptavidin was presumably affected by both TFO length and composition and the length of the linker between the TFO and the biotin, and varied from 40 to 80% (22, 23).

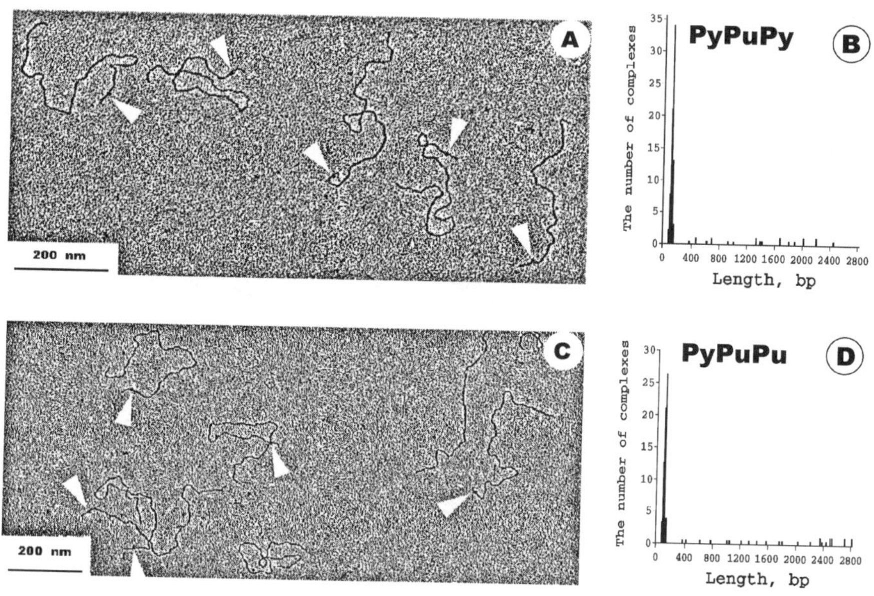

Figure 1. EM imaging of the regular triplexes formed by biotin-5'-dTCTCCTCCTCCTCCTTTT-3' (a) and biotin-5'-dAGGAGGAGGAGA-3' (b) oligonucleotides with plasmid DNA. The complexes are seen as streptavidin "beads", indicated by white arrowheads on the micrographs (A) and (C), respectively. (B) and (D) represent histograms of streptavidin positions obtained from the complexes formed by the oligonucleotides (a) and (b), respectively. Peaks are positioned at 119 ± 15 bp (B) and 128 ± 15 bp (D) from the nearest DNA end, pyrimidine·purine tract is located at 120-137 bp from the same DNA end. © 1993 by Academic Press Limited, London. Modified reproduction, with permission, from (22).

52

For another TFO, 5'-GGAGGAGGAGGAGGGGGAGG-3'-biotin, it was shown that efficiency of triplex formation and detection can attain 70% in the presence of NaCl, though in the presence of KCl this value dropped down to 20% (*24*). It was supposed that the negative effect of KCl on triplex formation was due to the self-association of the TFO in competitive structures such as duplexes and/or tetraplexes. In contrast, efficiency of triplex formation by the same TFO with supercoiled molecules was as high as 90% (Figure 2D), irrespective of the nature of the monovalent cations (*25*).

Similarly, triplex formation can be monitered by SFM. For instance, long structures believed to be triple-stranded DNA formed by either poly(dA) and poly(dT) or poly(dG) and poly(dC) can be revealed by increasing the apparent height of the filaments in respect to that of dsDNA (*26*). Sequence-specific interaction of short, 10-mer, 11-mer and 27-mer 3'-end-biotinylated pyrimidine TFOs with circular

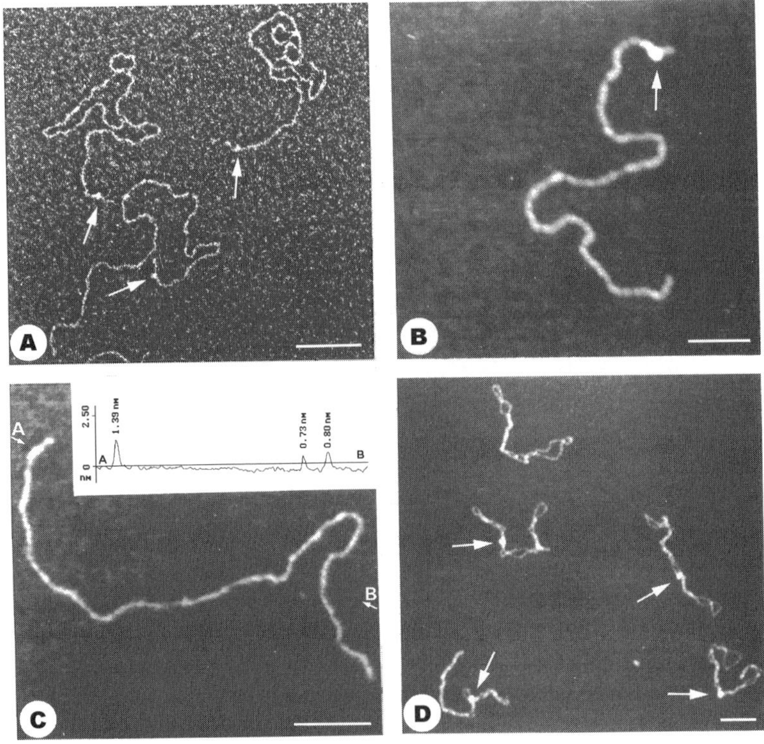

Figure 2. Visualization of the triplex formed by the biotin-5'-GGAGGAGGAGGAGGGGGAGG-3' oligonucleotide with the plasmid DNA in linear (**A**, **B** and **C**) and supercoiled (**D**) forms. (**A**) EM view; (**B**), (**C**) and (**D**) - SFM views. Triplex is detected *via* streptavidin shown by arrows on (**A**), (**B**) and (**D**). On (**C**) - streptavidin was omitted from the reaction and insert shows the profile of the cross section through the A-B line on the photograph. The scale markers indicate 100 nm. Reproduced, with permission, from (*25*).

DNA was easily detected using SFM *via* streptavidin-gold or streptavidin-alkaline phosphatase conjugates with ~10% efficiency (*17*). In order to increase triplex stability all TFOs used were 5'-conjugated with a psoralen. Streptavidin itself is also a relevant marker for triplex detection by SFM with linear (Figure 2B) and supercoiled molecules (Figure 2D). SFM imaging of the regular triplex without streptavidin marker shows that a few DNA molecules have extra corrugations of ~0.4 nm in height within the region encompassing the recognition sequence (Figure 2C). Moreover, SFM imaging of the triplex designed by the G, A-containing TFO tagged with a 74 nt-long oligonucleotide (Figure 3A) also confirms an increase in DNA height of 0.3-0.4 nm. Taken together these results demonstrate that triple-stranded DNA does have increased height relative to that of double-stranded DNA (*25*).

Pyrimidine·purine tracts several tens of bp-long cloned within plasmid DNA can participate in intermolecular triplex formation which is detected in EM by the occurrence of joints between two independent tracts, and not within them (*27, 28*). The yield of the joints is increased upon lowering the pH to 4 and extending the incubation time; however, once formed, they are stable at neutral pH. Both linear and circular DNA molecules are able to form joints. It is noteworthy that no plectonemic problems were caused by triplex formation between pyrimidine·purine tracts in circular DNAs.

Supercoiled DNA with a 60 bp-long pyrimidine·purine tract containing mirror repeats can form unusual conformations under acidic pH, viewed as molecules with a "stem" or "denaturation bubble" (*29*). The occurrence of these structures corresponded well with intramolecular protonated triplex formation (H form). "Stems" were often located at the apexes of the molecules which was interpreted as indicative of triplex-associated kink formation; however, local weakening or increasing of flexibility could also result in apically-positioned "stems".

In most cases TFOs play a negative role, inhibiting biochemical processes by preventing the normal interaction between the TFO-targeted DNA sequence and protein factors. However, oligonucleotides can be designed to simultaneously form a TFO and a hairpin with recognition sequences for regulatory proteins such as transcription factors (*30*). Oligonucleotide 5'-GGAGGAGGAGGAGGGGGAGG-3', capable of forming a stable triplex (*24*), was 3'-tagged to the 74 nt sequence 5'-TTT<u>GCTGACGCAGATGTCCTATATGGACATCCTGTTTTTTACAGGATGT CCATATTAGGACATCTGCGTCAGC</u>-3' forming a 33 bp-long hairpin (under-lined) by itself. The triplex is clearly detected in EM *via* a short DNA fragment uniquely located over the entire length of the plasmid DNA (Figure 3A). The measured length of the protruding DNA is equal to 29 ± 4 bp (mean ± S. D.) and coincides well with the hairpin length. At least 80% of DNA molecules have such protrusions, indicating that the efficiency of triplex formation cannot be less than this value. Protruding DNA is often seen in a perpendicular direction with respect to the DNA axis and is perceived as lying on the carbon film surface. The same hairpin-forming oligonucleotide 3'-conjugated with the same TFO but with an 8-nt-long linker displayed higher flexibility and was often seen parallel to the DNA axis.

54

SFM examination clearly demonstrates the occurrence of the protruding DNA in a correct position (Figure 3B, C). Likewise, protruding DNA is seen in a perpendicular direction with respect to the plasmid DNA axis. It was found that the short duplex protrudes in various directions either in the plane (Figure 3B) or out of the plane (Figure 3C). In the former case the length of the protruding DNA is close to the expected value of 10 nm, whereas in the latter case the apparent height of the protrusion is only ~1.5 nm – which means that the height of the protruding DNA *per se* is about 0.5-0.6 nm, and thus markedly less than the expected value of its length (10 nm). This discrepancy is probably due to thermal fluctuations of the hairpin which are significantly affected by the tip oscillations in the Tapping mode, resulting in a decrease in the apparent height of the protrusion.

Triplex can be formed in such a way that a T, G-containing oligonucleotide binds to oligopurine sequences which alternate between the two strands of duplex DNA (*1, 3, 6*). Indeed, oligonucleotides containing G and T may adopt either a parallel or an antiparallel orientation with respect to the oligopurine target, depending upon the

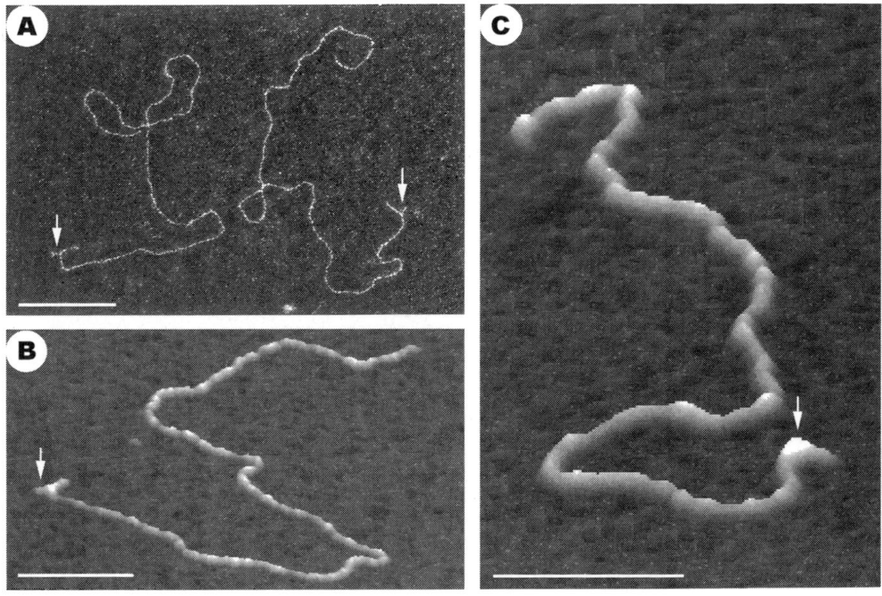

Figure 3. Images of the triplex formed by the oligonucleotide 5'-GGAGGAGGAGGAGGGGGAGG-3' tagged with a 74 nt-long oligonucleotide forming a 33 bp-long hairpin. (A) EM view, (B) and (C), SFM views. Protruded DNA indicated by arrows and seen as lying in the plane of imaging (A and B) or out of the plane (C). The scale markers indicate 100 nm. Reproduced, with permission, from (*25*).

sequence and, in particular, on the number of 5'-GpT-3' and 5'-TpG-3' steps (*31*). Following this rationale an oligonucleotide, 5'-GGTTTTTGTGT-3', was designed to form parallel triplets with the 5'-GGAAAA-3' segment and antiparallel ones with the 5'-AGAGA-3' sequence. Association of the oligonucleotide with the duplex DNA resulted in formation of a poorly stable triplex after long incubation. To augment the stability of the triplex, the oligonucleotide was conjugated with the intercalating chromophore oxazolopyridocarbazole (OPC; Figure 4A), and the alternate-strand triplex was clearly detected by EM (Figure 4B) using streptavidin as a marker (*32*).

IMAGING OF TRIPLEXES FORMED BY PNA

Homopyrimidine PNAs bind efficiently and highly selectively with the target sequence within long pieces of dsDNA *via* strand displacement and formation of PNA$_2$/DNA triplex (4,5). Interaction of PNA T$_{10}$-LysNH$_2$ with a (dA)$_{98}$·(dT)$_{98}$ target within plasmid DNA results in a full displacement of the non-complementary DNA strand, i.e., (dT)$_{98}$, and appearance of eye-looped molecules with different thicknesses for DNA strands within loops (Figure 5A, B). Together with the size and position of the loops, this demonstrated a highly specific interaction of the PNA with the target sequence (*33*).

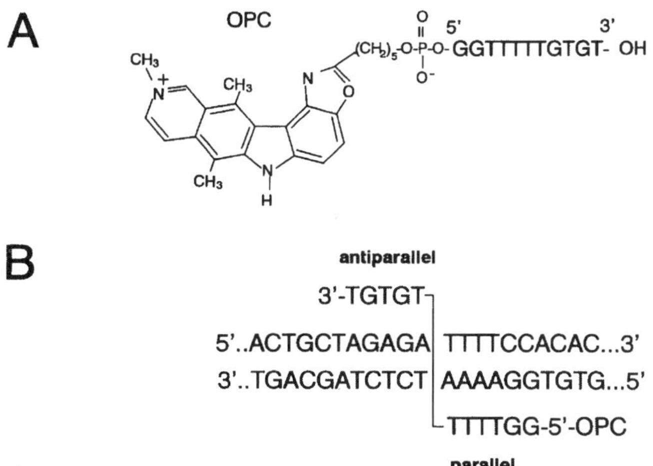

Figure 4. Structure of triple-helix forming oligonucleotides conjugated with the intercalating chromophore oxazolopyridocarbazole (**A**) and schematic presentation of the alternate-strand triplex (**B**). Reproduced, with permission, from (*32*).

56

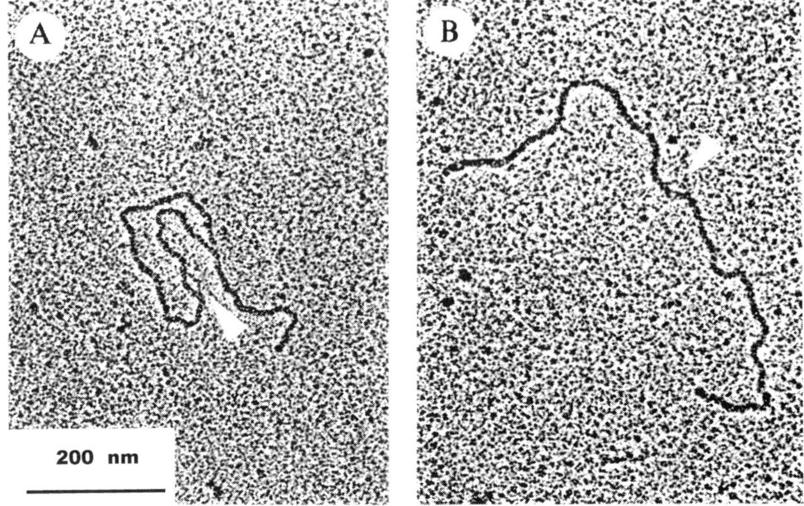

Figure 5. Electron microscopy visualization of DNA unwinding by PNA. Arrowheads point to the unwound regions (loops) resulted from the interaction of PNA T_{10}-LysNH$_2$ with (dA)$_{98}$·(dT)$_{98}$ insert. Copyright 1993, National Academy of Sciences, USA. Modified reproduction, with permission, from (*33*).

Short target sequences, e.g., 10 bp-long, can be detected in EM *via* their interaction with biotinylated derivatives of PNA decamers (Figure 6) and streptavidin (Figure 7A, 8A) as a marker (*34*). It is noteworthy that PNA, in addition to the triplex formed with the true target site corresponding to the major peak on the histogram of binding (Figure 7B), also forms stable complexes with the sites with one and two mismatches (seen as small satellite peaks on the histogram in Figure 7B). The latter become more pronounced upon increasing either PNA concentration or incubation time or both, consistent with the kinetic mechanism of PNA interaction with dsDNA (*35*). The stoichiometry of the (biotin-PNA)/DNA complex is thought to be similar to that of non-biotinylated PNA − i.e.,

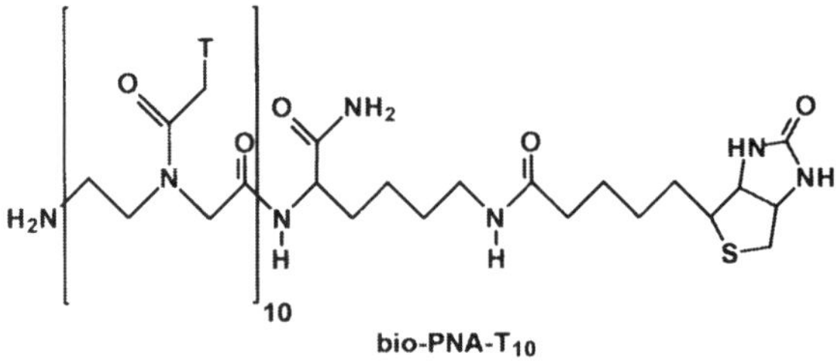

Figure 6. Chemical structure of the biotinylated PNA T_{10}-LysNH$_2$. © 1994, Oxford University Press. Reproduced, with permission, from (*34*).

PNA$_2$/DNA – due to the presence of two side-by-side streptavidin beads in the correct position on some DNA molecules (Figure 7C).

Interaction of PNA with the target site inserted into long linear or supercoiled plasmid DNA results in strong DNA bending, often visualized as kinks within or near the target sequence and detected both by EM and SFM (Figure 8). Kinks can be produced in the same manner as PNA decamers by bisPNAs, in which two PNA decamers connected *via* a flexible linker. The bending angle, i.e., the smallest angle between two DNA arms around the bending point, is ~130° (and independent of the presence of the streptavidin). Analysis of bending angle and the square of end-to-end distance distributions shows that bending is accompanied by increasing DNA flexibility. Special attempts were aimed at checking whether uranyl ions, needed for

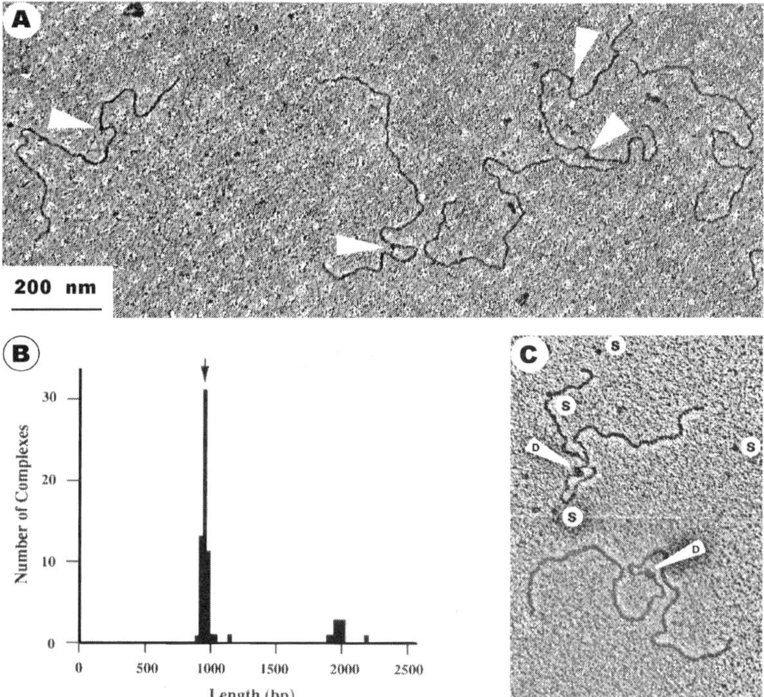

Figure 7. Electron-microscopic visualization of the triplexes formed by biotinylated PNA, bio-PNA-T$_{10}$, with linear DNA. (**A**) Micrograph of DNA/bio-PNA/streptavidin complexes. Streptavidin "beads" mark the position of PNA$_2$/DNA triplexes and are indicated by arrowheads. (**B**) Histogram of distribution of bound streptavidin molecules obtained from 66 DNA molecules. The arrow indicates the position of the target site. (**C**) Micrographs of DNA molecules bearing two side-by-side streptavidin molecules within the target sequence; S and D indicate, respectively, monomers and dimers of streptavidin. © 1994, Oxford University Press. Modified reproduction, with permission, from (*34*).

58

EM mounting onto carbon film, or magnesium ions, needed for SFM mounting onto mica, were responsible for kink formation. It was shown that neither magnesium ions, nor uranyl ions (separately or combined) are responsible for the kinks detected either by EM or SFM (*25*).

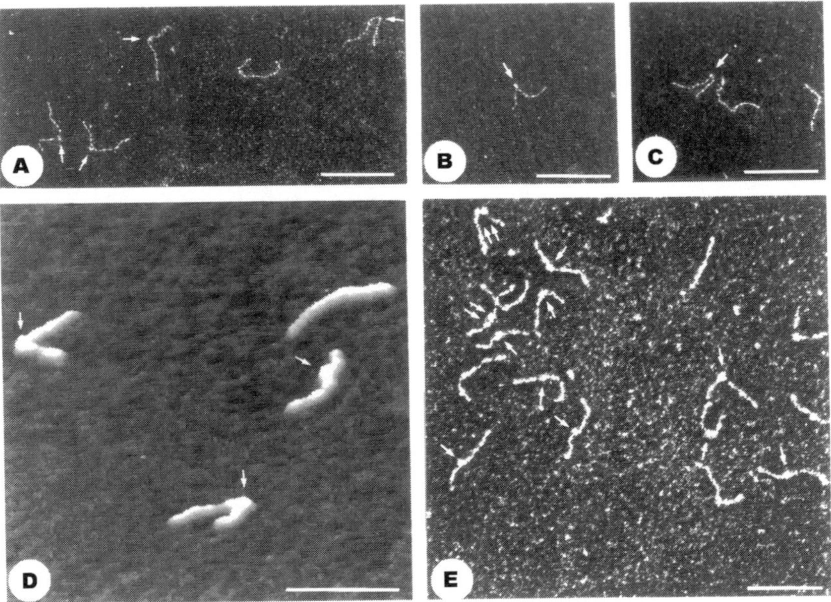

Figure 8. Images of PNA$_2$/DNA triplex formed by bisPNA H-TTJTTJTTTT-Lys-aha-Lys-aha-Lys-TTTTCTTCTT-LysNH$_2$ with DNA. **(A)**, **(B)**, **(C)** and **(E)**, EM views; **(D)** SFM view. Arrows indicate the presence of kinks in the sites of triplex formation. In **(E)** triplex was formed bio-bisPNA; arrows show streptavidin "beads". The scale markers indicate 100 nm. Modified reproduction, with permission, from (*25*).

PNA$_2$/DNA triplex formed by homopyrimidine PNA is not intrinsically curved (*36*) and is presumably less flexible relative to duplex DNA. Therefore DNA bending can occur at the junction(s) between DNA duplex and triplex, and coincides with previously reported anomalous gel mobilities of short DNA fragments strand-invaded by PNA T$_{10}$-Lys interpreted in favor of a DNA bend at or near the triplex (*37*). One cannot exclude increased flexibility at the junction region due to unstacked or partially stacked DNA bases.

SFM inspection of the PNA$_2$/DNA triplex showed a detectable increase in DNA height, ~35% of its original value, especially for molecules with kinks (Figure 8D). This can be explained by the slight increase in the geometrical diameter of the PNA$_2$/DNA triple helix (*36*), and by extrusion of the non-complementary DNA strand. However, major portions of the molecules without visible kinks did not have any extra corrugation of the double helix. PNA-associated deformation of the DNA helix cannot be excluded.

TRIPLEXES PROMOTED BY RecA PROTEIN

The RecA protein is an indispensable element for homologous recombination in prokaryotic organisms. It catalyses a specific DNA strand-exchange reaction between ssDNA and homologous dsDNA with the formation of a three-stranded nucleoprotein filament easily detected by EM (*38-41* and *herein*). The detailed mechanism of strand exchange as well as the structure of the complex are not yet clear. However, it is thought that homology recognition invokes a DNA triplex intermediate, albeit of a different nature from that formed by regular TFOs (*7, 8*).

Direct EM evidence of the capability of sequence-specific targeting of dsDNA by single-stranded DNA probes covered with RecA protein was demonstrated in (*42*). Two non-homologous ssDNA probes 222 and 446 nt-long displayed high efficiency in DNA targeting when used independently. However, the yield of homologously paired molecules when both probes were used simultaneously was much less than expected. Nevertheless, the results of these experiments are very promising. It is still expected that the length of ssDNA probes can be significantly diminished, allowing detection and presumably labeling of short target sites for any nucleotide sequence. In addition, this technique can be very useful for the location of interspersed repeated elements, such as SINE, LINE or endogenous retroviral sequences within long DNA fragments.

GOALS AND PERSPECTIVES

In brief, the major goal of triplex studies using microscopy techniques is the development of highly selective sequence-specific labeling methods for long pieces of linear and, especially, supercoiled DNA molecules. Certainly this can be achieved by detailed studies of the complexes as well as by biochemical and biophysical methods which constitute the particular aim of the experiments. Determining the distribution of particular sequences over long fragments of the genome is another very important aim of these studies. Studies will be concentrated on two different classes of triplexes, i.e., these formed with short target sequences and those formed with long (a few hundreds of bp) DNA sequences. It may be envisaged that new agents will be found for stabilizing short 6-8 nt-long oligonucleotides with non-exclusively homopurine DNA sequences. For the latter case RecA covered nucleotides will presumably be used.

REFERENCES

1. Beal, P. A. and Dervan, P. B. (1991). Second structural motif for recognition of DNA by oligonucleotide-directed triple-helix formation. *Science* **251**, 1360-1363.

2. Hélène, C. (1991). The anti-gene strategy: control of gene expression by triplex-forming-oligonucleotides. *Anticancer Drug Des.* **6**, 569-584.
3. Thuong, N. T. and Hélène, C. (1993). Sequence-specific recognition and modification of double-helical DNA by oligonucleotides. *Angew. Chem. Int. Ed. Engl.* **32**, 666-690.
4. Nielsen, P. E., Egholm, M., Berg, R. H. and Buchardt, O. (1991). Sequence selective recognition of DNA by strand displacement with a thymine-substituted polyamide. *Science* **254**, 1497-1500.
5. Nielsen, P. E., Egholm, M. and Burchardt, O. (1994). Peptide nucleic acid (PNA). A DNA mimic with a peptide backbone. *Bioconjugate Chem.* **5**, 3-7.
6. Frank-Kamenetskii, M. D. and Mirkin, S. M. (1995). Triplex DNA structures. *Annu. Rev. Biochem.* **64**, 65-95.
7. Zhurkin, V. B., Raghunathan, G., Ulyanov, N. B., Camerini-Otero, R. D. and Jernigan, R. L. (1994). A parallel DNA triplex as a model for the intermediate in homologous recombination. *J. Mol. Biol.* **239**, 181-200.
8. Kim, M. G., Zhurkin, V. B., Jernigan, R. L. and Camerini-Otero, R. D. (1995). Probing the structure of a putative intermediate in homologous recombination: the third strand in the parallel DNA triplex is in contact with the major groove of the duplex. *J. Mol. Biol.* **247**, 874-889.
9. Le Cam, E. and Delain, E. "Nucleic acids-ligands interactions." *In* Visualization of Nucleic Acids, Morel G. (Des), CRC Press Inc. Boca Raton, FL, USA. 331-356, 1995.
10. Bustamante, C. and Rivetti, C. (1996). Visualizing protein-nucleic acid interactions on a large scale with the scanning force microscope. *Annu. Rev. Biophys. Biomol. Struct.* **25**, 395-429.
11. Brack, C. (1981). DNA electron microscopy. *Crit. Rev. Biochem.* **10**, 113-169.
12. Delain, E. and Le Cam, E. "The spreading of nucleic acids." *In* Visualization of Nucleic Acids, Morel G. (ed.), CRC Press Inc. Boca Raton FL, USA. 35-56, 1995.
13. Podtelezhnikov, A. A., Kurakin, A. V., Vologodskii, A. V. and Cherny, D. I. (1994). Testing the quality of electron microscope mapping data for DNA molecules with sequence-specific ligands. *Micron* **25**, 439-446.
14. Delain, E., Fourcade, A., Poulin, J.-C., Barbin, A., Coulaud, D., Le Cam, E. and Paris, E. (1992). Comparative observation of biological specimens, especially DNA and filamentous actin molecules in atomic force, tunneling and electron microscopes. *Microsc. Microanal. Microsctruct.* **3**, 457-470.
15. Hansma, H. G., Laney, D. E., Bezanilla, M., Sinsheimer, R. L. and Hansma, P. K. (1995). Applications for atomic force microscopy of DNA. *Biophys. J.* **68**, 1672-1677.
16. Hansma, H. G., Bezanilla, M., Zenhauserm, F., Adrian, M and Sinsheimer, R. L. (1993). Atomic force microscopy of DNA in aqueous solutions. *Nucleic Acids Res.* **21**, 505-512.
17. Pfannschmidt, C., Schaper, A., Heim, G., Jovin, T. M. and Langowski, J. (1996). Sequence-specific labeling of supercoiled DNA by triple-helix formation and psoralen crosslinking. *Nucleic Acids Res.* **24**, 1702-1709.
18. Schaper, A., Starink, J. P. P. and Jovin, T. M. (1994). The scanning force microscopy of DNA in air and n-propanol using new spreading agents. *FEBS Lett.* **355**, 91-95.
19. Lyubchenko, Yu. L. and Sclyakhtenko, L. S. (1997). Visualization of supercoiled DNA with atomic force microscopy in situ. *Proc. Natl. Acad. Sci. USA* **94**, 496-501.

20. Bustamante, C., Rivetti, C. and Keller, D. J. (1997). Scanning force microscopy under aqueous solutions. *Curr. Opin. Struct. Biol.* **7**, 709-716.

21. Lee, J. S., Latimer, L. J., Haug, B. L., Pulleyblank, D. E., Skinner, D. M. and Burkholder, G. D. (1989). Triplex DNA in plasmids and chromosomes. *Gene* **82**, 191-199.

22. Cherny, D. I., Malkov, V. A., Volodin, A. A. and Frank-Kamenetskii, M. D. (1993). Electron microscopy visualization of oligonucleotide binding to duplex DNA *via* triplex formation. *J. Mol. Biol.* **230**, 379-383.

23. Cherny, D. I., Kurakin, A. V., Lyamichev, V. N., Frank-Kamenetskii, M. D., Zinkevich, V. E., Firman, K., Egholm, M., Burchardt, O., Berg, R. H. and. Nielsen, P. E. (1994). Electron microscopy studies of sequence specific recognition of duplex DNA by different ligands. *J. Molec. Recognition.* **7**, 171-176.

24. Svinarchuk, F., Cherny, D., Debin, A., Delain, E. and Malvy, C. (1996). A new approach to overcome potassium-mediated inhibition of triplex formation. *Nucleic Acids Res.* **24**, 3858-3865.

25. Cherny, D. I., Fourcade, A., Svinarchuk, F., Nielsen, P. E., Malvy, C. and Delain, E. (1998). Analysis of various sequence-specific triplexes by electron and atomic force microscopies. *Biophys. J.* **74**, 1015-1023.

26. Hansma, H. G., Revenko, I., Kim, K. and Laney, D. E. (1996). Atomic force microscopy of long and short double-stranded, single-stranded and triple-stranded nucleic acids. *Nucleic Acids Res.* **24**, 713-720.

27. Hampel, K. J., Burkholder, G. D. and Lee, J. S. (1993). Plasmid dimerization mediated by triplex formation between polypyrimidine-polypurine repeats. *Biochemistry* **32**, 1072-1077.

28. Lee, J. S., Ashley, C., Hampel, K. J., Bradley, R. and Scraba, D. G. (1995). A stable interaction between separated pyrimidine.purine tracts in circular DNA. *J. Mol. Biol.* **252**, 283-288.

29. Stokrova, J., Vojtiskova, M. and Palecek, E. (1989). Electron microscopy of supercoiled pEJ4 DNA containing homopurine.homopyrimidine sequences. *J. Biomol. Struct. Dyn.* **6**, 891-898.

30. Svinarchuk, F., Nagibneva, I., Cherny, D., Ait-Si-Ali, S., Pritchard, L. L., Robin, P., Malvy, C. and Harel-Bellan, A. (1997). Recruitment of transcription factors to the target site by triplex-forming oligonucleotides. *Nucleic Acids Res.* **25**, 3459-3464.

31. Sun, J. S., de Bizemont, T., Duval-Valentin, G., Montenay-Garestier, T. and Hélène, C. (1991). Extension of the range of recognition sequences for triple helix formation by oligonucleotides containing guanines and thymines. *C. R. Acad. Sci. Ser. III* **313**, 585-590.

32. Bouziane, M., Cherny, D. I., Mouscadet, J.-F. and Auclair, C. (1996). Alternate strand DNA triple-helix-mediated inhibition of HIV-1 U5 long terminal repeat integration in vitro. *J. Biol. Chem.* **271**, 10359-10364.

33. Cherny, D. I., Belotserkovskii, B. P., Frank-Kamenetskii, M. D., Egholm, M., Buchardt, O., Berg, R. H. and. Nielsen, P. E. (1993). DNA unwinding upon strand-displacement binding of a thymine-substituted polyamide to double-stranded DNA. *Proc. Natl. Acad. Sci. USA* **90**, 1667-1670.

34. Demidov, V. V., Cherny, D. I., Kurakin, A. V., Yavnilovich, M. V., Malkov, V. A., Frank-Kamenetskii, M. D., Sønnichsen, S. H. and. Nielsen, P. E. (1994). Electron microscopy mapping of oligopurine tracts in duplex DNA by peptide nucleic acid targeting. *Nucleic Acids Res.* **22**, 5218-5222.

35. Demidov, V. V., Yavnilovich, M. V. and Frank-Kamenetskii, M. D. (1997). Kinetic analysis of specificity of duplex DNA targeting by homopyrimidine peptide nucleic acid. *Biophys. J.* **72**, 2763-2769.

36. Betts, L., Josey, J. A., Veal, J. M. and Jordan, D. R. (1995). A nucleic acids triple helix formed by a peptide nucleic acid-DNA complex. *Science* **270**, 1838-1841.

37. Peffer, N. J., Hanvey, J. C., Bisi, J. E., Thomson, S. A., Hassman, C. F., Noble, S. A. and Babiss, L. E. (1993). Strand-invasion of duplex DNA by peptide nucleic acid oligomers. *Proc. Natl. Acad. Sci.* USA **90**, 10648-10652.

38. Stasiak, A. Z., Rosselli, W. and Stasiak, A. (1991). RecA-DNA helical filaments in genetic recombination. *Biochimie* **73**, 199-208.

39. Umlauf, S. W., Cox, M. M. and Inman, R. B. (1990). Triple-helical DNA pairing intermediates formed by recA protein. *J. Biol. Chem.* **265**, 16898-16912

40. Jain, S. K., Cox, M. M. and Inman, R. B. (1995). Occurrence of three-stranded DNA within a RecA protein filament. *J. Biol. Chem.* **270**, 4943-4949.

41. Morel, P., Cherny, D., Ehrlich, S. D. and Cassuto, E. (1997). Recombination-dependent repair of DNA double-strand breaks with purified proteins from Escherichia coli. *J. Biol. Chem.* **272**, 17091-17096.

42. Révet, B. M., Sena, E. P. and Zarling, D. A. (1993). Homologous DNA targeting with RecA protein-coated short DNA probes and electron microscope mapping on linear duplex molecules. *J. Mol. Biol.* **232**, 779-791.

5 CHEMICAL MODIFICATIONS OF TRIPLE HELIX FORMING OLIGONUCLEOTIDES

Ulysse Asseline

INTRODUCTION

Ever since its initial discovery, the triple-helix structure (*1*), because of possible applications in biotechnology, diagnostics, and therapeutics, has attracted considerable attention (*2-5*). It has been demonstrated that homopurine-homopyrimidine tracts of DNA can be targeted by third strand oligonucleotides which bind to the major groove of DNA, and held in place to purine bases by specific hydrogen bonds. A homopyrimidine third strand binds parallel to the purine strand of the double-stranded target forming T.A x T and C^+.G x C (where C^+ indicates protonated cytosine and the symbol x stands for Watson-Crick hydrogen bonds) isomorphous base triplets via Hoogsteen hydrogen bonding, whereas a purine third strand binds in an anti-parallel orientation forming G.G x C and A.A x T base triplets via reverse Hoogsteen hydrogen bonding. G and T containing oligonucleotides can also form triplexes. In this case, the orientation of the third strand is sequence-dependent.

Applications of the oligonucleotide directed triple-helix formation are, however, limited because the stability of triple-helical complexes formed with natural third strands, except for a few involving a G-rich purine sequence, is generally weak, and oligopurine tracts on the DNA target are required. Moreover, *in vivo* applications require triple-helix forming oligonucleotides that are not only able to form stable and specific triple-helical structures under physiological conditions but also to reach their target on DNA before being degraded by cellular nucleases. The delivery of triple helix forming oligonucleotides to the target is discussed in a companion chapter. Concerning nuclease resistance, triplex forming oligonucleotides have benefited from improvements made on the "antisense" oligonucleotides. Numerous chemical modifications of triplex forming oligonucleotides have been developed in order to realize the potentials of oligonucleotide-directed triple-helix formation. Several approaches have been used and will be discussed below.

MODIFICATION OF THE OLIGONUCLEOTIDE CHAIN FOR INCREASING STABILITY

Efforts to engineer oligonucleotides with a greater degree of nuclease resistance and improved binding properties include several modifications at different levels of the oligonucleotide chain: the phosphate-sugar backbone and the nucleic bases. Most of the modifications, first developed to improve the properties of "antisense" oligonucleotides, have also been incorporated into triple-helix forming oligonucleotides. Only a few of them have the ability to form triple-helical complexes with stabilities equivalent or superior to those of the triplex involving third strands made of natural nucleosides. Specific modifications of the triplex forming oligonucleotides have also been developed to overcome the limitations of triplex formation: (i) the pH dependence in the pyrimidine motif and the resulting difficulty in targeting G-C stretches due to the charge-charge repulsion between contiguous protonated cytosines, (ii) the non-isomorphism of triplets and the inhibition by monovalent cations in the purine and T.A x T/G.G x C motifs. Since many of these modifications have already been reviewed (2-5), only a few examples of those leading to triplex stabilization with maintenance of base-pairing specificity and recent new improvements will be reported.

Modification of the phosphate-sugar backbone

Modifications can be made in the oligonucleotide backbone by replacing non-bridging atoms as in the case of phosphoramidates (*6*), bridging atoms as in the case of N3'→P5' phosphoramidates (*7*), negatively charged phosphodiester linkages by a neutral internucleotide linkage as in the case of the "riboacetal" bond (*8*) or by a positively charged linkage as in the case of polycationic guanidinium oligonucleotides (*9*). In the case of peptide nucleic acids (PNA) (*10*) a more drastic change is made in that the phosphodiester-sugar backbone is replaced by a polyamine backbone. Pyrimidine oligonucleotide strands containing alternating phosphodiester and stereo-uniform cationic phosphoramidate linkages form triplexes whose stability is dependent on the configuration of the phosphoramidate linkages (*6*). One isomer binds to the duplex DNA target with an affinity superior to that of the corresponding phosphodiester oligonucleotide. The incorporation of the neutral "riboacetal" internucleotide linkage (*8*), which results in reduction of the negative charges and restricts the conformational freedom of the third-strand oligonucleotide analogue provides a strong stability increase for the triplex (4° C per unit). Short strands of guanidinium oligonucleotides bind to the duplex target with high affinity; but results with longer sequences are not yet known (*9*). N3'→P5' phosphoramidates are able to form stable triple-helical complexes under near-physiological conditions (*7*). PNA interact with DNA in a manner unlike that of classical triple helix forming oligonucleotides (*10*). The properties of these two classes of modified oligonucleotides are addressed in two chapters of this volume. Modifications of the sugar-containing third strand are also able to increase the stability of the triple helices. Among the groups substituted at the 2'-positions of oligopyrimidine third strands which have proven to stabilize the triplexes, the recently reported 2'-aminoethoxy group (11) provides the strongest enhancement in binding affinity. Replacement of the oxygen atom at the 4'-position of the sugar ring by a sulfur atom leads to a moderate stability increase when 4'-thio modifications are placed in contiguous stretches (12). Replacement of the ring oxygen of 2'-deoxycytidine by a carbon atom (inducing an increase of the pK_a value) leads to a more pronounced stability increase, whereas the carbocyclic analogue of thymidine decreases the triplex stability (13).

Modification of the nucleic base

One limitation to the formation of triplex with a pyrimidine third strand containing cytidine is its pH dependence, since formation of a C^+.G x C triplet requires protonation of the third strand cytidine residues on N3 in order to form two Hoogsteen-type hydrogen bonds. Consequently under physiological conditions the stability of triplexes involving a 2'-deoxycytidine containing pyrimidine third strand is weak, since the pK_a of dC is 4.3. Triplex stability can be increased by replacing third strand 2'-deoxycytidine by 5-methyl-2'-deoxycytidine residues (14). A more pronounced stabilization was observed using the 5-propynyl derivatives of 2'-deoxyuridine, whereas the presence of the 5-propynyl derivatives of 2'-deoxycytidine (less basic than 5-methyl-2'-deoxycytidine) decreased triplex stability (15). Unnatural cytidine analogues able to form two hydrogen bonds with G of the GC base-pair without protonation or able to undergo protonation under physiological conditions as well as unnatural purine analogues were incorporated in place of 2'-deoxycytidine in the third pyrimidine strand. Among the neutral nucleosides reported, pseudoisocytidine (16), 1-(2-deoxy-β-D-ribofuranosyl)-3-methyl-5-amino-1H-pyrazolo[4,3d]pyrimidin-7-one (17), and N^7-deoxyguanosine (18) have proven efficient for targeting GC stretches. The last two were shown to bind isolated GC base pairs with weaker affinity than did the 5-methyl-2'-deoxycytidine. Among the cytidine surrogates protonated at physiological pH, the recently reported 2-amino-(2'-deoxy-β-D-ribofuranosyl)pyridine exhibited interesting properties (19). Over a 6-8 pH range, third strand pyrimidine oligonucleotides containing this cytidine analogue bind strongly and with increased specificity to their target, as compared to those containing 5-methyl-2'-deoxycytidine. Furthermore, 2-amino-(2'-deoxy-β-D-ribofuranosyl)pyridine containing pyrimidine third strands are able to bind with targets containing stretches of GC at physiological pH, whereas those containing 5-methyl-2'-deoxycytidine do not. In order to overcome the problem of non-isomorphism in the case of T.A x T/G.G x C triplexes, purine analogues of thymidine have been incorporated in third strand oligonucleotides. Those involving 7-deaza-2'-deoxyxanthosine and 2'-deoxyguanosine were shown to strongly hybridize with the target sequence, in an anti-parallel orientation, under physiological conditions, whereas the oligonucleotides containing 2'-deoxyguanosine and thymidine did not bind (*20*). Triplex formation by purine or G and T containing third strands is pH independent but may be hampered by the tendency of guanine-rich oligonucleotides to form self-associated structures (G tetrads) under physiological conditions (e.g., 140 mM K⁺). Several analogues of guanine have also been incorporated into triple helix forming oligonucleotides to overcome this problem, but with limited success (*5* and references cited therein).

EXTENSION OF TARGETS ACCESSIBLE TO TRIPLEX FORMATION

Even though progress has been made in stabilizing triplexes by using nucleotide analogues or modified backbones in the third strand, their stability could be further increased by extending the range of recognition sequences, thus allowing the formation of longer triple-helical structures. Among several approaches explored to extend range of duplex DNA sequences accessible to triple-helix formation, the following two are most employed: the use of oligonucleotides capable of binding alternately to both strands of the duplex target and the design of base analogues able to specifically recognize inverted base-pair C.G and T.A in the polypurine-polypyrimidine target. Another strategy would be to develop oligonucleotides able to recognize all four base pairs of any DNA sequence.

Specific modifications for alternate strand recognition

The specific recognition of two contiguous homopurine sequences on alternate DNA strands can be achieved by using either oligonucleotides with constant backbone polarity employing a combination of purine and pyrimidine sequences or two oligopyrimidine third strands linked to 3'-3' or 5'-5' in order to fulfill the strand orientation requirements. The design of alternate third strand oligonucleotides requires linkers crossing the major groove adapted to each type of junction. Several linkers such as 1,3-propane diol or its oligomers (*21*), 1,2-dideoxy-D-ribose (*22*), have been used to tether the terminal phosphates of both pyrimidine third strands, while xylose residues have been used to link their terminal 2'-deoxyribose units (*23*). Alternate third strand oligonucleotides involving only a bridging phosphate between the two 3' ends of oligopyrimidine strands (*24*), as well as 3'-3' and 5'-5' alternate third strands involving base-to-base linkages (*25, 26*) via a flexible linker, have also been reported. These modified oligonucleotides bind to their double-stranded target with variable stabilities. Pronounced stability increases have been reported only with 3'-3' junctions (*23-25*). The junctions remain to be optimized to increase the triplex stability together with the participation of the bases close to the junction in the selectivity of hybridization. An alternate strategy for targeting purine on both strands of the DNA target consists in using two pyrimidine third strand oligonucleotides covalently linked at their 5'-end to a polyamide designed to bind sequence-specifically to the minor groove of DNA as a cooperative side-by-side antiparallel homodimer (*27*). Third strand binding parallel to the purine target is accompanied by dimerization of the polyamide in the minor groove, inducing cooperative binding of both conjugates associated with strong stabilization.

Use of a modified nucleoside for inverted base pair recognition in the polypyrimidine-polypurine target and general base pair recognition

When oligopurine targets are interrupted by a pyrimidine, a strong destabilization of the triplex is observed. This destabilization is similar to that observed when an abasic site which eliminates base-pairing and stacking interactions is introduced into the third strand (*28*). The replacement of nucleosides facing the base-pair inversion by azoles which allow the maintenance of stacking interactions with neighboring bases without recognition of any particular base induces stabilization of the triple-helical complexes (*29*). Unusual natural base triads which provide intermediate stability have also been selected (*30*). G can bind TA while T and C can bind CG with similar affinity. These base triplets involve a single hydrogen bond between the G or T and the pyrimidine of the inverted base pair which possesses only a single hydrogen bond forming site in the major groove. Increased stabilization and specific recognition require more than one hydrogen bond. Modified bases able to cross the major groove to form hydrogen bonds with both the pyrimidine and the purine on opposite strands have been reported. When incorporated into a pyrimidine third strand opposite a TA inverted base-pair, N^4-(3-carboxypropyl)-deoxycytidine better stabilizes the triplex than does G at the same position (*31*). Similarly, when N^4-(6-amino-2-pyridinyl) deoxycytidine is incorporated opposite the CG inverted base-pair, a stable triplex is formed at pH 7 (*32*). Much effort has been invested to extend this strategy to all four bases to allow the formation of a triplex on any double stranded target, but with limited success (*5* and *33* and references cited therein).

COVALENT LINKING OF STABILIZING AND REACTIVE MOLECULES TO THE THIRD STRAND OLIGONUCLEOTIDE

When added to triple-helical complexes, many ligands such as intercalators, minor groove binding agents or polycationic molecules have been shown to enhance their stabilities. For an exhaustive list in this field, the reader is referred to a companion chapter in this volume. A further stabilization increase can be obtained by covalently linking these molecules to the triplex forming oligonucleotides. However, this increase depends on many parameters, such as the size and nature of the linker between the two entities as well as the positions on the oligonucleotide and the ligand involved in the coupling. To provide triplex forming oligonucleotides with additional properties it is also possible to endow them with ligands able to induce irreversible modifications of the target sequence such as cleavage or covalent linking at specific sites. In this case, the best efficiency is achieved when both strands of the target DNA are modified. A few examples of these triplex forming oligonucleotide-ligand conjugates will be reported. Concerning the DNA cleaving reagents the reader is also referred to a companion chapter in this volume devoted to artificial nucleases.

Covalent linking of intercalators, groove binders and other stabilizing ligands to the third strand

When attached to the 5'-end of a third strand oligonucleotide, 2-methoxy-6-chloro-9-amino-acridine, oxazolopyridocarbazole and daunorubicine derivatives have been shown to strongly stabilize the triplex formed with the DNA target by intercalation at the triplex-duplex junction (*34* and references cited therein, *35*). New families of intercalators with a higher affinity for the triple-helical rather than the double-stranded complex are under investigation in different laboratories. Among them, benzopyridoindole and benzopyridoquinoxaline derivatives have been covalently linked to pyrimidine third strands via flexible linkers of varying sizes at either the 5' ends or internucleotide positions (*36, 37*). Under near physiological conditions the strongest stability increases were observed with a benzo[h]pyridoquinoxaline derivative conjugated to the 5' end, and with a benzo[f]pyridoquinoxaline or a benzo[e] pyridoindole derivative linked to internucleotide positions of the oligonucleotides, respectively. Hoechst 33258, a minor groove binder, has also been covalently linked to the 5'-end of a pyrimidine third strand (*38*). Complex stabilization by simultaneous major and minor groove binding has been observed. In the same way, the covalent coupling to a pyrimidine third strand of a minor groove binding ligand composed of polyamides containing N-methylimidazole and N-methylpyrrole amino acids, combined side-by-side, leads to an increased stability of the triple-helical complex (*39*). When linked to the 5'-ends (*40*) or at the N^4 position of internal cytosine (*41*) of triple helix forming oligonucleotides, spermine stabilizes the triplex as do cationic peptides linked to the 5'-ends of pyrimidine third strands (*42*).

Covalent linking of alkylating and cleaving reagents to the third strand

Even though certain triple-helix forming oligonucleotides modified at the level of the oligonucleotide chain and covalently linked to intercalators are able to form stable triplexes under physiological conditions, they can be displaced by replication or transcription machinery. One way to improve the potential of the triple helix forming oligonucleotides in regulating gene expression is to attach to them reagents able to induce irreversible modifications of the target sequence such as cleaving or crosslinking (*2* and references cited therein). A variety of reactive groups have been used in order to cleave the DNA target or to link the triple helix forming oligonucleotides to their specific binding sites. These reagents can be divided into three groups. The first group concerns metal chelates Fe-EDTA, Fe-Porphyrin, Cu-Orthophenanthroline and more recently Fe-Bleomycin (*43*), which induce cleavage of the target via oxidative degradation of the deoxyribose ring. The second group includes compounds such as psoralen and azido-containing compounds which are able to induce a covalent bond between the triple helix forming oligonucleotide and the target under irradiation. These reagents which do not require cofacteurs are not reactive until activated by light. Psoralen-oligonucleotide conjugates resistant to nucleases have been used to demonstrate that the DNA target sequence is accessible in the cell nuclei (*44*). The last group involves alkylating compounds which are able to react spontaneously with the target via electrophilic alkylation of the bases. Among them the most attractive for *in vivo* applications are those that are chemically stable when tethered to a single strand oligonucleotide but very reactive after hybridization of the oligonucleotide with its target. When incorporated into a pyrimidine third strand, ethano-5-methyl-2'-deoxycytidine has been shown to alkylate, with high yield (>95%) but slowly, a G residue facing the modified C in the purine target sequence under physiological salt conditions (*45*). More recently the N-5-methyl-cyclopropapyrroloindole residue, a structural analogue of cyclopropapyrroloindole the reactive subunit of the potent antibiotic CC-1065 which alkylates adenines at their N-3 position, has been linked at both the 5'-and 3'-positions of a purine third strand. This conjugate has proven capable of alkylating, efficiently (>85%) and quickly, in a physiological buffer, both strands of the target duplex region immediately adjacent to its binding site on the double-stranded target (*46*). By analogy with the RNase H mediated activity of antisense oligonucleotides, an alternative way for biological applications would be the recruitment of a cellular enzyme which would be directed to perform a sequence-specific cleavage of the target DNA. When covalently linked to the 3'-end of a pyrimidine third strand, camptothecin, an inhibitor of topoisomerase I, has been proven to induce *in vitro* sequence-specific cleavage of one strand of the double-stranded DNA target by topoisomerase I (*47*).

HAIRPIN AND CIRCULAR OLIGONUCLEOTIDES FOR TRIPLEX FORMATION WITH A SINGLE-STRANDED TARGET

A single-stranded oligopurine sequence can be recognized by two oligopyrimidine strands tethered via a flexible hexaethylene linker, providing the strand orientation requirement is maintained, to form a triple-helix complex more stable than that obtained when the two pyrimidine sequences are separated (*48*). It is possible to increase the stability of such a system by lengthening the Watson-Crick part to create a strong binding site (a duplex-triplex junction) for an intercalator attached to the 5' end of the Hoogsteen pyrimidine strand. When the intercalating agent is a psoralen it is possible to photo-induce covalent attachment of all three stands providing a TpA sequence is present at the triplex-duplex junction on the oligopurine containing strand (*49*). Circular oligonucleotides have also proven to form very stable triplexes with an oligopurine sequence. Enhanced stability can be achieved by cross-linking both strands of the circular oligonucleotide (*50*).

CONCLUSION

The formation and stability of triple-helical complexes depend on several parameters: the electrostatic interactions among the three strands, the number of hydrogen bonds between the third strand and the duplex, triplet isomorphism, stacking and hydrophobic interactions, and the hydration states of the target and triplex forming oligonucleotides, whose relative contributions are difficult to evaluate. A great deal of work has been carried out in many laboratories to engineer chemically modified triplex forming oligonucleotides with improved properties. Progress has been made concerning triplex stabilization by chemical modification of the oligonucleotide third strands or by coupling stabilizing ligands to them. Reactive agents capable of inducing irreversible modifications of the DNA target have been linked to the triple-helix forming oligonucleotides. Much effort has been expended to extend the duplex DNA sequences accessible to triple-helix formation. Although modified oligonucleotides are able to recognize pyrimidine interruptions in the polypurine strand or two short purine tracts on alternate strands of the DNA, a general solution to the recognition of DNA by triplex forming oligonucleotides remains a challenge to chemists and is not likely to be available in the near future.

REFERENCES

1. Fesenfeld, G., Davies, D. R. and Rich, A. (1957). Formation of a three-stranded polynucleotide molecule. *J. Am. Chem. Soc.* **79**, 2023-2024.
2. Thuong, N. T. and Hélène, C. (1993). Sequence-specific recognition and modification of double-helical DNA by oligonucleotides. *Angew. Chem. Int. Ed.* **32**, 666-690.
3. Soyfer, V. N. and Potoman, V. (1996). Triple-helical nucleic acids. *Springer-Verlag, New York.*
4. Sun, J.-S., Garestier, T. and Hélène C. (1996). Oligonucleotide directed triple helix formation. *Current Opinion in Structural Biology.* **6**, 327-333.
5. Doronina, S. O. and Behr, J.-P. (1997). Towards a general triple helix mediated DNA recognition. *Chem. Soc. Rev.* 63-71.
6. Chaturvedy, S., Horn, T. and Letsinger R. L. (1996). Stabilization of triple-stranded oligonucleotide complexes: use of probes containing alternating phosphodiester and stereo-uniform cationic phosphoramidate linkages. *Nucleic Acids Res.* **24**, 2318-2323.

7. Gryaznov, S. M., Lloyd, D. H., Chen, J.-K., Schultz, R. N., DeDionisio, L. A., Ratmeyer, L. and Wilson, W. D. (1995). Oligonucleotide N3'→P5' phosphoramidates. *Proc. Natl. Acad. Sci. USA.* **92**, 5798-5802.

8. Jones, R. J., Swaminathan, S., Milligan, J. F., Wadwani, S., Froehler, B. C. and Matteucci, M. D. (1993). Oligonucleotides containing a covalent conformationally restricted phosphodiester analog for high-affinity triple helix formation: The riboacetal internucleotide linkage. *J. Am. Chem. Soc.* **115**, 9816-9817.

9. Blasko, A., Dempcy, R. O., Minyat, E. E. and Bruice, T. C. (1996). Association of short-strand DNA oligomers with guanidinium-linked nucleosides. A kinetic and thermodynamic study. *J. Am. Chem. Soc.* **118**, 7892-7899.

10. Dueholm, K. L. and Nielsen, P. E. (1997). Chemistry, properties and applications of PNA. *New J. Chem.* **21**, 19-31.

11. Cuenoud, B., Casset, F., Hüsken, D., Natt, F., Wolf, R. M., Altmann, K.-H., Martin, P. and Moser, K. H. (1998). Dual recognition of double-stranded DNA by 2'-aminoethoxy-modified oligonucleotides. *Angew. Chem. Int. Ed. Engl.* **37**, 1288-1291.

12. Jones, G. D., Altmann, K.-H., Hüsken, D. and Walker, R. T. (1997). Duplex and triplex-forming properties of 4'-thio-modified oligodeoxynucleotides. *Bioorganic & Medicinal Chemistry Lett.* **7**, 1275-1278.

13. Froehler, B. C. and Ricca, D. J. (1992). Triple-helix formation by oligonucleotides containing the carbocyclic analogs of thymidine and 5-methyl-2'-deoxycytidine. *J. Am. Chem. Soc.* **114**, 8320-8322.

14. Posvic, T. J. and Dervan, P. B. (1989). Triple helix formation by oligonucleotides on DNA extended to the physiological pH range. *J. Am. Chem. Soc.* **111**, 3059-3061.

15. Froehler, B. C., Wadwani, S., Terhorst, T. J. and Gerrard, S. R. (1992). Oligonucleotides containing C-5 propyne analogs of 2'-deoxyuridine and 2'-deoxycytidine. *Tetrahedron Lett.* **33**, 5307-5310.

16. Ono, A., Ts'o, P. O. P. and Kan, L.-S. (1992). Triplex formation of an oligonucleotide containing 2'-O-methylpseudoisocytidine with a DNA duplex at neutral pH. *J. Org. Chem.* **57**, 3225-3230.

17. Priestley, E. S. and Dervan, P. B. (1995). Sequence composition effects on the energetics of triple helix formation by oligonucleotides containing a designed mimic of protonated cytosine. *J. Am. Chem. Soc.* **117**, 4761-4765.

18. Brunar, H. and Dervan, P. B. (1996). Sequence composition effects on the stabilities of triple helix formation by oligonucleotides containing N[7]-deoxyguanosine. *Nucleic Acids Res.* **24**, 1987-1991.

19. Hildbrand, S., Blaser, A., Parel, S. P. and Leumann, C. J. (1997). 5-Substituted 2-aminopyridine C-Nucleosides as protonated cytidine equivalents: increasing efficiency and selectivity in DNA triple-helix formation. *J. Am. Chem. Soc.* **119**. 5499-5511.

20. Milligan, J. F., Krawczyk, S. H., Wadwani, S. and Matteucci, M. D. (1993). An anti-parallel triple helix motif with oligodeoxynucleosides containing 2'-deoxyguanosine and 7-deaza-2'-deoxyxanthosine. *Nucleic Acids Res.* **21**, 327-333.

21. Ono, A., Chen, C.-N. and Kan, L.-S. (1991). DNA triplex formation of oligonucleotide analogues consisting of linker groups and octamer segments that have opposite sugar-phosphate backbone polarities. *Biochemistry* **30**, 9914-9921.

22. Horne, D. A. and Dervan, P. B. (1990). Recognition of mixed-sequence duplex DNA by alternate-strand triple-helix formation. *J. Am. Chem. Soc.* **112**, 2435-2437.
23. Froehler, B. C., Terhorst, T., Shaw, J.-P. and McCurdy, S. N. (1992). Triple-helix formation and cooperative binding by oligonucleotides with a 3'—3' internucleotide junction. *Biochemistry* **31**, 1603-1609
24. De Napoli, L., Messere, A., Montesarchio, D., Pepe, A., Piccialli, G. and Varra, M. (1997). Synthesis and triple helix formation by alternate strand recognition of oligonucleotides containing 3'-3' phosphodiester bonds. *J. Org. Chem.* **62**, 9024-9030.
25. Zhou, B.-W., Marchand, C., Asseline, U., Thuong, N. T., Sun, J.-S., Garestier, T. and Hélène, C. (1995). Recognition of alternating oligopurine/oligopyrimidine tracts of DNA by oligonucleotides with base-to-base linkages. *Bioconjugate Chem.* **6**, 516-523.
26. Asseline, U. and Thuong, N. T. (1994). 5'-5' Tethered oligonucleotides via nucleic bases: A potential new set of compounds for alternate strand triple-helix formation. *Tetrahedron Lett.* **35**, 5221-5224.
27. Szewczyk, J. W., Baird, E. E. and Dervan, P. B. (1996). Cooperative triple-helix formation via a minor groove dimerization domain. *J. Am. Chem. Soc.* **118**, 6778-6779.
28. Horne, D. A. and Dervan, P. B. (1991). Effects of an abasic site on triple helix formation characterized by affinity cleaving. *Nucleic Acids Res.* **19**, 4963-4965.
29. Durland, R. H., Rao, T. S., Bodepudi, V., Seth, D. M., Jayaraman, K. and Revankar, G. R. (1995). Azole substituted oligonucleotides promote antiparallel triplex formation at non-homopurine duplex target. *Nucleic Acids Res.* **23**, 647-653.
30. Mergny, J. L., Sun, J.-S., Rougée, M., Montenay-Garestier, T., Barcelo, F., Chomilier, J. and Hélène, C. (1991). Sequence specificity in triple-helix formation: experimental and theoretical studies of the effect of mismatches on triplex stability. *Biochemistry* **30**, 9791-9798.
31. Verma, S. and Miller, P. S. (1996). Interactions of cytosine derivatives with T.A interruptions in pyrimidine.purine.pyrimidine DNA triplexes. *Bioconjugate Chem.* **7**, 600-605.
32. Huang, C.-Y., Bi, G. and Miller, P. S. (1996). Triplex formation by oligonucleotides containing novel deoxycytidine derivatives. *Nucleic Acids Res.* **24**, 2606-2613.
33. Lehmann, T. E., Greenberg, W. A., Liberles, D. A., Wada, C. K. and Dervan, P. B. (1997). Triple-helix formation by pyrimidine oligonucleotides containing nonnatural nucleosides with extended aromatic nucleobases: intercalation from the major groove as a method for recognizing C.G and T.A base pairs. *Helvetica Chemica Acta* **80**, 2002-2022.
34. Asseline, U., Thuong, N. T. and Hélène, C. (1997). Synthesis and properties of oligonucleotides covalently linked to intercalating agents. *New J. Chem.* **21**, 5-17.
35. Garbesi, A., Bonazzi, S., Zanella, S., Capobianco, M. L., Giannini, G. and Arcamone, F. (1997). Synthesis and binding properties of conjugates between oligodeoxynucleotides and daunorubicin derivatives. *Nucleic Acids Res.* **25**, 2121-2128.
36. Silver, G. C., Sun, J. S., Nguyen, C. H., Boutorine, A. S., Bisagni, E. and Hélène, C. (1997). Stable triple-helical DNA complexes formed by benzopyridoindole and benzopyrido-quinoxaline-oligonucleotide conjugates. *J. Am. Chem. Soc.* **119**, 263-268.

37. Silver, G. C., Nguyen, C. H., Boutorine, A. S., Bisagni, E., Garestier, T. and Hélène, C. (1997). Conjugates of oligonucleotides with triplex-specific intercalating agents. Stabilization of triple-helical DNA in the promoter region of the gene for the α-subunit of interleukin 2 (IL-2Rα). *Bioconjugate Chem.* **8**, 15-22.

38. Robles, J., Rajur, S. B. and McLaughlin, L. W. (1996). A parallel-stranded DNA triplex tethering a Hoechst 33258 analogue results in complex stabilization by simultaneous major groove and minor groove binding. *J. Am. Chem. Soc.* **118**, 5820-5821.

39. Szewczyk, J. W., Baird, E. E. and Dervan, P. B. (1996). Sequence-specific recognition of DNA by a major and minor groove binding ligand. *Angew. Chem. Int. Ed. Engl.* **35**, 1487-1489.

40. Tung, C.-H., Breslauer, K. J. and Stein, S. (1993). Polyamine-linked oligonucleotides for DNA triple helix formation. *Nucleic Acids Res.* **21**, 5489-5494.

41. Rajeev, K. G., Jadhav, V. R. and Ganesh K. N. (1997). Triplex formation at physiological pH: comparative studies on DNA triplexes containing 5-Me-dC tethered at N^4 with spermine and tetraethyleneoxyamine. *Nucleic Acids Res.* **25**, 4187-4193.

42. Tung, C.-H., Breslauer, K. J, and Stein, S. (1996). Stabilization of DNA triple-helix formation by appended cationic peptides. *Bioconjugate Chem.* **7**, 529-531.

43. Sergeyev, D. S., Godovikova, T. S. and Zarytova, V. F. (1995). Catalytic site-specific cleavage of a DNA-target by an oligonucleotide carrying bleomycin A5. *Nucleic Acids Res.* **23**, 4400-4406.

44. Giovannangeli, C., Diviacco, S., Labrousse, V., Gryaznov, S., Charneau, P. and C. Hélène. (1997). Accessibility of nuclear DNA to triplex-forming oligonucleotides: the integrated HIV-1 provirus as a target. *Proc. Natl. Acad. Sci. USA.*, **94**, 79-84.

45. Shaw, J.-P., Milligan, J. F., Krawczyk, S. H. and Matteucci, M. (1991). Specific, high-efficiency, triple-helix-mediated cross-linking to duplex DNA. *J. Am. Chem. Soc.* **113**, 7765-7766.

46. Lukhtanov, E. A., Mills, A. G., Kutyavin, I. V., Gorn, V. V., Reed, M. W. and Meyer, R. B. (1997). Minor groove DNA alkylation directed by major groove triplex forming oligodeoxyribonucleotides. *Nucleic Acids Res.* **25** 5077-5084.

47. Matteucci, M., Lin, K.-Y., Huang, T., Wagner, R., Sternbach, D. D., Mehrotra, M. and Besterman, J. M. (1997). Sequence-specific targeting of duplex DNA using a camptothecin-triple helix forming oligonucleotide conjugate and topoisomerase I. *J. Am. Chem. Soc.* **119**, 6939-6940.

48. Giovannangeli, C., Montenay-Garestier, T., Rougée, M., Chassignol, M., Thuong, N. T. and Hélène, C. (1991). Single-Stranded DNA as a Target for triple-helix formation. *J. Am. Chem. Soc.* **113**, 7775-7777.

49. Giovannangeli, C., Thuong, N. T. and Hélène, C. (1993). Oligonucleotide clamp arrest DNA synthesis on a single-stranded DNA target. *Proc. Natl. Acad. Sci. USA.* **90**, 10013-10017.

50. Kool, E. T. (1997). Design of triplex-forming oligonucleotides for binding DNA and RNA: optimizing affinity and selectivity. *New J. Chem.* **21**, 33-45.

6 TRIPLE HELIX FORMATION WITH MODIFIED OLIGONUCLEOTIDES

Sergei M. Gryaznov

INTRODUCTION

Significant interest in synthetic oligonucleotides is largely determined by their ability to form specific complexes – duplexes and/or triplexes – with nucleic acid targets of interest, primarily DNA and RNA molecules. The high specificity of these interactions is governed mainly by the formation of proper Watson-Crick and Hoogsteen hydrogen bonds between heterocyclic bases of targeted nucleic acids and the oligonucleotide ligands. This type of selective recognition of genetic information carriers can potentially open up promising new opportunities in modern rational drug design and drug discovery, as well as in the creating of powerful molecular biological and biochemical tools allowing regulation of gene expression and studying of gene functions. However, several very important problems need to be resolved before oligonucleotides may become pharmaceutical agents. Among these are increase of thermodynamic stability of the complexes formed by the oligomers with the targets, specificity of interaction with chosen molecules, hydrolytic stability and bioavailability of oligonucleotides in cells and in model animal systems as well as in human tissues and organs. Additionally, the chemical structure of the oligonucleotides and the choice of suitable and functionally meaningful molecular targets, as well as administration or delivery methods may play a crucial role in the success of oligonucleotide based therapeutic approaches.

This chapter will concentrate on properties, comparative analysis and potential utilities of sugar phosphate backbone modified oligonucleotides as triplex forming compounds, targeted to the double-stranded DNA molecules.

OLIGONUCLEOTIDE PHOSPHOROTHIOATES

Oligonucleotides containing internucleoside phosphorothioate linkages constitute the most extensively studied class of oligonucleotide analogues to date (Figure 1). This modification was originally introduced into oligonucleotides in order to protect their quite labile internucleoside phosphodiester linkages against hydrolysis by extra- and intracellular nucleases (*1*). Substitution of phosphorothioate linkages for the phosphodiester counterparts leads to creation of a new stereocenter at each modified internucleoside phosphorus atom and, as anticipated, has significantly stabilized oligonucleotides against enzymatic degradation (*2*). Moreover, efficient, economical and convenient synthetic protocols for preparation of oligonucleotide phosphorothioates, based primarily on phosphoramidite and H-phosphonate methods were developed and commercialized (*3, 4*). Finally, the first generation of antisense oligonucleotide-based drug candidates under evaluation in clinical trials are oligonucleotide phosphorothioates (*5, 6*).

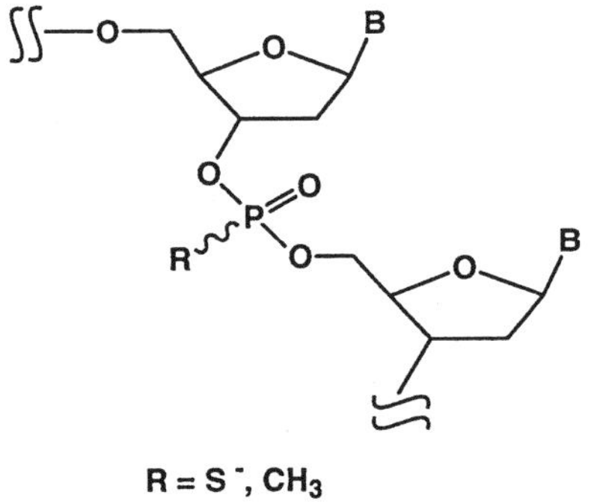

$$R = S^-, CH_3$$

Figure 1. Structure of oligonucleotide phosphorothioates (R = S) and oligonucleotide methylphosphonates (R = CH$_3$).

Duplexes formed by mixed-base stereorandom phosphorothioate oligomers containing mixtures of both Rp and Sp isomers or by stereopure compounds containing only Rp or Sp isomers, with complementary single-stranded DNA and RNA molecules are generally less thermodynamically stable than the ones formed by the parent phosphodiesters. The reduction in duplex melting temperatures, T_m, is typically 0.5-1° C per linkage (*7, 8*).

Formation of triple-stranded complexes by the oligonucleotide phosphorothioates as a third strand binding to phosphodiester duplexes has also been extensively studied. It was demonstrated that incorporation of stereorandom internucleoside phosphorothioate linkages into pyrimidine T, C-containing oligonucleotide 11-mers significantly destabilized the triplexes formed by these compounds with d(A, G)-containing duplex. Reduction of the triplexes' T_m was approximately 2° C per substitution, and formation of the triplex with fully modified oligonucleotide was observed only at pH 5, but not at pH 6, as demonstrated by gel-shift experiments and by melting curves (*9*). Similar observations were reported for the polynucleotide poly d(T, C), containing stereopure Rp phosphorothioate linkages (*10*). This compound does not form triplexes with the phosphodiester complement poly d(A, G), whereas polypurine all-Rp phosphorothioate polynucleotide poly d(A, G) did form a 1:2 triplex with two poly d(T, C) phosphodiester strands even at pH 8 (*10*). Analogously, d(T, C)-containing 16-mer oligonucleotide phosphorothioates, containing either stereorandom internucleoside linkages, or all-Rp isomers did not formed triplexes with d(A, G) oligomers, as judged by DNase I footprint titration experiments (*11*). It was also reported previously that the oligonucleotide containing a single phosphorothioate linkage in Sp configuration formed less stable duplexes than the parent unmodified oligomer (*12*). In contrast, d(T, G) containing 19-mer phosphorothioates with 75% of guanosine nucleosides and containing either a stereorandom mixture of isomers or all-Rp isomers formed stable triplexes with phosphodiester d(A, G) duplexes, being aligned in parallel to the polypurine strand (*11*). Stability of these triple-stranded complexes was similar to the one formed by the phosphodiester d(A, G) counterpart. In corroboration of these results, it was reported that the stereorandom oligonucleotide phosphorothioate 13-mer d(G$_4$AG$_5$AG$_2$) produced very stable and specific triplexes with polypurine-polypyrimidine duplexes, being directed in antiparallel orientation to the purine strand (*13*). Under near-physiological conditions, the melting temperature of the triplex was about 65° C and close to the T$_m$ of the duplex target. Thermal stability of this phosphorothioate oligomer formed triplex was practically the same as that for the isosequential phosphodiester compound. Also, the *in vivo* dimethylsulfate treatment generated footprint indicates that the triplex preformed by this purine-containing phosphorothioate and electroporated into the cells is quite stable inside the cells for at least 24 hours (*13*).

In summary, the incorporation of stereorandom or stereopure phosphorothioate linkages into the pyrimidine d(T, C)-containing oligonucleotide appears to destabilize triplexes formed by these compounds, whereas mostly purine-containing d(A, G) or d(T, G) phosphorothioates – whether as stereorandom or as all-Rp isomers – form stable triplexes, with the third strand running either antiparallel or parallel to the duplexes purine strands, respectively.

OLIGONUCLEOTIDE METHYLPHOSPHONATES

Oligonucleotide methylphosphonates (Figure 1) were introduced almost two decades ago as non-ionic uncharged analogues of phosphodiester oligomers which are resistant to enzymatic degradation (*14, 15*).

Similar to phosphorothioate derivatives, substitution of one non-bridging oxygen atom of internucleoside phosphodiester by a methyl group resulted in creation of a new stereocenter at the phosphorus atom. Stereorandom methyl phosphonates form duplexes with complementary DNA and RNA strands. The thermodynamic stability of these complexes generally, with the exception of oligothymidilates, is similar to that for the unmodified oligonucleotides at physiologically relevant buffers and ionic strength. Also, the T$_m$ of these duplexes is essentially salt concentration independent

(*15*). Stereopure oligonucleotide methylphosphonates with Rp configuration form more stable duplexes than do their Sp counterparts as well as isosequential phosphodiesters (*15*).

It was originally reported that the alternating methylphosphonate-phosphodiester decathymidilates containing only one unassigned isomer form 2T:1A triple-stranded complexes with poly r(A), and another isomer forms a very unstable 2T:1A triplex with poly d(A) (*16*). Moreover, oligothymidylates with alternating phoshodiester-methylphosphonate groups, containing a random mixture of both Rp and Sp isomers in every methylphosphonate position, form 2T:1A triplexes with oligo d(A) with a T_m 22° C lower than that of the unmodified phosphodiester oligomer. Triplex formation was not observed for the uniformly modified stereorandom 19-mer oligothymidylate methylphosphonate with either a natural oligo d(A):oligo d(T) duplex, or with a 0.5 mole equivalent of oligo d(A) strand (*17*). In contrast, oligo-adenylate with alternating phosphodiester–stereorandom methylphosphonate groups formed a stable 2T:1A triplex with oligothymidylates containing either natural phosphodiester or alternating phosphodiester–methylphosphonate groups, where the increase in the melting temperature of the complexes was 25° C or 16° C, respectively, relative to the all-phosphodiester complexes (*17*). In line with these observations, the later reports demonstrate formation of stable 2Pu:1Py[1] triple-stranded structures between oligopyrimidine phosphodiester RNA – r(T, C) and DNA – d(T, C) strands and stereorandom oligopurine d(A, G)-containing methylphosphonates (*18, 19*), as shown by gel-shift and thermal dissociation experiments. Melting temperatures of the triplexes with RNA targets ranged from 40° C to 54° C at pH 7.2 and close to physiological ionic strength of the buffer. Interestingly, such triplexes were not formed by the isosequential phosphodiester d(A, G) oligopurines (*19*). Also, it was reported that these triple-stranded complexes specifically inhibited mRNA translation *in vitro* at 1 μM of oligonucleotide methylphosphonates (*19*).

ZWITTERIONIC OLIGONUCLEOTIDE WITH INTERNUCLEOSIDE PHOSPHODIESTERAMIDATE LINKAGES

Formation of the triplexes requires association of three negatively charged strands. Thus stability of these complexes is very dependent on the ionic strength of the environment, and a high concentration of salts usually promotes triplex formation. In order to reduce interstrand repulsion between the triplex-forming strands, oligonucleotides containing positively charged groups (Figure 2, X = O) were designed and studied (*20, 21*).

Oligonucleotides containing 2'-deoxy or 2'-methoxy pyrimidines and alternating internucleoside phosphodiester and N-(dimethylaminopropyl) phosphodiester amidate linkages were prepared (*21*). At pH 7 these linkages are anionic and cationic,

---

[1] Pu, purine; Py, pyrimidine

respectively. These d(T/U, C) zwitterionic amidate pentadecanucleotides containing only one unassigned isomer form stable 2Py:1Pu triplexes, T_m 24-42° C with polypurine/polypyrimidine phosphodiester duplexes in buffers with 0.1 M NaCl, pH 7.0. Under the same experimental conditions, no triplex formation was observed for the other isomer of zwitterionic phosphoramidates, and the complexes formed by the isosequential phosphodiester compounds were also significantly less stable: the T_m was reduced by 23-30° C. Interestingly, zwitterionic phosphoramidates with the isomer, which did not form 2T:1A triplexes with oligo d(A)/oligo d(T) duplex, apparently formed very unusual 2A:1T triple-stranded complexes with poly d(A) at 1 M NaCl, as demonstrated by thermal dissociation experiments and circular dichroism spectroscopy (21).

Moreover, stable 2Pu:1Py triplexes were also reported for the alternating zwitterionic oligopurine phosphoramidates with a stereorandom mixture of isomers at every cationic internucleoside position (22). The zwitterionic d(A, G) phosphoramidates were aligned in antiparallel orientation to the purine strands of the duplexes, and the triplex stability was significantly higher than that for the phosphodiester counterparts, as judged by gel-shift analysis. It is important to note that triplex formation by the zwitterionic compounds was potassium concentration independent, whereas for the isosequential phosphodiesters this process was

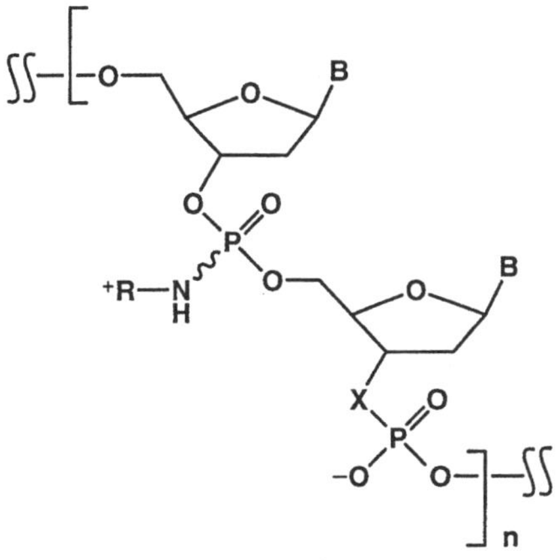

Figure 2. Structure of zwitterionic oligonucleotide phosphodiesters (X = O) and zwitterionic N3'→P5' phosphoramidates (X = NH).

practically completely inhibited by 150 mM KCl, very likely due to the self-association of the purine-rich oligonucleotides (*23*). Interestingly, alternating oligonucleotide phosphoramidates containing neutral N-(methoxyethyl)amine substituents at every other internucleoside phosphorus atom along with negatively charged phosphodiesters formed noticeably less stable complexes than did the zwitterionic or fully cationic compounds (*22*).

In line with these results, the zwitterionic oligodeoxyadenylate 11-mer with alternating negatively charged N3'→P5' phosphoramidates and positively charged stereorandom N-(dimethylaminopropyl)phosphoramidate internucleoside groups (Figure 2, X = NH) formed stable 2T:1A or 2r(U):1A triplexes with poly d(T) or with poly r(U) (*24*). Formation of duplexes by this compound with complementary polypyrimidines was not observed even at a 1:1 ratio between the strands. The melting temperatures of these complexes were 45.4° C and 41.2° C in low ionic strength buffers. For comparison, under similar experimental conditions, negatively charged natural all-phosphodiester oligoadenylate formed less stable complexes with poly d(T), T_m 22° C, or with poly r(U), $T_m < 0°$ (*24*).

OLIGONUCLEOTIDE N3'→P5' PHOSPHORAMIDATES

Uniformly modified oligodeoxyribonucleotides containing internucleoside N3'→ P5' phosphoramidate linkages (Figure 3) have been described recently (*25*). These compounds contain 3'-amino-2'-deoxynucleosides, instead of the natural 3'-hydroxy counterparts, and the aminonucleosides are linked together in oligonucleotides *via* phosphoramidate monoester linkages, unlike phosphodiester linkages for the native DNA and RNA. Similarly to the phosphodiesters, these oligonucleotide

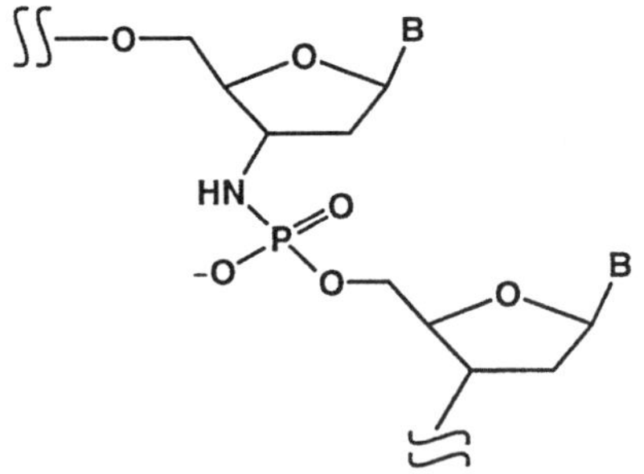

Figure 3. Structure of oligonucleotide N3'→P5' phosphoramidates.

phosphoramidates are negatively charged at neutral pH, and their internucleoside phosphorus atoms are achiral.

The oligonucleotide N3'→P5' phosphoramidates form very stable duplexes with complementary DNA and RNA strands. Increases in the duplex melting temperature reached 0.75-1.2° C and 2.3-2.6° C per modified nucleoside unit for the complexes formed with DNA and RNA, respectively, relatively to the isosequential phosphodiesters (*25, 26*). Compared to phosphodiester oligonucleotides, these compounds also have different chromatographic characteristics under reversed phase (RP), ion exchange (IE) and gel electrophoresis conditions, which are indicative of the increased hydration and rigidity of the sugar-phosphate backbone (*27, 28*). It was demonstrated by a high resolution NMR study that the 3'-aminonucleosides 2'-deoxyfuranose conformation is predominantly C3'-endo, or N-type, for the 3'-aminonucleosides and for single- and double-stranded N3'→P5' phosphoramidates, unlike the 3'-hydroxy-2'-deoxynucleosides, which prefer C2'-endo, or S-type, conformations (*29*). These findings were confirmed by X-ray analysis of crystals of a self complementary Dickerson-Drew dodecamer CGCGAATTCGCG containing only N3'→P5' phosphoramidate linkages (*30*). The phosphoramidate duplex adopts an RNA-like A-type of double helix, despite being made from 2'-deoxynucleosides (*29, 30*). Additionally, X-ray analysis revealed a more extensive hydration pattern of the phosphoramidate sugar-phosphate backbone, compared to that for phosphodiesters, and also formation of – uncharacteristic for the cognate phosphodiesters – hydrogen bonds between 3'-aminogroups, water and salt ions in the crystal (*30*).

The pyrimidine-containing oligonucleotide N3'→P5' phosphoramidates form very stable triplexes with phosphodiester duplexes, as demonstrated by various techniques, including thermal dissociation, gel-shift, cross-linking, endonuclease restriction arrest, and FTIR spectroscopy experiments. Thus, the triplex melting temperature of the phosphoramidate decathymidylate with a phosphodiester duplex was 47.2° C under near physiological pH and salt conditions (*25*). Even better complex stability was observed for the d(T, C)-containing phosphoramidate CTTCTTCCTTA, where the triplex T_m was 62° C. Parent phosphodiester oligonucleotides failed to form triplexes under similar conditions (*26*). Another T, C-containing phosphoramidate oligonucleotide, TTTCCTCCTCT, formed a much more stable triplex than did isosequential phosphodiester linked DNA, or RNA, or 2'- OMe oligomers: triplex T_m values were 45, <10, 26 and 26° C, respectively (*31*). Additionally, similarly to the phosphodiesters, N5-methylation of the cytosine bases of the phosphoramidates increased T_m values for the phosphoramidate triplex up to 61° C (*31*).

It was reported that among natural nucleic acids the pyrimidine-containing phosphodiester RNA oligomers form the most stable triplexes with DNA duplexes, with the third strand aligned in parallel orientation relatively to the targeted purine strand (*32*). This was attributed in part to the N-type of ribonucleoside sugar puckering, which may facilitate and stabilize Hoogsteen base pairing. In this respect, it is important to note that the oligonucleotide phosphoramidates also having RNA-like N-type nucleoside conformation still form much more stable triplexes than

RNA phosphodiesters do. Moreover, 2'-OMe phosphodiesters which also adopt C3'-endo or N-type sugar conformations form less stable triplexes than the isosequential and conformationally close N3'→P5' phosphoramidates (*31*). Interestingly, it was reported that 2'-fluoro C, U-containing phosphodiester oligonucleotides formed the least stable triplexes relative to the 2'-deoxy, 2'-OMe or 2'-ribo counterparts. Melting temperature for their triplexes at pH 6.1 was 20.7, 28.3, 36.8 and 37.7° C, respectively (*33*). The reduction in 2'-fluoro phosphodiester oligonucleotide triplex stability takes place despite predominance of the C3'-endo RNA-like nucleoside sugar conformation for 2'-fluoronucleosides (*33, 34*). This indicates that some additional to nucleosides sugar puckering factors and combination of those factors such as, for instance, sugar phosphate backbone hydration and rigidity, spatial positioning of the nucleoside bases, and change in basicity of heterocyclic bases caused by sugar ring substitutions and modifications, may contribute to the elevated or diminished triplex stability. Oligopyrimidine N3'→P5' phosphoramidates may have an optimal combination of triplex–favoring properties: N-type sugar puckering and increased hydration of the backbone due to the presence of 3'-aminogroups which can donate and accept hydrogen bonds (*30*). Additionally, the oligonucleotide phosphoramidate backbone is apparently more rigid than the phosphodiester backbone due to the partial electron conjugation between a free pair of 3'-nitrogen and phosphorus atoms. The X-ray structural data for the phosphoramidate duplex crystal revealed an almost planar position of the three substituents at the 3'-nitrogen atoms, which may indicate partial double bond character for the internucleoside P-N link (*30*).

In line with RNA purine-rich oligomers, d(A, G)-containing phosphoramidates seem do not form stable triplexes with antiparallel orientation to the duplexes polypurine strand (*31, 35*). Preferred C3'-endo nucleoside sugar puckering by both types of oligonucleotides very likely disfavors formation of reverse Hoogsteen base pairs between two oligopurine strands.

Oligonucleotide N3'→P5' phosphoramidates having blocks of thymidine and guanosine nucleotides apparently form more stable triplexes than do T, C-containing compounds with the same dsDNA target, being oriented in parallel to the duplex oligopurine strand. Thus, the T_m for the triplex formed by a 15-mer phosphoramidate, $T_4CT_4G_6$, with dsDNA at pH 7 was 62° C in comparison with 40° C for a $T_4CT_4C_6T$ oligomer. The melting temperature for the triplexes formed by the cognate T, G- and T, C-containing phosphodiesters was 32° C and <10° C, respectively (*31, 35*). The triplexes formed by these T, C- and T, G-containing phosphoramidates were able to effectively protect the dsDNA target site from cleavage by the restriction enzyme Dra I (*31*). Moreover, triplex formation by the T, G-phosphoramidate resulted in an efficient inhibition of transcription, both *in vitro* and in cell culture, of the HIV-1 *nef* gene at submicromolar concentrations of the oligomer (*36, 37*). The high target recognition specificity by the triplex forming oligonucleotide phosphoramidates was confirmed by the UV light induced cross-linking of posphoramidate-psoralen conjugates to the predicted sites on the dsDNA targets (*36*).

Interestingly, pentadecathymidine phosphoramidate formed only a 2T:1A triplex with 2'-deoxy pentadecaadenylate, even if the molar ratio between the purine and pyrimidine strands was 1:1, as demonstrated by gel shift analysis, thermal dissociation experiments and FTIR spectroscopy (*38*). In addition, it was demonstrated by gel shift experiments that the phosphoramidate pentadecathymidylate effectively displaced a phosphodiester dT_{15} strand from a $dT_{15}:dA_{15}$ duplex to form a 2T:1A triplex with two dT-phosphoramidate and one phosphodiester oligoadenylate oligomer (*38*).

In summary, oligonucleotide N3'→P5' phosphoramidates form extremely stable triplexes with dsDNA targets, and they are both resistant to enzymatic digestion and cell permeable. This indicates that phosphoramidate oligonucleotides could be used as efficient modulators of gene expression *in vitro* and *in vivo*.

CONCLUSIONS

The sugar-phosphate backbone modified oligonucleotide analogues described herein can form triple-stranded complexes with purine-containing phosphodiester DNA and RNA oligomers. Stability, structure, strand composition and strand orientation for these triplexes is largely determined by the chemical nature and spatial positioning of the modifications. Among the compounds discussed, the pyrimidine-containing oligonucleotide N3'→P5' phosphoramidates may form one of the most stable complexes studied to date.

ACKNOWLEDGMENTS

I would like to thank Krisztina Pongracz for technical assistance.

REFERENCES

1. Eckstein, F. (1966). Nucleoside phosphorothioates. *J. Am. Chem. Soc.* **88**, 4292-4293.
2. Eckstein, F. (1985). Nucleoside phosphorothioates. *Annual Rev. Biochem.* **54**, 367-402.
3. Stec, W. J., Zon, G., Egan, W. and Stec, B. J. (1984). Automated solid phase synthesis, separation and stereochemistry of phosphorothioate analogues of oligodeoxyribonucleotides. *J. Am. Chem. Soc.* **106**, 6077-6079.
4. Zon, G. and Stec, W. J. (1991). *In* Eckstein, F. (Ed.) Oligonucleotides and Analogues: A Practical Approach, IRL Press, Oxford, UK, 87-108.
5. Glaser, V. (1996). Oligonucleotide therapies move toward efficacy trials to treat HIV, CMV, cancer. *Genetic Engineering News* **16**, February 1, 1-21.
6. Roush, W. (1997). Antisense aims for a renaissance. *Science* **276**, 1192-1193.
7. Kebler-Herzog, L., Zon, G., Uznanski, B., Whittier, G. and Wilson, W. D. (1991). Duplex stabilities of phosphorothioate, methylphosphonate, and RNA analogs of two DNA 14-mers. *Nucleic Acids Res.* **19**, 2979-2986.

8. Koziolkiewicz, M., Krakowiak, A., Kwinkowski, M., Boczkowska, M. and Stec, W. J. (1995). Stereodifferentiation – the effect of P chirality of oligonucleoside phosphorothioates) on the activity of bacterial RNase H. *Nucleic Acids Res.* **23**, 5000-5005.

9. Xodo, L., Alunni-Fabbroni, M., Manzini, G. and Quadrifoglio, F. (1994). Pyrimidine phosphorothioate oligonucleotides form triple-stranded helices and promote transcription inhibition. *Nucleic Acids Res.* **22**, 3322-3330.

10. Latimer, L., Hampel, K. and Lee, J. S. (1989). Synthetic repeating sequence DNAs containing phosphorothioates: nuclease sensitivity and triplex formation. *Nucleic Acids Res.* **17**, 1549-1561.

11. Hacia, J. G., Wold, B. and Dervan, P. B. (1994). Phosphorothioate oligonucleotide-directed triple helix formation. *Biochemistry* **33**, 5367-5369.

12. Kim, S.-G., Tsukahara, S., Yokoyama, S. and Takaku, H. (1992). The influence of oligodeoxyribonucleotide phosphorothioate pyrimidine strands on triplex formation *FEBS Lett.* **314**, 29-32.

13. Svinarchuk, F., Debin, A., Bertrand, J.-R. and Malvy, C. (1996). Investigation of the intracellular stability and formation of a triple helix formed with a short purine oligonucleotide targeted to the murine c-*pim-1* proto-oncogene promoter. *Nucleic Acids Res.* **24**, 295-302.

14. Miller, P. S., Yano, J., Yano, E., Carrol, C., Jayraman, K. and Ts'o, P. O. P. (1979). Nonionic nucleic acid analogues. Synthesis and characterization of deoxyribonucleoside methylphosphonates. *Biochemistry* **18**, 5134-5143.

15. Miller, P. S. (1991). Oligonucleotide methylphosphonates as antisense reagents. *Bio/Technology* **9**, 358-361.

16. Miller, P. S., Dreon, N., Pulford, S. M. and McParland K. B. (1980). Oligothymidylate analogues having stereoregular, alternating methylphosphonate/phosphodiester backbone. *J. Biol. Chem.* **255**, 9659-9665.

17. Kibler-Herzog, L., Zon, G., Mizan, S. and Wilson, W. D. (1993). Stabilities of duplexes and triplexes of $dA_{19}+dT_{19}$ with alternating methylphosphonate and phosphodiester linkages. *Anti-Cancer Drug Design* **8**, 65-79.

18. Trapane, T., Hogrefe, R. I., Reynolds, M. A., Kan, L.-S. and Ts'o, P. O. P. (1996). Interstrand complex formation of purine oligonucleotides and their nonionic analogues: the model system of $d(AG)_8$ and its complement, $d(CT)_8$. *Biochemistry* **35**, 5495-5508.

19. Reynolds, M. A., Arnold, L. J., Jr., Almazan, M. T., Beck, T. A., Hogrefe, R. I., Metzler, M. D., Stoughton, S. R., Tseng, B. Y., Trapane, T., Ts'o, P. O. P. and Woolf, T. M. (1994). Triple-strand-forming methylphosphonate oligodeoxynucleotides targeted to mRNA efficiently block protein synthesis. *Proc. Natl. Acad. Sci. USA* **91**, 12433-12437.

20. Letsinger, R. L., Singman, C. N., Histand, G. and Salunkhe, M. (1988). Cationic oligonucleotides. *J. Am. Chem. Soc.* **110**, 4470-4471.

21. Chaturvedi, S., Horn. T. and Letsinger, R. L. (1996). Stabilization of triple-stranded oligonucleotide complexes: use of probes containing alternating phosphodiester and stereo-uniform cationic phosphoramidate linkages. *Nucleic Acids Res.* **24**, 2318-2323.

22. Dagle, J. M. and Weeks, D. L. (1996). Positively charged oligonucleotides overcome potassium-mediated inhibition of triplex DNA formation. *Nucleic Acids Res.* **24**, 2143-2149.

23. Cheng, A.-J. and Van Dyke, M. W. (1993). Monovalent cation effects on intermolecular purine-purine-pyrimidine triple-helix formation. *Nucleic Acids Res.* **21**, 5630-5635.

24. Mignet, N. and Gryaznov, S. M. (1998). Zwitterionic oligodeoxyribonucleotide N3'→P5' phosphoramidates: synthesis and properties. *Nucleic Acids Res.* **26**, 431-438.

25. Gryaznov, S. and Chen, J.-K. (1994). Oligodeoxyribonucleotide N3'→P5' phosphoramidates: synthesis and hybridization properties. *J. Am. Chem. Soc.* **116**, 3143-3144.

26. Gryaznov, S. M., Lloyd, D. H., Chen, J.-K., Schultz, R. G., DeDionisio, L. A., Ratmeyer, L. and Wilson, W. D. (1995). Oligonucleotide N3'→P5' phosphoramidates. *Proc. Natl. Acad. Sci. USA* **92**, 5798-5802.

27. Chen, J.-K., Schultz, R. G., Lloyd, D. H. and Gryaznov, S. M. (1995). Synthesis of oligodeoxyribonucleotide N3'→P5' phosphoramidates. *Nucleic Acids Res.* **23**, 2661-2668.

28. Gryaznov, S. M. (1997). Synthesis and properties of the oligonucleotide N3'→P5' phosphoramidates. *Nucleosides & Nucleotides* **16**, 899-905.

29. Ding, D., Gryaznov, S. M., Lloyd, D. H., Chandrasekaran, S., Yao, S., Ratmeyer, L., Pan, Y. and Wilson. W. D. (1996). An oligodeoxyribonucleotide N3'→P5' phosphoramidate duplex forms an A-type helix in solution. *Nucleic Acids Res.* **24**, 354-360.

30. Tereshko, V., Gryaznov, S. and Egli, M. (1998). Consequences of replacing the DNA 3'-oxygen by an amino group: high resolution crystal structure of a fully modified N3'→P5' phosphoramidate DNA dodecamer duplex. *J. Am. Chem. Soc.* **120**, 269-283.

31. Escudé, C., Giovannangeli, C., Sun, J.-S., Lloyd, D. H., Chen, J.-K., Gryaznov, S. M., Garestier, T. and Hélène, C. (1996). Stable triple helices formed by oligonucleotide N3'→P5' phosphoramidates inhibit transcription elongation. *Proc. Natl. Acad. Sci. USA* **93**, 4365-4369.

32. Roberts, R. W. and Crothers, D. M. (1992). Stability and properties of double and triple helices: dramatic effects of RNA or DNA backbone composition. *Science* **258**, 1463-1466.

33. Shimizu, M., Konishi, A., Shimada, Y., Inoue, H. and Ohtsuka, E. (1992). Oligo(2'-O-methyl)ribonucleosides. Effective probes for duplex DNA. *FEBS Lett.* **302**, 155-158.

34. Guschlbauer, W. and Jankowski, K. (1980). Nucleoside conformation is determined by the electronegativity of the sugar substituent. *Nucleic Acids Res.* **8**, 1421-1433.

35. Semerad, C. L. and Maher, L. J., III. (1994). Exclusion of RNA strand from a purine motif triple helix. *Nucleic Acids Res.* **22**, 5321-5325.

36. Giovannangeli, C., Perrouault, L., Escudé, C., Gryaznov, S. and Hélène, C. (1996). Efficient inhibition of transcription elongation *in vitro* by oligonucleotide phosphoramidates targeted to proviral HIV DNA. *J. Mol. Biol.* **261**, 386-398.

37. Giovannangeli. C., Diviacco, S., Labrousse, V., Gryaznov, S., Charneau, P. and Hélène, C. (1997). Accessibility of nuclear DNA to triplex-forming oligonucleotides: the integrated HIV-1 provirus as a target. *Proc. Natl. Acad. Sci. USA* **94**, 79-84.

38. Zhou-Sun, B.-W., Sun, J.-S., Gryaznov, S. M., Liquier, J., Garestier, T., Hélène, C. and Taillandier, E. (1997). A physico-chemical study of triple helix formation by an oligothymidylate with N3'→P5' phosphoramidate linkages. *Nucleic Acids Res.* **25**, 1782-1787.

7 OLIGONUCLEOTIDES AS ARTIFICIAL DNases

Valentina Zarytova and Asya Levina

SUMMARY

Triplex forming oligonucleotides (TFO) bearing reactive groups which can cause site-specific damage or direct cleavage of double-stranded DNA targets are the subject of this chapter.

INTRODUCTION

The ability of oligonucleotides of certain sequences to form triplex structures with dsDNA open the wide opportunity for so called antigene strategy (*1-5*). Triplex forming oligonucleotides (TFOs) provide a possibility to achieve purine and mixed purine-pyrimidine sequences of DNA through Hoogsteen or reverse Hoogsteen bond formation in the major groove of the double helix. Such triplex forming oligonucleotides equipped with appropriate modifying moieties can deliver this group to the sequence of interest and damage it. The major advantage of using modified oligonucleotides is the more diverse and longer recognition sequences compared to small DNA-binding drugs and other chemicals. Such oligo-conjugates are 10^6 times more specific than restriction enzymes (*6, 7*). The site-specific cleavage of double-helical DNA is a potentially powerful new approach in the design of artificial nucleases; it is essential for chromosomal mapping, gene isolation, DNA sequencing, and selective inhibition of gene expression and, thereby, for designing potential drugs to be used in anticancer and antiviral chemotherapy.

OLIGONUCLEOTIDE CONJUGATES AS CHEMICAL DNases

Reactive substituents linked to TFO can be divided into several groups: metal chelates, alkylating agents, photosensitizers, radiolabels, nonspecific enzymes, and others. They were initially developed for the antisense strategy, i.e. for targeting RNA and ssDNA. In this chapter we will discuss the literature data concerning modification of only dsDNA in triplexes.

TFO conjugates used for this purpose consist usually of 15-20 nucleotide units and contain reactive groups tethered to either the 3'- or 5'-terminal phosphates or else to the C5 position of deoxyuridine. The length and structure of the linker between TFO and substituents are very important and must be optimized in each case. The main mechanisms of damaging DNA include oxidation of d-ribose, covalent photo-induced cross-linking or direct cleavage, and electrophilic alkylation of the bases.

Oligonucleotides bearing metal chelates

Action of this class of oligonucleotide conjugates is based on the ability of metal-chelates to produce active oxygen-containing species which attack d-ribose and mediate cleavage of the DNA backbone in the presence of reducing agents.

EDTA·Fe(II)-oligo conjugates (*6-16*). Oligonucleotides equipped with an EDTA·Fe(II) moiety at the C5-position of thymidine (T*) produce sequence-specific double strand breaks with efficiency ranging from 5 to 25% at their target sites within genomes as large as that of bacteriophage λ (48.5 kbp) (*6-12*).

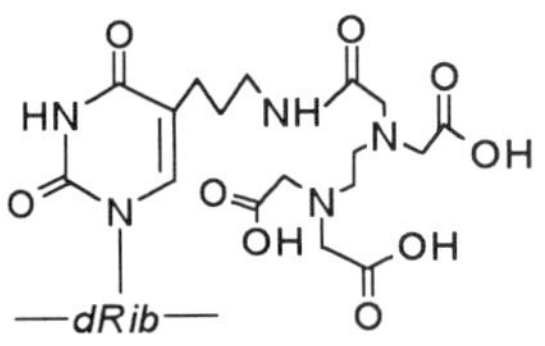

Pyrimidine oligonucleotides coupled 3'-3' via a 1,2-dideoxy-D-ribose linker possess the requisite orientation for binding of 5'(Pu)$_m$(Py)$_n$3' sequences within the major groove of a double helix (*12*). This probe produced site-specific double-strand cleavage (14%) at the target site of plasmid DNA (4.05 kbp). This result is an example of a larger class of potential *multimeric crossover oligonucleotides*.

Affinity cleavage analysis with EDTA-oligo probes was used for detection different helical conformations of RNA and DNA hybrid triplexes (*13*). The same technique was used for analyzing the structure of the protein-associated joint molecule formed between a RecA·EDTA-oligo filament and dsDNA (*14*). It was shown that the filament containing TFO bound antiparallel to its complementary sequence, which indicated its location in the minor groove.

Two oligonucleotides (one of them contained T*) that formed a Y-shaped complex with dsDNA were used for cleavage in this complex with a yield higher than that observed by a single oligo-EDTA targeted to the same site (*15*).

Phenanthroline-oligo conjugates (*17-25*). Circular and linearized SV40 DNA was cleaved at a single binding site with high efficiency (70%) by using the phenanthroline-polypyrimidine conjugate in the presence of Cu(II) (*17,18*). Location of the cleavage sites on the two strands strongly suggests that phenanthroline intercalates within the double helix at the boundary of the triplex helix, and that cleavage occurs via reactions in the minor groove, even though the oligonucelotide binds to the major groove of DNA.

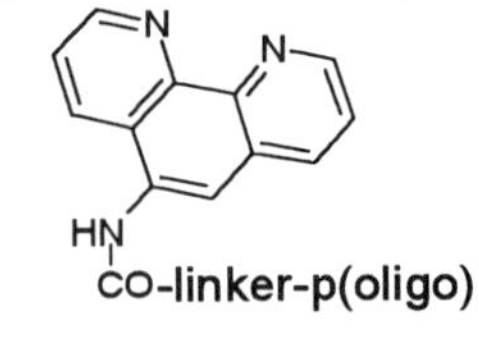

Sequence-specific 10% double-stranded scission of the linearized plasmid pGM820 was accomplished with phenanthroline containing RNA about 110 bases long within the R-loop (*21, 22*). R-loop directed scission is the first method for DNA cleavage applicable to *any* sequence (without limitation of being a polyPy·polyPu sequence).

The triple helices containing canonical base triplets showed, as expected, relatively higher cleavage efficiency (60%) as compared with other triplets (*23*). Cleavage of the promoter region of the human thrombomodulin gene was performed by phenanthroline linked 2'-O-Me-RNAs, which were found to work as preferntial purine-strand cutters (*24*).

Oligonucleotide conjugates with other metal chelates. The complex of antitumor antibiotic bleomycin A_5 with Fe(II) is known to induce degradation of a DNA backbone after formation of an "active" oxygen-coordinated form (*26*). A bleomycin derivative of $(pT)_{16}$ was used for efficient (70%) site-specific cleavage of the dsDNA fragment (30 bp) (*26*).

Fe-bleomycin-oligo

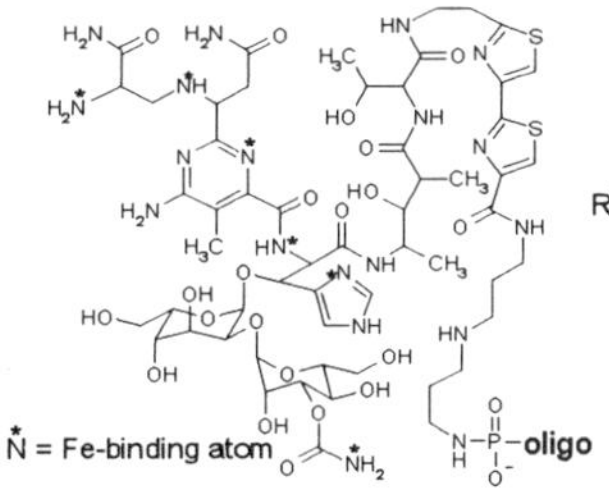

Mn-porphyrin~oligo

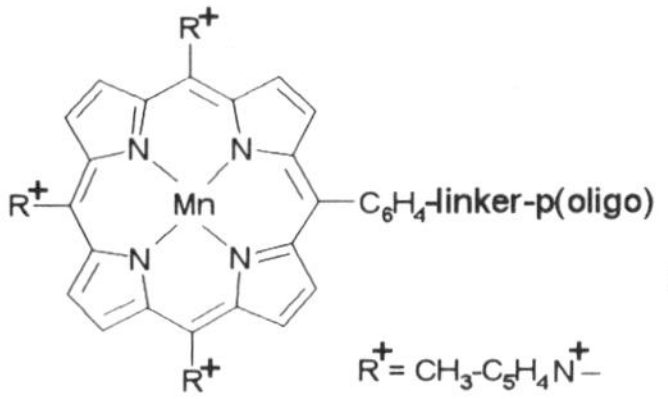

Cu-dispheral~oligo

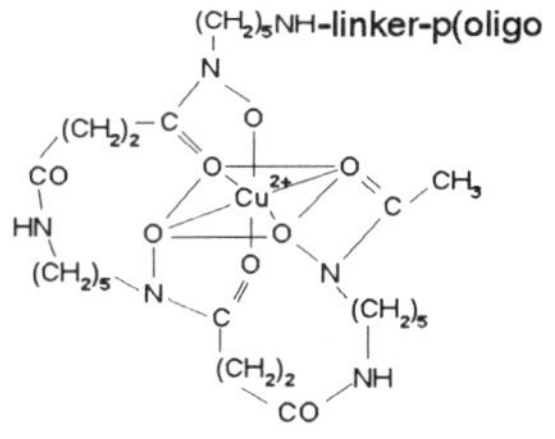

A high yield (80%) of cleavage of a 29-mer dsDNA present in the HIV-1 genome was obtained by using the oligo-Mn(II)-porphyrin conjugate when the porphyrin residue was attached to the 5'-end *via* a long linker containing spermine (*27*). The positive charge of cationic metalloporphyrin endows it with a high affinity for nucleic acids. The synthetic 32-mer duplex was cleaved with reasonable site-specificity and efficiency by Cu(II)-metallodispheral-oligo conjugate (*28*).

Oligonucleotides bearing alkylating groups

This class of substituents englobes the electrophilic compounds that alkylate preferentially the most nucleophilic center in DNA, N7 of the guanine base. Alkaline (piperidine) treatment of alkylated DNA products results in depurination and strand cleavage at the modified sites.

The first group used for design of reactive oligonucleotide derivatives was an alkylating moiety, an aromatic 2-chloroethylamino group (RCl) (*29*). RCl reacts with nucleophiles through an intermediate aziridinium ion that attacks N7 of guanines (*5*). The oligonucleotide-RCl conjugates were employed for investigation of triplex structures (*30-33*). Sequence-specific alkylation (60-70%) of dsDNA fragments of immunoglobin gene (*33, 34*), *c-fos* gene (*35*), human interferon gene (*36*), and bovine papilloma vector expressing human interferon-*g* (*37*) has been performed. The efficiency of alkylation of a superhelical DNA enhances with an increase of the plasmid's negative superhelicity (*34, 38*). The quantitative modification of the G bases was achieved with repeated treatment of DNA with gradually increasing concentrations of the reagent (*37*).

RCl~oligo

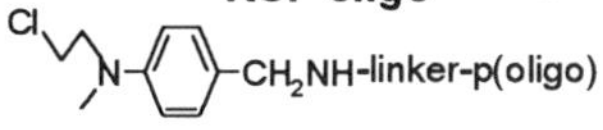

chlorambucil~oligo

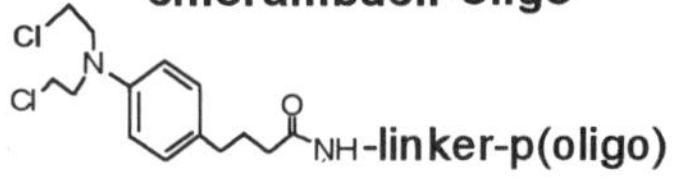

TFO bearing the clinically used chlorambucil

and other nitrogen mustards are able, like RCl, to alkylate guanines through aziridinium ion formation (*39-45*). One strand of dsDNA target (40-mer) is alkylated site-specifically with high efficiency (77%); 3',5'-bis-chlorambucil-oligo conjugates alkylate both strands of dsDNA (*39*). The triplex selective intercalator coralyne (*40-43*) or an analog of guanine in the TFO structure (*41*) enhances both triplex stability and the efficiency of alkylation (85-90%). Chlorambucil-containing oligonucleotides were successfully used for site-directed modification of a native chromosomal gene for the chemokine receptor in intact human cells (*41*) and human genomic DNA (DQb1*0302 allele) (*42*). Buffer composition and structure of nitrogen mustard influence alkylation rate (*43*). In contrast to the TFO conjugates, the complex RecA protein with a chlorambucil-oligo conjugate (*42*) was less efficient (50%), but allowed modification of *any* target sequence, which makes this method applicable to whole genomic DNA.

Oligonucleotides bearing an aziridine residue in the heterocyclic base can site-specifically and highly-efficiently (95%) cross-link with the nucleophilic N7 of guanine in a triplex (*46*).

aziridine~oligo

Electrophilic properties of the N-halo-acetamido group tethered to TFO were exploited for sequence-specific alkylation of dsDNA (*47-52*). The rate constant ratios were estimate for halo-acetamido-oligo conjugates: $k_I/k_{Br}=0.2$; $k_{Cl}/k_{Br}=0.06$, so the Br-acetamide derivatives appeared to be the most efficient(*46*). Double-strand cleavage (60-90%) at a single site of

Br-acetamido~oligos adjacent inverted purine tracts of dsDNA of different sizes: restriction fragments, plasmids and yeast chromosome III was carried out after alkylation with Br-acetamido-oligonucleotides (*49-51*). Bromoacetamido-oligonucleotides, which can form sequence-specific complexes composed of the adjacent triplex and D-loop regions, were used for trapping D-loops in a supercoiled dsDNA (*52*).

A very promising alkylating reagent, MCPI, a structural analog of the reactive subunit of the potent antibiotic CC-1065, has been recently proposed to modify dsDNA (*53*). MCPI moiety reacts in the DNA minor groove,

MCPI~oligo alkylating adenines at the N3 position. MCPI-oligo conjugate (24 nt) quantitatively alkylated dsDNA (65 bp). The required long linker between the TFO and MCPI was chosen to traverse one strand of the target duplex from the major groove-bound TFO and deliver the reaction group to the minor groove. The proposed types of conjugates seem to be advantageous reagents because they are very stable in aqueous solution even in the presence of strong nucleophiles and react very efficiently once bound to the target DNA.

Oligonucleotides bearing photoactive groups

Photoreactive oligonucleotide derivatives provide a unique opportunity to modify target DNA at any moment by initiating reactions with irradiation. Proflavine (*54*),

azidoproflavine~oligo **proflavine~oligo** **azidophenacyl~oligo** **ellipticine*~oligo**

R = linker-p(oligo)

azidophenacyl (*55*), azidoproflavine (*56*), and ellipticine* (*57, 58*) groups, being attached to homopyrimidine oligonucleotides, caused a photosensitized reaction on either strand of the dsDNA within a triplex structure. The cross-linked bases could be identified after piperidine treatment of the DNA fragment that led to the cleavage reaction. Although the above conjugates did not result in a high efficiency of DNA cleavage, they played a significant role in studying triplex structures formed by either oligo-β- or oligo-α-deoxynucleotides.

A 14-mer oligonucleotide carrying the spherical fullerene residue has been used for the direct cleavage of a dsDNA (24 bp) and a 41-mer hairpin (*59*). In both

fullerene~oligo **chlorin~oligo**

CO-O-linker-p(oligo)

NHCO-linker-p(oligo)

cases, upon irradiation at 310 nm this conjugate modified the target with high site-specificity, nearly exclusively at the G bases, which is consistent with singlet oxygen mediation. The same profile of site and base selectivity was observed with an oligonucleotide-chlorin conjugate (*60*). Upon irradiation at 428 or 668 nm, piperidine labile sites at G positions were produced in a sequence-specific manner with a total efficiency 30-50%.

The dsDNA fragments (190 bp and 110 bp) were subjected to photomodification at 320-420 nm by TFOs containing ethydium and azidoethydium residues (*61*). Both formation of covalent adducts and hidden modification (detected after the piperidine treatment) were observed (80%). Sequence-specific photomodification (50%) of a 27-mer dsDNA fragment was carried out upon irradiation at 303-365 nm using a perfluoroarylazido-TFO conjugate (*62*). Azidoethydium and azidoperfluoroaryl groups, due to their high quantum yield of photomodification (0.3-0.5), allow one to employ a relatively low intensity of irradiation and/or significantly shorter irradiation time.

arylazido~oligos

Very efficient and widely used photoactive psoralen-containing oligonucleotides are described in another chapter.

[125]I-Labeled oligonucleotides

TFOs bearing ^{125}I at the C5 position of cytidine were used to deliver radionuclides to target DNA sequences so that ^{125}I decay produced sequence-specific double-strand breaks of dsDNA (up to 25%) within 10 bp around the decay site (*63-66*). Synthetic dsDNAs (*65, 66*), linearized plasmid pCR 12713 Nef (*63*), and a stretch of the human HPRT gene (*63*) were used as the targets. It is commonly suggested that the main targets in the DNA molecule are deoxyriboses and phosphate groups.

Oligonucleotide-nuclease conjugates

The polypyrimidine oligonucleotide-staphylococcal nuclease adduct is capable of binding to supercoiled duplex DNA (*67-70*). This conjugate directly incorporates into duplex DNA by strand invasion *via* D-loop formation. Oligo-nuclease adduct site-specifically cleaves pUC19 and pDP20 plasmids with 60-75% efficiency (*67-69*). Semisynthetic nucleases hydrolyze the phosphodiester bonds of DNA and generate products with intact termini, which are suitable for subsequent enzymatic manipulation.

The nuclease domain of the conjugate seems to play a dual role involving both substrate recognition (this is along with the oligonucleotide domain) and substrate cleavage (*70*). Moreover, nuclease increases stability of the complex dsDNA·oligo-nuclease. Detailed understanding of how the nuclease accomplishes this may lead to the design of conjugates which are more amenable to *in vitro* and *in vivo* applications.

Special cases

An attractive option for biological application would be the recruitment of a cellular enzyme which would be directed to perform a sequence-specific cleavage or modification of dsDNA. Such a system using the topoisomerase inhibitor, camptothecin, has been demonstrated in a recent study (*71*). After incubation of the camptothecin-oligo conjugate with a DNA restriction fragment in the presence of topoisomerase I, a pronounced trapping (30%) of the cleavable complex containing one-strand breaks had occurred in a sequence-specific manner.

A new class of cleaving molecules, named deoxyribozymes, has recently been reported (*72*), which are able to cleave single-stranded DNAs. The authors have found an effective means for specific cleavage (100%) of one strand of an extended dsDNA using repetitive cycles of thermal denaturation and reannealing.

CONCLUSION

TFOs carrying reactive moieties can damage target DNAs, making their interaction irreversible, in contrast to that with unmodified oligonucleotides. Because of this property, such conjugates may be more promising agents as potential drugs, which can be targeted at particular genes to put them out of action irreversibly. TFO conjugates also represent a very useful tool for studying different aspects of molecular and cell biology.

Extensive use of TFO-conjugates as artificial endonucleases depends on the stringency of the hybridization reaction and the efficiency of modification within triplexes. Most of the substituents do not prevent triplex formation and even slightly strengthen it; sometimes intercalators (attached or not to TFO conjugates) are used to increase triplex stability and, thereby, the efficiency of modification.

The extent of cleavage of target DNAs is a crucial point. TFO-conjugates bearing some metal-chelate cleavers and alkylating groups seem to be the most efficient. They produce 80-100% cleavage, which makes the corresponding oligonucleotide conjugates very efficient tailored artificial DNases.

REFERENCES

1. Gee, J. E. and Miller, D. M. (1992). Structure and application of intermolecular DNA triplexes. *Am. J. Med. Sci.* **304**, 366-372.
2. Helene, C. (1991). The anti-gene strategy: control of gene expression by triplex-forming-oligonucleotides. Anti-Cancer Drug Des. **6**, 569-584.
3. Radhakrishnan, I. and Patel, D. J. (1994). DNA triplexes: solution structures, hydration sites, energetics, interactions, and function. *Biochemistry* **33**, 11405-11416.
4. Helene, C. And Garestier, T. "Oligonucleotide-directed recognition of double-helical DNA. In: Chemical synthesis." In *Chemical Synthesis*, Chatgilialoglu, C. and Snieckus, V., ed. Netherlands: Kluwer Academic Publishers. 1996.
5. Knorre, D. G., Vlassov, V. V., Zarytova, V. F., Lebedev, A. V. and Fedorova, O. S. *Design and targeted reactions of oligonucleotide derivatives*. Boca Raton-Ann Arbor-London-Tokio: CRC Press, 1994.
6. Strobel, S. A. and Dervan, P. B. (1989). Cooperative site-specific binding of oligonucleotides to DNA. *J. Am. Chem. Soc.* **111**, 7286-7287.
7. Strobel, S. A. and Dervan, P. B. (1990). Site-specific cleavage of a yeast chromosome by oligonucleotide-directed triple-helix formation. *Science* **249**,73-75.
8. Moser, H. E. and Dervan, P. B.(1987). Sequence-specific cleavage of double helical DNA by triple helix formation .*Science* **238**, 645-650.
9. Povsic, T. J. and Dervan, P. B. (1989). Triple helix formation by oligonucleotide on DNA extended to the physiological pH range. *J. Am. Chem. Soc.* **111**, 3059-3062.
10. Maher, L. J., III, Wold, B. and Dervan, P. B. (1989). Inhibition of DNA binding proteins by oligonucleotide-directed triple helix formation. *Science* **245**, 725-730.
11. Griffin, L. C. and Dervan, P. B. (1989). Recognition of thymine-adenine base pairs by guanine in a pyrimidine triple helix motiff. *Science* **245**, 967-971.
12. Horne, D. A. and Dervan, P. B. (1990). recognition of mixed-sequence duplex DNA by alternate-strand triple-helix formation. *J. Am. Chem. Soc.* **112**, 2435-2437.
13. Han, H. and Dervan, P. B. (1994). Different conformational families of pyrimidine·purine· pyrimidine triple helices depending on backbone composition. *Nucl. Acids Res.* **22**, 2837-2844.
14. Baliga, R., Singleton, J. W. and Dervan, P. B. (1995). RecA oligonucleotide filaments bind in the minor groove of double-stranded DNA. *Proc. Natl. Acad. Sci USA* **92**, 10393-10397
15. Distefano, M. D., Shin, J. A. and Dervan, P. B. (1991). Corporative binding of oligonucleotides to DNA by triple helix formation: dimerisation via Watson-Crick hydrogen bonds. *J. Am. Chem. Soc.* **113**, 5901-5908.
16. Boidot-Forget, M, Chassignol, M, Takasugi, M, Thuong, N. T, and Helene, C. (1988)

Site-specific cleavage of single-stranded and double-stranded DNA sequences by oligodeoxyribonucleotides covalently linked to an intercalating agent and an EDTA-Fe chelate. *Gene* **72**, 361-371

17. François, .J. C., , Saison-Behmoaras, T., Chassignol, M., Thuong, N. T. and Helene, C. (1988). Artificial nucleases: specific cleavage of the double helix of DNA by oligonucleotides linked to copper-phenanthroline complex. *C. R. Acad. Sci. III* **307**, 849-854

18. Francois, J. C., Saison-Behmoaras, T., Barbier, C., Chassignol, M., Thuong, N. T. and Helene, C. (1989). Sequence-specific recognition and cleavage of duplex DNA via triple-helix formation by oligonucleotides covalently linked to a phenanthroline-copper chelate. *Proc. Natl .Acad. Sci. USA* **86**, 9702-9706

19. Francois, J.C., Saison-Behmoaras, T., Chassignol, M., Thuong, N.T and Helene, C. (1989). Sequence-targeted cleavage of single- and double-stranded DNA by oligothymidylates covalently linked to 1,10-phenanthroline. *J. Biol. Chem.* **264**, 5891-5898.

20. Francois, J. C. and Helene, C. (1995). Recognition and cleavage of single-stranded DNA containing hairpin structures by oligonucleotides forming both Watson-Crick and Hoogsteen hydrogen bonds. *Biochemistry* **34**, 65-72

21. Chen, C. H., Gorin, M. B. and Dervan, P. B. (1993). Sequence-specific scission of DNA by the chemical nuclease activity of 1,10-phennathroline-copper (I) targeted by RNA. *Proc. Natl. Acad. Sci. USA* **90**, 4206-4210.

22. Sigman, D. S., Chen, C. H. and Gorin, M. B. (1993) Sequence-specific scission of DNA by RNAs linked to a chemical nuclease. *Nature* **363**, 474-475.

23. Shimizu, M, Inoue, H. and Ohtsuka, E. (1994). Detailed study of sequence-specific DNA cleavage of triplex-forming oligonucleotides linked to 1,10-phenanthroline. *Biochemistry* **33**, 606-613.

24. Shimizu, M., Morioka, H., Inoue, H. and Ohtsuka, E. (1996). Triplex-mediated cleavage of DNA by 1,10-phenanthroline-linked 2'-O-methyl RNA. *FEBS Lett.* **384**, 207-210.

25. Tsukahara, S., Suzuki, J., Ushijima, K., Takai, K. and Takaku, H. (1996). Nonenzymatic sequence-specific cleavage of duplex DNA via triple-helix formation by homopyrimidine phosphorothioate oligonucleotides. *Bioorg. Med. Chem.* **4**, 2219-2224.

26. Sergeev, D. S. and Zarytova, V. F. (1996). Interaction of bleomycin and its oligonucleotide derivatives with nucleic acids. *Rus. Chem. Rev.* **65**, 355-378.

27. Bigey, P., Pratviel, G. and Meunier, B. (1995). Cleavage of double-stranded DNA by 'metalloporphyrin-linker-oligonucleotide' molecules: influence of the linker. *Nucl. Acids Res.* **32**, 3894-3900.

28. Joshi, R. R. and Ganesh, K. N. (1994) Duplex and triplex directed DNA cleavage by oligonucleotide-Cu(II)/Co(III) metallodesferal conjugates. *Biochim. Biophys. Acta* **1201**, 454-460.

29. Belikova, A. M., Zarytova, V. F. and Grineva, N. I. (1967). Synthesis of ribonucleotides and diribonucleotide phosphates containing 2-chloroethylamine and nitrogen mustard residues. *Tetrahedron Lett.* **37**, 3557-3562.

30. Fedorova, O. S., Knorre. D. G., Podust, L. M. and Zarytova, V. F. (1988). Complementary addressed modification of double-stranded DNA within a ternary complex. *FEBS Lett.* **228**, 273-276.

31. Knorre, D. G., Zarytova, V. F., Podust, L. M. and Fedorova O. S. (1988). Complementary addressed modification of double stranded DNA in triple-stranded complexes. *Dokl. Akad. Nauk. SSSR* **300**, 1006-1009.

32. Podust, L. M., Gaidamakov, S. A., Abramova, T. V., Vlasov, V. V. and Gorn, V. V. (1989). Sequence specificity modification of double-stranded DNA with an alkylating

derivative of oligodeoxyribonucleotide pT(CT)₆. *Bioorg. Khim.* **15**, 363-369.

33. Vlassov, V. V., Gaidamakov, S. A., Zarytova, V. F., Knorre, D. G., Levina, A. S., Nikonova, A. A., Podust, L. M. and Fedorova, O. S. (1988). Sequence-specific chemical modification of double-stranded DNA with alkylating oligodeoxyribonucleotide derivatives. *Gene* **72**, 313-322.
34. Vlasov, V. V., Gaidamakov, S. A., Nikonova, A. A. and Levina, A. S. (1989). Complementary addressed modification of plasmid DNA. **Mol. Biol. (Mosk)** **23**, 556-561.
35. Mastiugin, V. E., Lavrovskii, Ya. V. and Vlasov, V. V. (1996). Site specific alkylation of dual-stranded DNA of the murine c-fos gene promotor region and modulation of transcription by derivatives of pyrimidine oligonucleotides, containing residues of 2-chloroethylamine. *Mol. Biol. (Mosk)* **30**, 293-306.
36. Brossalina, E. B., Demchenko, E. N. and Vlassov, V. V. (1993). Specificity of interaction of pyrimidine oligonucleotides with DNA at acidic pH in the presence of magnesium ions: affinity modification study. Antisense Res. Dev. **3**, 357-365.
37. Brossalina, E. B., Demchenko, E. N., Vlassov, V. V. and Mamaev, S. V. (1991). Sequence-specific alkylation of dsDNA with derivatives of pyrimidine oligonucleotides conjugated to 2-chloroethylamine groups. *Antisense Res. Dev.* **1**, 229-242.
38. Gaidamakov, S. A., Tsirel'nikov, N. I. and Vlasov, V. V. (1988). Complementary addressed modification of single- and double-stranded DNA by alkylating derivatives of oligonucleotides isolated by partial DNA fragmentation. *Mol. Gen. Mikrobiol. Virusol.* **7**, 42-47
39. Kutyavin, I. V., Gamper, H. B., Gall, A. A. and Meyer, R. B. (1993). Efficient, specific interstrand cross-linking of double-stranded DNA by a chlorambucil-modified, triple-forming oligonucleotide. *J. Am. Chem. Soc.* **115**, 9303-9304.
40. Lampe, J. N., Kutyavin, I. V., Rhinehart, R., Reed, M. W., Meyer, R. B. and Gamper, H. B. (1997). Factors influencing the extent and selectivity of alkylation within triplexes by reactive G/A motif oligonucleotides. *Nucl. Acids Res.* **25**, 4123-4131.
41. Belousov, E. S., Afonina, I. A., Kutyavin, I. V., Gall, A. A., Reed, M. W., Gamper, H. B., Wydro, R.M. and Meyer, R. B. (1998). Triplex targeting of a native gene in permeabilized intact cells: covalent modification of the gene for the chemokine receptor CCR5. *Nucl. Acids Res.* **26**, 1324-1328.
42. Belousov, E. S., Afonina, I. A., Podyminogin, M. A., Gamper, H. B., Reed, M. W., Wydro, R. M. and Meyer, R. B. (1997). Sequence-specific targeting and covalent modification of human genomic DNA. *Nucl. Acids Res.* **25**, 3440-3444.
43. Reed, M. W., Lukhtanov, E. A., Gorn, V. V., Kutyavin, I. V., Gall, A., Wald, A. and Meyer R. B. (1998). Synthesis and reactivity of aryl nitrogen mustard-oligodeoxyribonucleotide conjugates. *Bioconjugate Chem.* **9**, 64-71.
44. Podyminogin, M. A., Meyer, R. B. and Gamper, H. B. (1996). RecA-catalyzed, sequence-specific alkylation of DNA by cross-linking oligonucleotides. Effects of length and nonhomologous base substitutions. *Biochemistry* **35**, 7267-7274.
45. Podyminogin, M. A., Meyer, R. B. and Gamper, H. B. (1995). Sequence-specific covalent modification of DNA by cross-linking oligonucleotides. Catalysis by RecA and implication for the mechanism of synaptic joint formation. *Biochemistry* **34**, 13098-13108.
46. Shaw, J.-P. Milligan, J. F., Krawczyk, S. H. and Matteucci, M. (1991). Specific, high-efficiency, triple-helix-mediated cross-linking to duplex DNA. *J. Am. Chem. Soc.* **113**, 7765-7766.
47. Povsic, T. J. and Dervan, P. B. (1990). Sequence-specific alkylation of double-helical DNA by oligonucleotide-directed triple-helix formation. *J. Am. Chem. Soc.* **112**, 9428-9430.

96

48. Taylor, M. J. and Dervan, P. B. (1997). Kinetic analysis of sequence-specific alkylation of DNA by pyrimidine oligodeoxyribonucleotide-directed triple-helix formation. *Bioconjugate Chem.* **8**, 354-364.
49. Dervan, P. B. (1992). Reagents for the site-specific cleavage of megabase DNA. *Nature* **359**, 87-88.
50. Povsic, T. J., Strobel, S. A. and Dervan, P. B. (1992). Sequence-specific double-strand alkylation and cleavage of DNA mediated by triple-helix formation. *J. Am. Chem. Soc.* **114**, 5934-5941.
51. Grant, K. B. and Dervan, P. B. (1996). Sequence-specific alkylation and cleavage of DNA mediated by purine motif triple helix formation. *Biochemistry* **35**, 12313-12319.
52. Gamper, H. B., Hou, Ya-M., Stamm, M. R., Podyminogin, M. A. and Meyer, R. B. (1998). Strand-invasion of supercoiled DNA by oligonucleotides wiyh triplex guide sequence. *J. Am. Chem. Soc.* **120**, 2182-2183.
53. Lukhtanov, E. A., Mills, A. G., Kutyavin, I. V., Gorn, V. V., Reed, M. W., Meyer, R. B. (1997). Minor groove DNA alkylation directed by major groove triplex forming oligodeoxyribonucleotides. *Nucl. Acids Res.* **25**, 5077-5084.
54. Praseuth, D., Le Doan, T., Chassignol, M., Decout, J. L., Habhoub, N., Lhomme, J., Thuong, N. T. and Helene, C. (1988). Sequence-targeted photosensitized reactions in nucleic acids by oligo-alpha-deoxynucleotides and oligo-beta-deoxynucleotides covalently linked to proflavin. *Biochemistry* **27**, 3031-3038.
55. Praseuth, D., Perrouault, L., Le Doan, T., Chassignol, M., Thuong, N. and Helene, C. (1988). Sequence-specific binding and photocrosslinking of alpha and beta oligodeoxynucleotides to the major groove of DNA via triple-helix formation. *Proc. Natl. Acad. Sci.* USA **85**, 1349-1353.
56. Le Doan, T., Perrouault, L., Praseuth, D., Habhoub, N., Decout, J. L., Thuong, N. T., Lhomme, J. and Helene, C. (1987). Sequence-specific recognition, photocrosslinking and cleavage of the DNA double helix by an oligo-[alpha]-thymidylate covalently linked to an azidoproflavine derivative. *Nucl. Acids Res.* **15**, 7749-7760.
57. Perrouault, L., Asseline, U., Rivalle, C., Thuong, N. T., Bisagni, E., Giovannangeli, C., Le Doan, T. and Helene, C. (1990). Sequence-specific artificial photo-induced endonucleases based on triple helix-forming oligonucleotides. *Nature* **344**, 358-360.
58. Le Doan, T., Perrouault, L., Asseline, U., Thuong, N. T., Rivalle, C., Bisagni, E. and Helene, C. (1991). Recognition and photo-induced cleavage and cross-linking of nucleic acids by oligonucleotides covalently linked to ellipticine. *Antisense Res. Dev.* **1**, 43-54.
59. Boutorine, A. S., Tokuyama, H.,Takasugi, M., Isobe, H., Nakamura, E. and Helene, C. (1994). Fullerene-oligonucleotide conjugates: photoinduced sequence-specific DNA cleavage. *Angew. Chem. Int. Ed. Engl.* **33**, 2462-2465.
60. Boutorine, A. S., Brault, D., Takasugi, M., Delgado, O. and Helene, C. (1996). Chlorin-oligonucelotide conjugates: Synthesis, properties, and red light-induced photochemical sequence-specific DNA cleavage in duplexes and triplexes. *J. Am Chem. Soc.* **118**, 9469-9476.
61. Koshkin, A. A., Kropachev, K. Yu., Mamaev, S. V., Bulychev, N. V., Lokhov, S. G., Vlassov, V. V. and Lebedev, A. V. (1994). Ethydium and azidoethydium oligonucleotide derivatives: synthesis, complementary complex formation and sequence-specific photomodification of the single-stranded and double-stranded target oligo- and polynucleotides. *J. Mol. Rec.* **7**, 177-188.
62. Levina, A. S., Tabatadze, D. R., Dobrikov, M. I., Shishkin, G. V. and Zarytova, V. F. (1996). Sequence-specific photomodification of single and double stranded DNA fragments by oligonucleotide arylazide derivative. *Antisense Nucl. Acid Drug Dev.* **6**, 127-132.
63. Panyutin, I. G. and Neumann, R. D. (1994). Sequence-specific DNA double-strand

breaks induced by triplex forming ^{125}I labeled oligonucleotides. *Nucl. Acids Res.* **22**, 4979-4982.

64. Panyutin, I. G. and Neumann, R. D. (1996). Sequence-specific DNA breaks produced by triplex-directed decay of iodine-125. *Acta Oncol.* **35**, 817-823.

65. Panyutin, I. G. and Neumann, R. D. (1997). Radioprobing of DNA: distribution of DNA breaks produced by decay of ^{125}I incorporated into a triplex-forming oligonucleotide correlates with geometry of the triplex. *Nucl. Acids Res.* **25**, 883-887.

66. Karamychev, V. N., Panyutin, I. G., Reed, M. W. and Neumann, R. D. (1997). Effect of radionuclide linker structure on DNA cleavage by 125I-labeled oligonucleotides. *Antisense Nucl. Acid Drug Dev.* **7**, 549-557.

67. Corey, D., Pei, D. and Schultz, P. G. (1989). Sequence-selective hydrolysis of duplex DNA by an oligonucelotide-directed nuclease. *J. Am. Chem. Soc.* **111**, 8523-8525.

68. Pei, D., Corey, D. R. and Schultz, P. G. (1990). Site-specific cleavage of duplex DNA by a semisynthetic nuclease via triple-helix formation. *Proc. Natl. Acad. Sci. USA* **87**, 9858-9862.

69. Pei, D. and Schultz, P. G. (1991). Engineering protein specificity: gene manipulation with semisynthetic nucleases. *J. Am. Chem. Soc.* **113**, 9398-9400.

70. Corey, D. R., Munoz-Medellin, D. and Huang, A. (1995). Strand invasion by oligonucleotide-nuclease conjugates. *Bioconjugate Chem.* **6**, 93-100.

71. Matteucci, M., Lin, K.-Y., Huang, T., Wagner, R., Stembach, D. D., Mehrota, M. and Besterman, J. M. (1997). Sequence-specific targeting of duplex DNA using a camptothecin-triple helix forming oligonucleotide conjugate and topoisomerase I. *J. Am. Chem. Soc.* **119**, 6939-6940.

72. Carmi, N., Balkhi, S. R. and Breaker, R. R. (1998). Cleaving DNA with DNA. *Proc. Natl. Acad. Sci. USA* **95**, 2233-2237.

THE BIOLOGY OF TRIPLE HELICES

8 POTENTIAL MECHANISMS OF ACTION

L. James Maher

SUMMARY

Triple helix forming oligonucleotides (TFOs) are oligonucleotides or oligonucleotide analogs that can participate in triple-helical nucleic acid complexes. Because of their potential sequence-selectivity, rational mechanisms can be conceived for application of these agents to the artificial regulation of target nucleic acids in molecular or cellular systems. Two classes of possible TFO mechanisms are discussed. First, antigene TFOs might be targeted against duplex DNA, with the goal of altering DNA replication or transcription by at least seven possible mechanisms. Second, antisense TFOs might participate in three-stranded complexes with RNA targets. Several mechanisms might be envisioned for the possible application of antisense TFOs in blocking RNA processing, reverse transcription, and translation. It remains to be seen whether serious constraints on oligonucleotide-directed triple helix formation in biological contexts can be overcome to allow the potential mechanisms discussed here to be practically implemented.

INTRODUCTION

Oligonucleotides and oligonucleotide analogs can display biological activity. The allure of oligonucleotide-based therapies is the promise of rationally-designed, mechanism-based inhibitors of specific genetic information. A long-term challenge in this field involves understanding the biological activity of oligonucleotide-based compounds. Do such compounds function through intended rational mechanisms? Unintended mechanisms (sequence-dependent or sequence-independent) may be very important, and must be carefully considered. Only when intended mechanisms have been repeatedly proven will it be possible to substantiate the claim that oligonucleotides provide a rational drug design paradigm.

TFOs are oligonucleotides or oligonucleotide analogs that can participate in triple-helical nucleic acid complexes. This chapter will briefly review examples of *intended* mechanisms by which TFOs might be hoped to act in complex biological systems. Two classes of TFOs are considered here (Figure 1). Antigene TFOs are targeted against duplex DNA, whereas antisense TFOs are designed to participate in three-stranded complexes with RNA targets. The present review is not exhaustive, but attempts to provide representative literature examples documenting TFO mechanisms. Recent reviews of TFO strategies may also be of interest (this volume and *1-5*).

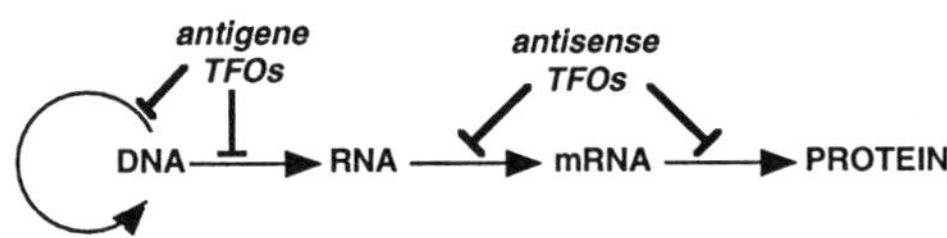

Figure 1. The central dogma of molecular biology showing possible sites of action of TFOs. Antigene TFOs bind to duplex DNA (left). In principle, the resulting structures might interfere with DNA replication or transcription. Antisense TFOs recognize single-stranded or duplex RNA sequences (right). In principle, antisense TFOs might interfere with RNA processing or translation.

MECHANISMS OF ANTIGENE TFOs

DNA cleavage

Initial studies of oligonucleotide-directed triple helix formation took advantage of *in vitro* affinity cleaving techniques (*6, 7*). TFOs were covalently tethered to either direct cleaving agents such as EDTA•Fe(II), or to indirect cleaving agents such as azidoproflavin that break the sugar-phosphate backbone upon piperidine treatment. Several other approaches to oligonucleotide-directed cleavage of duplex DNA have subsequently been demonstrated. These include the tethering of 1,10-phenanthroline (*8*), alkylating agents (*9*), nucleases (*10*), or drugs that recruit cleaving enzymes (*11*). One disadvantage of these techniques is the potential heterogeneity of cleavage products due to the promiscuity of the tethered cleaving agent. In some cases it is possible to endow restriction endonucleases with the specificity of TFOs by oligonucleotide-directed methylation protection (*12*). This approach has proven useful for a variety of molecular biology applications *in vitro* (*13*). Although valuable for *in vitro* analysis and manipulation, no direct DNA cleavage method has been mediated within cells by modified TFOs. As with other irreversible covalent damage that might be generated by modified TFOs, it seems likely that consequences to the genome might be catastrophic to the cell. There is concern that the exquisite sequence-specificity of the reagent might therefore be lost due to toxicity.

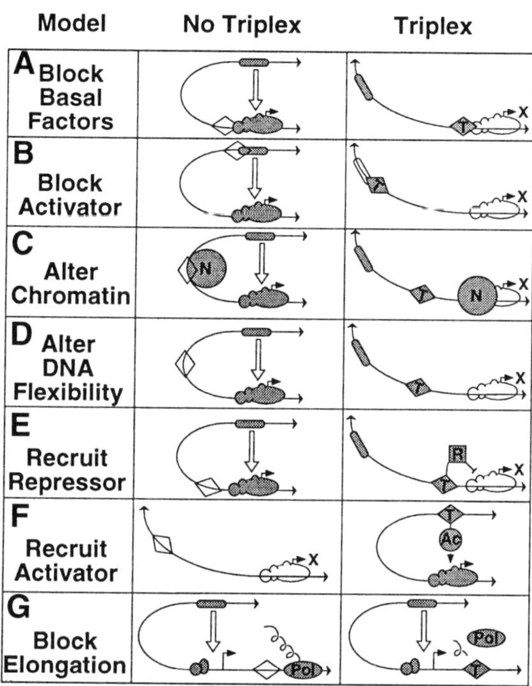

Figure 2. Possible mechanisms of action of antigene TFOs. **A** and **B**, disruption of basal transcription factors (**A**) or transcription activator proteins (**B**) by direct steric occlusion. **C**, Alteration of nucleosome positioning along the DNA template, perhaps inhibiting a local promoter. **D**, Alteration of DNA flexibility, thereby inhibiting formation of nucleoprotein structures required for transcription activation. **E** and **F**, bifunctional TFOs might simultaneously bind duplex DNA and present affinity domains to recruit endogenous protein repressors (**E**) or activators (**F**). **G**, TFO binding to duplex DNA downstream of the transcription start point might arrest RNA polymerase. Shading indicates occupation of the binding site. Bent arrow indicates active transcription. "X" indicates no transcription. Diamond, TFO binding site. Oval, transcription activator binding site. Irregular shape, basal transcription factors and RNA polymerase. "T", occupation of duplex DNA by TFO. "R", endogenous repressor protein. "Ac", endogenous activator protein. "N", nucleosome structure. "Pol", RNA polymerase. Coil indicates nascent RNA transcript.

Inhibition of protein binding at promoters

After the demonstration of oligonucleotide-directed triple helix formation, an obvious issue was the extent to which triple-helical complexes might disrupt protein-DNA interactions. Early studies using simple *in vitro* systems demonstrated that preformed triple-helical structures overlapping protein recognition sites could block the binding of restriction endonucleases (*12, 14, 15*), restriction methylases (*12, 15*), transcription factors (*12*), and nonspecific nucleases (*12*).

The ability of certain preformed triple-helical complexes to block protein binding to duplex DNA raises the possibility that such complexes might alter the function of gene promoters. Figure 2 suggests some possible mechanisms whereby such artificial regulation might occur. TFOs with binding sites that overlap positions occupied by the basal transcription machinery could inhibit recruitment of RNA polymerase (Figure 2A). TFO inhibition of transcription activators by competition for DNA binding sites is a second possibility (Figure 2B). There is evidence that direct binding site overlap may be required (*16*). Several *in vitro* studies have confirmed the ability of triple-helical complexes to block DNA binding by transcription factor Sp1 (*12, 17-21*), and extended these results to transcription factor NF-κB (*22-26*) and the progesterone receptor (*27*). In other cases, triple helix formation has been suggested to reduce promoter activity in the absence of documented transcription factor inhibition (*28-32*). TFOs have also been shown to inhibit retroviral integrase activity (*33*). Obviously, promoter activation might be achieved if TFOs were targeted to *cis* elements bound by transcriptional repressors.

A critical issue in possible TFO mechanisms is the extent to which oligonucleotide-directed triple helix formation is plausible within the context of chromatin. If triple helix formation were to occur in the absence of nucleosomes (e.g., just after DNA replication), it is possible that the triple-helical complex might alter subsequent nucleosome positioning (Figure 2C; *34, 35*). This effect might either increase or decrease transcription from local promoters. Externally administered TFOs have been shown to bind intended sequences within cultured cells at detectable levels (*36, 37*). This suggests that at least some triplex target sites in endogenous genomic DNA are accessible to oligonucleotide binding. On the other hand, pre-existing chromatin has been shown to act as a severe impediment to triplex formation (*38*).

Alteration of DNA physical properties

Triple-helical DNA is characterized by a high negative charge density and altered helical repeat parameters relative to double-helical DNA (*39*). These changes in geometry suggest that the inherent flexibility of the double helix might be reduced when in triple-helical form. Evidence supporting this view was first obtained in gel mobility experiments using circular permutation assays (*40*). For DNA fragments of equal mass and charge, stiffer molecules are expected to exhibit greater gel mobility when the region of stiffness is centrally located within the fragment. DNA fragments with centrally-located triplexes were observed to migrate more rapidly than fragments where the triplexes were nearer one terminus. Changes in DNA stiffness could influence the formation of nucleoprotein structures important for transcription initiation (Figure 2D).

In addition to the potential for stiffening of DNA, other local structural perturbations have been detected in the vicinity of some triplexes as measured by chemical hyperreactivity *(41-43)*. Such perturbations might reflect irregularities in base stacking or very small bends at the triplex-duplex junction *(44)*.

A clever design for inducing substantial DNA bending using TFOs has been independently demonstrated by two groups *(45-49)*. This concept is presented schematically in Figure 3A. Pairs of isolated triplex target sequences are bound by binary TFOs composed of two recognition domains separated by a restraining tether. If the free energy release associated with triplex formation is sufficient to accommodate the cost of DNA bending, short tethers can result in the capture of bent DNA conformations. It remains to be seen whether such designs might be implemented in the context of physiological nucleoprotein complexes.

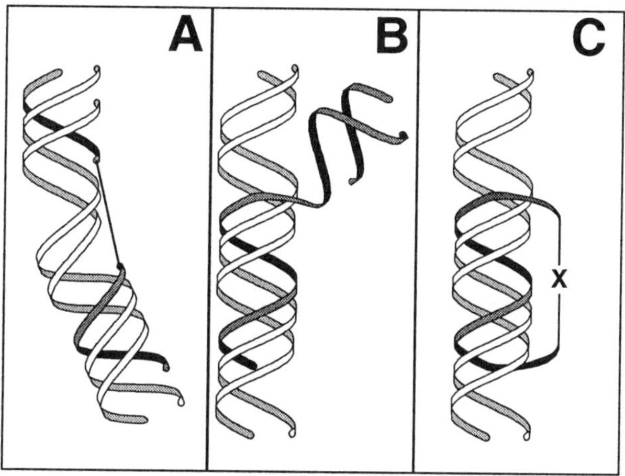

Figure 3. Novel mechanisms for antigene TFOs. **A**, DNA bending by tethered TFOs. If the free energy release upon triplex formation is sufficient, a binary TFO with restraining tether can capture and stabilize a bent conformation of the target DNA. **B**, Tethering a protein binding site to duplex DNA. The shaded TFO contains a duplex domain to be recognized by a sequence specific DNA binding activity such as a transcription activator or repressor protein. Recruitment of such a factor to the vicinity of the triplex might activate or repress local gene expression. **C**, Formation of a locked knot on duplex DNA. "X" indicates site of ring closure after binding of the TFO. Though not covalently linked to the DNA duplex target, the locked knot would form an obstacle to polymerases and helicases.

Recruiting proteins

Another tantalizing mechanism for artificial regulation of transcription by TFOs would be use of a modified triple helix to recruit accessory proteins to the vicinity of a promoter. These concepts are depicted in Figure 2E-F. If a TFO is modified to contain an appended protein recognition domain (such as the duplex DNA sequence recognized by a DNA binding protein), perhaps this appended domain might recruit the desired factor to the vicinity of the promoter. Such a complex is shown schematically in Figure 3B. A related concept (tethering an RNA to duplex DNA *via* a protein) has proven capable of recruiting transcription activators to an artificial yeast promoter *in vivo* (*50*). We (*51*) and others (*52*) have explored TFO recruitment of accessory proteins. Although the concept remains intriguing, it suffers from a severe constraint due to the relatively low affinity of typical TFOs for duplex DNA, compared to the affinities of accessory proteins for the appended domain on the TFO. Thus, to saturate the target sequence in duplex DNA, a large molar excess of TFO is required. If the TFO contains an appended domain intended to recruit accessory proteins, the vast majority of these domains will be present on oligonucleotides not bound to the duplex DNA target. This unbound fraction will titrate the accessory protein. Although these constraints can be avoided in contrived *in vitro* systems by providing excess accessory protein, or by removing unbound TFO, the applicability of these designs within cells is unclear. A clear advantage would be gained if recruitment of the accessory protein to the DNA was somehow made dependent on prior triplex formation.

Blocking transcription elongation

A principal limitation to blocking protein/DNA interactions by triple helix formation is the requirement for appropriate target sequences that overlap the binding sites of key transcription regulators. Despite the fact that homopurine sequences are statistically overrepresented in mammalian genomes (*53*), they are still too rare to occur in useful positions within most promoters. An obvious strategic improvement could be gained if triple-helical complexes could be made to act at a distance from promoters by blocking the elongation phase of RNA synthesis downstream from the initiation site (Figure 2G). The transcribed region of a gene is typically much larger than its promoter, improving the probability of potential triplex target sites. Moreover, the precise position of such sites would likely be of less importance.

Early studies examined the ability of triple-helical complexes to inhibit transcription initiation and/or elongation by bacteriophage RNA polymerases. These polymerases are extremely fast, highly processive, and able to elongate through a variety of stable protein/DNA complexes. In experiments studying triple-helical complexes at sites of 21 base pairs or less, no transcription inhibition was observed unless the triplex overlapped the bacteriophage promoter (*54-57*). The results of subsequent studies with longer triplex complexes (*58*), or under different conditions (*59*) suggest that elongation pausing or termination can be observed in some cases.

For mammalian RNA polymerase II, partial triplex inhibition of transcription elongation has been observed (*60*). Not surprisingly, there is evidence that more robust triple-helical complexes [such as those stabilized by appended intercalators

(*61*) or deoxyribose analogs (*62*)] induce more efficient elongation arrest. Thus, there is reason for optimism that identification of potential targets for TFOs will be aided by complexes capable of arresting RNA polymerase elongation. Interestingly, primer extension on single-stranded DNA templates by DNA polymerases can also be inhibited by some site-specific triple helices (*63*).

It might be argued that covalent attachment of TFOs to their targets will create structures that more efficiently terminate transcription elongation. In particular, certain photoactive agents such as psoralen can covalently cross-link the two strands of the double helix. The utility of such covalent chemistries must be evaluated with consideration of other issues such as toxicity to the entire chromosome (see above), and DNA repair (see below).

Powerful DNA helicases are capable of disrupting some triple-helical structures (*64*). It is intriguing to consider how triplexes might be made more effective at blocking DNA unwinding without covalent attachment to the DNA. A possibility is shown in Figure 3C. In this case, a TFO is designed to form an intramolecular cross-link *after* triplex formation on the duplex target. The resulting structure would topologically knot the target sequence, forming a kind of molecular lock. We (*65*) and others (*66*) have made preliminary studies of such structures, termed either Klemheist knots or pseudorotaxanes. In the latter case, a precircularized TFO was observed to thread onto a short DNA duplex in some cases. Useful application of this potential locking technique will require development of a method to engage the circularization of the TFO only after it has bound duplex DNA.

Induction of error-prone DNA repair

Recent elegant studies with shuttle vector systems have shown that photocrosslinking of psoralen-modified TFOs to chromosomal targets within intact cells can lead to modest mutagenesis near the target site (*36, 67-71* and reviewed elsewhere in this volume). Mutations presumably arise through induction of an error-prone, transcription-coupled repair response (*72*). Of particular note was the observation that TFOs lacking psoralen also induced detectable mutagenesis, albeit at reduced levels (*72*). These interesting results have several implications. At the most fundamental level, these studies are among the very few that document triple helix formation at intended chromosomal targets after exogenous delivery of oligonucleotides (see also *37, 73*). Mechanistically, these experiments show that triplex formation can induce some degree of damage repair response. On one hand, such a response ultimately acts to reverse triplex inhibition of gene expression by removing the TFO. On the other hand, the assembly of a transient repair complex may contribute to temporary gene repression (*74*). Although mutagenesis is very inefficient (not exceeding 2% of targeted chromosomes), and the spectrum of induced mutations cannot be tightly controlled, TFO-directed mutagenesis has been discussed as a practical tool (*36, 67, 75*). It will remain to be seen whether TFOs modified with other chemical cross-linking agents such as N_4,N_4-ethano-5-methyl-deoxycytidine (*76*) will give rise to similar patterns of mutagenesis. The kinetics of DNA repair may depend on TFO length and the biological system under study (*69, 77*).

Other Mechanisms

Peptide nucleic acids (PNAs) are oligonucleotide analogs that show interesting promise as antigene agents whose initial DNA recognition mechanism involves triple helix formation (*78-81* and reviewed elsewhere in this volume). In this regard, possible mechanisms of action of antigene PNAs are similar to those reviewed above.

On the other hand, PNA reagents are uniquely capable of DNA strand invasion under appropriate ionic conditions and at appropriate DNA sequences (*78, 80* but see also *82*). A particularly interesting observation is that the single-stranded displacement loop extruded from duplex DNA upon PNA strand invasion acts as an artificial promoter, recognized as a spontaneous initiation site by both prokaryotic and eukaryotic RNA polymerases (*83*). This result is consistent with promoter activity of single-stranded DNA loops observed in other systems (*84-86*). If PNA strand invasion could be made to occur under physiological conditions, the promoter activity of the resulting single-stranded DNA loop might be useful in two ways. First, such an artificial promoter might be used to increase expression of a target gene. Second, if configured to direct convergent transcription toward a target promoter, the resulting transcriptional interference might contribute to inhibition of expression of the target gene.

MECHANISMS OF ANTISENSE TFOs

As shown in Figure 1, TFOs can also be directed toward single-stranded targets such as RNA. A homopurine or homopyrimidine target sequence in RNA is first recognized by a complementary (antisense) oligonucleotide by Watson-Crick base pairing. A third domain that recognizes the resulting duplex by Hoogsteen or reverse-Hoogsteen hydrogen bonding can then be added to generate a triple helix involving the target RNA strand. A particularly effective strategy has been to combine the second and third strands of the TFO into either a circle (reviewed in *87*) or a hairpin structure (*88, 89*) such that the third strand is held in high local concentration near the nascent duplex. Related studies have targeted TFOs to RNA or DNA secondary structures by triple helix formation (*90, 91*). It should be noted that the homopurine/homopyrimidine sequence requirement inherent in triple helix formation remains as a constraint for antisense TFOs.

Possible mechanisms of antisense TFOs are similar to those proposed for antisense oligonucleotides and analogs. It has been suggested that antisense TFOs might bind with higher affinity and sequence specificity due to the ensemble of both Watson-Crick and Hoogsteen (or reverse-Hoogsteen) hydrogen bonds.

Figure 4A shows that antisense TFOs might direct degradation of target RNAs by endogenous nucleases such as RNase H. Perhaps surprisingly (in light of steric constraints), RNase H activity appears to recognize and degrade RNA in a triple-helical complex with an antisense TFO (*89*), although it remains unclear whether the nuclease is attacking the duplex or triplex form of the complex. In another study, antisense TFOs inhibited RNase H activity induced by a DNA oligonucleotide hybridized to an RNA target *in vitro* (*92*). The ability of this hybridized DNA oligonucleotide to act as a primer for reverse transcriptase was inhibited when a naturally-occurring polypurine tract in the HIV-1 RNA genome was targeted (*92*). Figure 4B suggests that targeting antisense TFOs to pre-mRNA sequences

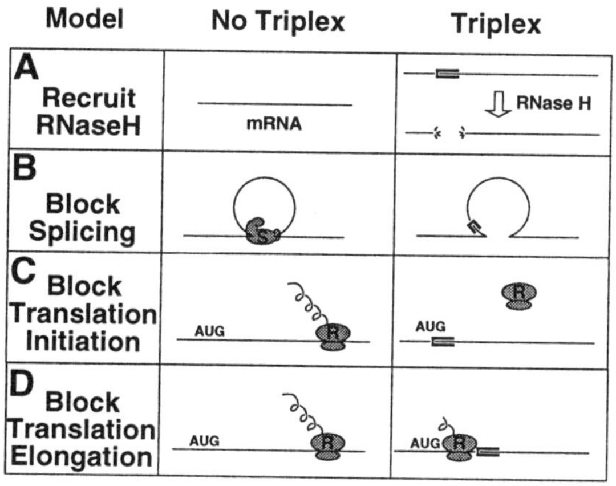

Model	No Triplex	Triplex
A Recruit RNaseH	mRNA	RNase H
B Block Splicing		
C Block Translation Initiation	AUG	AUG
D Block Translation Elongation	AUG	AUG

Figure 4. Possible mechanisms of action of antisense TFOs. **A**, In some cases an antisense TFO may recruit RNase H activity to degrade the mRNA target, but this mechanism is not consistently observed (*89, 92*). **B**, Normal RNA splicing (left) might be inhibited by a TFO that interacts with RNA at or near splice recognition sequences (right). **C** and **D**, The translation of mRNA might be inhibited by TFO binding through arrest of either translation initiation (**B**) or translation elongation (**C**). "AUG", translation initiation codon. "R", ribosome. TFO is shown as a hairpin structure.

recognized by splicing factors might inhibit specific splicing reactions. Panels C and D of Figure 4 illustrate the possibility that antisense TFOs might be made to act by sequestering translation initiation signals (Figure 4C), or, if very tightly bound to RNA, by arresting an elongating ribosome (Figure 4D).

FUTURE DIRECTIONS

Several novel mechanisms can be imagined for TFOs in the future. An obvious improvement will be TFO conjugates carrying multiple appended functions, e.g.,

110

conferring the ability to efficiently cross membranes, be delivered to the nucleus, resist degradation, and bind tightly to the target sequence. Another useful advance would be the development of antigene mechanisms (other than covalent cross-linking) that are triggered only by triple helix formation. For example, could one engineer a TFO that displays a transcription factor binding domain only when the TFO is bound to DNA?

The prospect of synthetic, sequence-specific ligands for duplex DNA and RNA is exciting. This review has surveyed some of the possible molecular mechanisms of action available to oligonucleotides and analogs that recognize nucleic acids by triple helix formation. It remains to be seen whether the many obstacles to application of TFOs *in vivo* can be overcome.

ACKNOWLEDGMENTS

The dedication of present and past members of the Maher laboratory is acknowledged, together with the encouragement of P. Dervan and B. Wold. K. Gates provided a helpful discussion of knot nomenclature. Work in the author's laboratory is supported by the Mayo Foundation, and by grants from GM47814 and GM54411 from the NIH. The author is a Harold W. Siebens Research Scholar.

REFERENCES

1. Maher, L. J. (1996). Prospects for the therapeutic use of antigene oligonucleotides. *Cancer Invest.* **14**, 66-82.
2. Thuong, N. T. and Hélène, C. (1993). Sequence-specific recognition and modification of double-helical DNA by oligonucleotides. *Angew. Chem. Ed. Engl.* **32**, 666-690.
3. Maher, L. J. (1992). DNA triple helix formation: an approach to artificial gene repressors? *BioEssays* **14**, 807-815.
4. Hélène, C. (1991). Rational design of sequence-specific oncogene inhibitors based on antisense and antigene oligonucleotides. *Eur. J. Cancer* **27**, 1466-1471.
5. Hélène, C. (1991). The anti-gene strategy: control of gene expression by triplex-forming-oligonucleotides. *Anti-Cancer Drug Design* **6**, 569-584.
6. Moser, H. E. and Dervan, P. B. (1987). Sequence-specific cleavage of double helical DNA by triple helix formation. *Science* **238**, 645-650.
7. Le Doan, T., Perrouault, L., Praseuth, D., Habhoub, N., Décout, J.-L., Thuong, N. T., L'homme, J. and Hélène, C. (1987). Sequence-specific recognition, photocrosslinking and cleavage of the DNA double helix by an oligo-[α]-thymidylate covalently linked to an azidoproflavine derivative. *Nucl. Acids Res.* **15**, 7749-7760.
8. Shimizu, M., Morioka, H., Inoue, H. and Ohtsuka, E. (1996). Triplex-mediated cleavage of DNA by 1,10-phenanthroline-linked 2'-O-methyl RNA. *FEBS Lett.* **384**, 207-210.
9. Povsic, T. J., Strobel, S. A. and Dervan, P. B. (1992). Sequence-specific double-strand alkylation and cleavage of DNA mediated by triple-helix formation. *J. Am. Chem. Soc.* **114**, 5934-5941.

10. Pei, D. and Schultz, P. G. (1991). Engineering protein specificity: Gene manipulation with semisynthetic nucleases. *J. Am. Chem. Soc.* **113**, 9398-9400.

11. Matteucci, M., Lin, K. Y., Huang, T., Wagner, R., Sternbach, D. D., Mehrotra, M. and Besterman, J. M. (1997). Sequence-specific targeting of duplex DNA using a Camptothecin-triple helix forming oligonucleotide conjugate and topoisomerase I. *J. Am. Chem. Soc.* **119**, 6939-6940.

12. Maher, L. J., Wold, B. and Dervan, P. B. (1989). Inhibition of DNA binding proteins by oligonucleotide-directed triple Helix formation. *Science* **245**, 725-730.

13. Strobel, S. A. and Dervan, P. B. (1992). Triple helix-mediated single-site enzymatic cleavage of megabase genomic DNA. *Meth. Enzymol.* **216**, 309-321.

14. François, J.-C., Saison-Behmoaras, T., Thuong, N. T. and Hélène, C. (1989). Inhibition of restriction endonuclease cleavage *via* triple helix formation by homopyrimidine oligonucleotides. *Biochemistry* **28**, 9617-9619.

15. Hanvey, J. C., Shimizu, M. and Wells, R. D. (1989). Site-specific inhibition of *Eco*RI restriction/modification enzymes by a DNA triple helix. *Nucl. Acids Res.* **18**, 157-161.

16. Huang, C.-C., Nguyen, D., Martinez, R. and Edwards, C. A. (1992). Triple-helix formation is compatible with an adjacent DNA-protein complex. *Biochemistry* **31**, 993-998.

17. Gee, J. E., Blume, S., Snyder, R. C., Ray, R. and Miller, D. M. (1992). Triplex formation prevents Sp1 binding to the dihydrofolate reductase promoter. *J. Biol. Chem.* **267**, 11163-11167.

18. Mayfield, C., Ebbinghaus, S., Gee, J., Jones, D., Rodu, B., Squibb, M. and Miller, D. (1994). Triplex formation by the human Ha-*ras* promoter inhibits Sp1 binding and *in vitro* transcription. *J. Biol. Chem.* **269**, 18232-18238.

19. Kim, H. G. and Miller, D. M. (1998). A novel triplex-forming oligonucleotide targeted to human cyclin D1 (*bcl-1*, proto-oncogene) promoter inhibits transcription in HeLa cells. *Biochemistry* **37**, 2666-2672.

20. Bianchi, N., Rutigliano, C., Passadore, M., M., T., Pippo, L., Mischiati, C., Feriotto, G. and Gambari, R. (1997). Targeting of the HIV-1 long terminal repeat with chromomycin potentiates the inhibitory effects of a triplex-forming oligonucleotide on Sp1-DNA interactions and *in vitro* transcription. *Biochem. J.* **326**, 919-927.

21. Tu, G. C., Cao, Q. N. and Israel, Y. (1995). Inhibition of gene expression by triple helix formation in hepatoma cells. *J. Biol. Chem.* **270**, 28402-28407.

22. Orson, F. M., Thomas, D. W., McShan, W. M., Kessler, D. J. and Hogan, M. E. (1991). Oligonucleotide inhibition of IL2Rα mRNA transcription by promoter region collinear triplex formation in lymphocytes. *Nucl. Acids Res.* **19**, 3435-3441.

23. Grigoriev, M., Praseuth, D., Robin, P., Hemar, A., Saison-Behmoaras, T., Dautry-Varsat, A., Thuong, N. T., Hélène, C. and Harel-Bellan, A. (1992). A triple helix-forming oligonucleotide-intercalator conjugate acts as a transcriptional repressor *via* inhibition of NF-κB binding to interleukin-2 receptor alpha-regulatory sequence. *J. Biol. Chem.* **267**, 3389-3395.

24. Grigoriev, M., Praseuth, D., Guieysse, A. L., Robin, P., Thuong, N. T., Hélène, C. and Harel-Bellan, A. (1993). Inhibition of interleukin-2 receptor alpha-subunit gene expression by oligonucleotide-directed triple helix formation. *C. R. Acad. Sci. Sér. III*, **316**, 492-495.

25. Grigoriev, M., Praseuth, D., Guieysse, A. L., Robin, P., Thuong, N. T., Hélène, C. and Harel-Bellan, A. (1993). Inhibition of gene expression by triple helix-directed DNA cross-linking at specific sites. *Proc. Natl. Acad. Sci. USA* **90**, 3501-3505.

26. Kochetkova, M. and Shannon, M. F. (1996). DNA triplex formation selectively inhibits granulocyte-macrophage colony-stimulating factor gene expression in human T cells. *J. Biol. Chem* **271**, 14438-14444.

27. Ing, N. H., Beekman, J. M., Kessler, D. J., Murphy, M., Jayaraman, K., Zendegui, J. G., Hogan, M. E., O'Malley, B. W. and Tsai, M.-J. (1993). *In vivo* transcription of a progesterone-responsive gene is specifically inhibited by a triplex-forming oligonucleotide. *Nucl. Acids Res.* **21**, 2789-2796.

28. Cooney, M., Czernuszewicz, G., Postel, E. H., Flint, S. J. and Hogan, M. E. (1988). Site-specific oligonucleotide binding represses transcription of the human c-*myc* gene *in vitro*. *Science* **241**, 456-459.

29. Postel, E. H., Flint, S. J., Kessler, D. J. and Hogan, M. E. (1991). Evidence that a triple-forming oligodeoxyribonucleotide binds to the c-*myc* promoter in HeLa cells, thereby reducing c-*myc* mRNA levels. *Proc. Natl. Acad. Sci. USA* **88**, 8227-8231.

30. Noonberg, S. B., Scott, G. K., Hung, C. A., Hogan, M. E. and Benz, C. C. (1994). Inhibition of transcription factor binding to the *HER2* promoter by triplex-forming oligodeoxyribonucleotides. *Gene* **149**, 123-126.

31. Kovacs, A., Kandala, J. C., Weber, K. T. and Guntaka, R. V. (1996). Triple helix-forming oligonucleotide corresponding to the polypyrimidine sequence in the rat alpha1(I) collagen promoter specifically inhibits factor binding and transcription. *J. Biol. Chem.* **271**, 1805-1812.

32. Lavrovsky, Y., Stoltz, R. A., Vlassov, V. V. and Abraham, N. G. (1996). c-*fos* protooncogene transcription can be modulated by oligonucleotide-mediated formation of triplex structures *in vitro*. *Eur. J. Biochem.* **238**, 582-590.

33. Mouscadet, J.-F., Carteau, S., Goulaouic, H., Subra, F. and Auclair, C. (1994). Triplex-mediated inhibition of HIV DNA integration *in vitro*. *J. Biol. Chem.* **269**, 21635-21638.

34. Westin, L., Blomquist, P., Milligan, J. F. and Wrange, O. (1995). Triple helix DNA alters nucleosomal histone-DNA interactions and acts as a nucleosome barrier. *Nucl. Acids Res.* **23**, 2184-2191.

35. Espinás, M. L., Jiménez-García, E., Martínez-Balbás, A. and Azorín, F. (1996). Formation of triple-stranded DNA at d(GA.TC)n sequences prevents nucleosome assembly and is hindered by nucleosomes. *J. Biol. Chem.* **271**, 31807-31812.

36. Chan, P. P. and Glazer, P. M. (1997). Triplex DNA: Fundamentals, advances, and potential applications for gene therapy. *J. Mol. Med.* **75**, 267-282.

37. Giovannangeli, C., Diviacco, S., Labrousse, V., Gryaznov, S., Charneau, P. and Hélène, C. (1997). Accessibility of nuclear DNA to triplex-forming oligonucleotides: The integrated HIV-1 provirus as a target. *Proc. Natl. Acad. Sci. USA* **94**, 79-84.

38. Brown, P. M. and Fox, K. R. (1996). Nucleosome core particles inhibit DNA triple helix formation. *Biochem. J.* **319**, 607-611.

39. Shin, C. and Koo, H. S. (1996). Helical periodicity of GA-alternating triple-stranded DNA. *Biochemistry* **35**, 968-972.

40. Maher, L. J., Dervan, P. B. and Wold, B. (1992). Analysis of promoter-specific repression by triple-helical DNA complexes in a eukaryotic cell-free transcription system. *Biochemistry* **31**, 70-81.

41. Hartman, D. A., Kuo, S.-R., Broker, T. R., Chow, L. T. and Wells, R. D. (1992). Intermolecular triplex formation distorts the DNA duplex in the regulatory region of human papillomavirus Type-11. *J. Biol. Chem.* **267**, 5488-5494.

42. Olivas, W. M. and Maher, L. J. (1994). DNA recognition by alternate strand triple helix formation: affinities of oligonucleotides for a site in the human p53 gene. *Biochemistry* **33**, 983-991.

43. Olivas, W. M. and Maher, L. J. (1996). Binding of DNA oligonucleotides to sequences in the promoter of the human *bcl-2* gene. *Nucl. Acids Res.* **24**, 1758-1764.

44. Chomilier, J., Sun, J.-S., Collier, D. A., Garestier, T., Hélène, C. and Lavery, R. (1992). A computational and experimental study of the bending induced at a double-triple helix junction. *Biophys. Chem.* **45**, 143-152.

45. Kessler, D. J., Pettitt, B. M., Cheng, Y.-K., Smith, S. R., Jayaraman, K., Vu, H. M. and Hogan, M. E. (1993). Triple helix formation at distant sites: hybrid oligonucleotides containing a polymeric linker. *Nucl. Acids Res.* **21**, 4810-4815.

46. Liberles, D. A. and Dervan, P. B. (1996). Design of artificial sequence-specific DNA bending ligands. *Proc. Natl. Acad. Sci. USA* **93**, 9510-9514.

47. Akiyama, T. and Hogan, M. E. (1996). The design of an agent to bend DNA. *Proc. Natl. Acad. Sci. USA* **93**, 12122-12127.

48. Akiyama, T. and Hogan, M. E. (1996). Microscopic DNA flexibility analysis - Probing the base composition and ion dependence of minor groove compression with an artificial DNA bending agent. *J. Biol. Chem.* **271**, 29126-29135.

49. Akiyama, T. and Hogan, M. E. (1997). Structural analysis of DNA bending induced by tethered triple helix forming oligonucleotides. *Biochemistry* **36**, 2307-2315.

50. SenGupta, D. J., Zhang, B. L., Kraemer, B., Pochart, P., Fields, S. and Wickens, M. (1996). A three-hybrid system to detect RNA-protein interactions *in vivo*. *Proc. Natl. Acad. Sci. USA* **93**, 8496-8501.

51. Soukup, G. A. and Maher, L. J. *unpublished observations*

52. Svinarchuk, F., Nagibneva, I., Cherny, D., Ait-Si-Ali, S., Pritchard, L. L., Robin, P., Malvy, C. and Harel-Bellan, A. (1997). Recruitment of transcription factors to the target site by triplex-forming oligonucleotides. *Nucl. Acids Res.* **25**, 3459-3464.

53. Behe, M. (1987). The DNA sequence of the human β-globin region is strongly biased in favor of long strings of contiguous purine or pyrimidine residues. *Biochemistry* **26**, 7870-7885.

54. Maher, L. J. (1992). Inhibition of T7 RNA polymerase initiation by triple-helical DNA complexes: a model for artificial gene repression. *Biochemistry* **31**, 7587-7594.

55. Ross, C., Samuel, M. and Broitman, S. L. (1992). Transcriptional inhibition of the bacteriophage T7 early promoter region by oligonucleotide triple helix formation. *Biochem. Biophys. Res. Commun.* **189**, 1674-1680.

56. Skoog, J. U. and Maher, L. J. (1993). Repression of bacteriophage promoters by DNA and RNA oligonucleotides. *Nucl. Acids Res.* **21**, 2131-2138.

57. Skoog, J. U. and Maher, L. J. (1993). Relief of triple-helix-mediated promoter inhibition by elongating RNA polymerases. *Nucl. Acids Res.* **21**, 4055-4058.

58. Rando, R. F., DePaolis, L., Durland, R. H., Jayaraman, K., Kessler, D. J. and Hogan, M. E. (1994). Inhibition of T7 and T3 RNA polymerase directed transcription elongation *in vitro*. *Nucl. Acids Res.* **22**, 678-685.

59. Alunni-Fabbroni, M., Manfioletti, G., Manzini, G. and Xodo, L. E. (1994). Inhibition of T7 RNA polymerase transcription by phosphate and phosphorothioate

triplex-forming oligonucleotides targeted to a R/Y site downstream from the promoter. *Eur. J. Biochem.* **226**, 831-839.

60. Young, S. L., Krawczyk, S. H., Matteucci, M. D. and Toole, J. J. (1991). Triple helix formation inhibits transcription elongation *in vitro. Proc. Natl. Acad. Sci. USA* **88**, 10023-10026.

61. Giovannangeli, C., Perrouault, L., Escudé, C., Thuong, N. and Hélène, C. (1996). Specific inhibition of *in vitro* transcription elongation by triplex-forming oligonucleotide-intercalator conjugates targeted to HIV proviral DNA. *Biochemistry* **35**, 10539-10548.

62. Giovannangeli, C., Perrouault, L., Escudé, C., Gryaznov, S. and Hélène, C. (1996). Efficient inhibition of transcription elongation *in vitro* by oligonucleotide phosphoramidates targeted to proviral HIV DNA. *J. Mol. Biol.* **261**, 386-398.

63. Samadashwily, G. M. and Mirkin, S. M. (1994). Trapping DNA polymerases using triplex-forming oligodeoxyribonucleotides. *Gene* **149**, 127-136.

64. Maine, I. P. and Kodadek, T. (1994). Efficient unwinding of triplex DNA by a DNA helicase. *Biochem. Biophys Res. Commun.* **204**, 1119-1124.

65. Borgeson, C., Ferber, M. and Maher, L. J. *unpublished observations*

66. Ryan, K. and Kool, E. T. (1998). Triplex-directed self-assembly of an artificial sliding clamp on duplex DNA. *Chem. & Biol.* **5**, 59-67.

67. Havre, P. A., Gunther, E. J., Gasparro, F. P. and Glazer, P. M. (1993). Targeted mutagenesis of DNA using triple helix-forming oligonucleotides linked to psoralen. *Proc. Natl. Acad. Sci. USA* **90**, 7879-7883.

68. Gasparro, F. P., Havre, P. A., Olack, G. A., Gunther, E. J. and Glazer, P. M. (1994). Site-specific targeting of psoralen photoadducts with a triple helix-forming oligonucleotide: Characterization of psoralen monoadduct and crosslink formation. *Nucl. Acids Res.* **22**, 2845-2852.

69. Wang, G. and Glazer, P. M. (1995). Altered repair of targeted psoralen photoadducts in the context of an oligonucleotide-mediated triple helix. *J. Biol. Chem.* **270**, 22595-22601.

70. Wang, G., Levy, D. D., Seidman, M. M. and Glazer, P. M. (1995). Targeted mutagenesis in mammalian cells mediated by intracellular triple helix formation. *Mol. Cell. Biol.* **15**, 1759-1768.

71. Faruqi, A. F., Seidman, M. M., Segal, D. J., Carroll, D. and Glazer, P. M. (1996). Recombination induced by triple-helix-targeted DNA damage in mammalian cells. *Mol. Cell. Biol.* **16**, 6820-6828.

72. Wang, G., Seidman, M. M. and Glazer, P. M. (1996). Mutagenesis in mammalian cells induced by triple helix formation and transcription-coupled repair. *Science* **271**, 802-805.

73. Guieysse, A. L., Praseuth, D., Grigoriev, M., Harel-Bellan, A. and Hélène, C. (1996). Detection of covalent triplex within human cells. *Nucl. Acids Res.* **24**, 4210-4216.

74. Macaulay, V. M., Bates, P. J., McLean, M. J., Rowlands, M. G., Jenkins, T. C., Ashworth, A. and Neidle, S. (1995). Inhibition of aromatase expression by a psoralen-linked triplex-forming oligonucleotide targeted to a coding sequence. *FEBS Lett.* **372**, 222-228.

75. Sandor, Z. and Bredberg, A. (1994). Repair of triple helix directed psoralen adducts in human cells. *Nucl. Acids Res.* **22**, 2051-2056.

76. Shaw, J.-P., Milligan, J. F., Krawczyk, S. H. and Matteucci, M. (1991). Specific, high-efficiency, triple-helix-mediated cross-linking to duplex DNA. *J. Am. Chem. Soc.* **113**, 7765-7766.

77. Musso, M., Wang, J. C. and Van Dyke, M. W. (1996). *In vivo* persistence of DNA triple helices containing psoralen-conjugated oligodeoxyribonucleotides. *Nucl. Acids Res.* **24**, 4924-4932.

78. Nielsen, P. E. (1991). Sequence-selective DNA recognition by synthetic ligands. *Bioconj. Chem.* **2**, 1-12.

79. Knudsen, H. and Nielsen, P. E. (1996). Antisense properties of duplex- and triplex-forming PNAs. *Nucl. Acids Res.* **24**, 494-500.

80. Boffa, L. C., Morris, P. L., Carpaneto, E. M., Louissaint, M. and Allfrey, V. G. (1996). Invasion of the CAG triplet repeats by a complementary peptide nucleic acid inhibits transcription of the androgen receptor and TATA-binding protein genes and correlates with refolding of an active nucleosome containing a unique AR gene sequence. *J. Biol. Chem.* **271**, 13228-13233.

81. Nielsen, P. E. (1997). Peptide nucleic acid (PNA) - From DNA recognition to antisense and DNA structure. *Biophys. Chem.* **68**, 103-108.

82. Gamper, H. B., Hou, Y. M., Stamm, M. R., Podyminogin, M. A. and Meyer, R. B. (1998). Strand invasion of supercoiled DNA by oligonucleotides with a triplex guide sequence. *J. Am. Chem. Soc.* **120**, 2182-2183.

83. Mollegaard, N. E., Buchardt, O., Egholm, M. and Nielsen, P. E. (1994). Peptide nucleic acid-DNA strand displacement loops as artificial transcription promoters. *Proc. Natl. Acad. Sci. USA* **91**, 3892-3895.

84. Aiyer, S. E., Helmann, J. D. and deHaseth, P. L. (1994). A mismatch bubble in double-stranded DNA suffices to direct precise transcription initiation by *E. coli* RNA polymerase. *J. Biol. Chem.* **269**, 13179-13184.

85. Daubendiek, S. L., Ryan, K. and Kool, E. T. (1995). Rolling-circle RNA synthesis: Circular oligonucleotides as efficient substrates for T7 RNA polymerase. *J. Am. Chem. Soc.* **117**, 7818-7819.

86. Daubendiek, S. L. and Kool, E. T. (1997). Generation of catalytic RNAs by rolling transcription of synthetic DNA nanocircles. *Nature Biotech.* **15**, 273-277.

87. Kool, E. T. (1996). Circular oligonucleotides: New concepts in oligonucleotide design. *Annu. Rev. Biophys. Biomol. Struct.* **25**, 1-28.

88. Giovannangeli, C., Montenay-Garestier, T., Rougée, M., Chassignol, M., Thuong, N. T. and Hélène, C. (1991). Single-stranded DNA as a target for triple-helix formation. *J. Am. Chem. Soc.* **113**, 7775-7777.

89. Kandimalla, E. R. and Agrawal, S. (1994). Single-strand-targeted triplex formation: stability, specificity and RNase H activation properties. *Gene* **149**, 115-121.

90. Brossalina, E., Pascolo, E. and Toulmé, J.-J. (1993). The binding of an antisense oligonucleotide to a hairpin structure *via* triplex formation inhibits chemical and biological reactions. *Nucl. Acids Res.* **21**, 5616-5622.

91. Brossalina, E., Demchenko, E., Demchenko, Y., Vlassov, V. and Toulmé, J. J. (1996). Triplex-forming oligonucleotides trigger conformation changes of a target hairpin sequence. *Nucl. Acids Res.* **24**, 3392-3398.

92. Volkmann, S., Jendis, J., Frauendorf, A. and Moelling, K. (1995). Inhibition of HIV-1 reverse transcription by triple-helix forming oligonucleotides with viral RNA. *Nucl. Acids Res.* **23**, 1204-1212.

9 OLIGONUCLEOTIDE UPTAKE AND DELIVERY IN TISSUE CULTURE CELLS

Scot W. Ebbinghaus, Nadarajah Vigneswaran,
Charles M. Mayfield, David T. Curiel,
and Donald M. Miller

INTRODUCTION

Triplex forming oligonucleotides (TF-ODNs) hold promise for the specific modulation of gene expression (*1*), and recent data indicate that intermolecular triplex formation can take place in living cells and in chromatin (*2-6*). However, TF-ODNs must overcome significant cellular barriers in order to arrive at the target gene in the nucleus. Oligonucleotides designed for triple helix anti-gene applications must cross the cell membrane, escape the endosome, and resist degradation by nucleases to be available for triplex formation in the nucleus. Oligonucleotides (ODNs) do not passively diffuse across cell membranes, and available data indicate uptake into cells by adsorptive endocytosis, fluid phase pinocytosis, or endocytosis after binding to specific receptors on the surface of cells, such as the integrin protein, Mac-1 (*7*). Uptake of ODNs from cell culture media varies widely by cell type (*8, 9*) and may be insufficient for the desired biological effects of the ODN because of low uptake efficiency and sequestration in endosomal vesicles (*7*).

In this chapter, we will review the critical issue of triplex formation in cells. The available data suggest that triplex formation in chromatin is possible if the membrane barriers to triplex formation are removed. We will present our data with a novel and highly sensitive immmunohistochemical method to analyze the uptake and subcellular distribution of ODNs in cultured cells and tissue sections. Finally, we will present our data on the delivery of phosphodiester TF-ODNs to the nucleus of tumor cells in culture with the adenovirus-polylysine vector.

INTERMOLECULAR TRIPLEX DNA IN CHROMATIN

There are several reports that demonstrate a reduction in target gene expression and other biological effects after treating cells in culture with ODNs designed for triplex

118

formation (*1*). While these data demonstrate the potential for the triplex DNA antigene strategy to be effective, one must interpret these reports cautiously. ODNs often have both sequence-specific and non-sequence-specific effects on cells that are independent of binding to their intended targets, and it is therefore of critical importance to demonstrate that the effects of ODNs designed for triplex formation actually exert their cellular effects because of intracellular triplex formation with the target gene. In 1996, three reports convincingly demonstrated intermolecular triplex formation with a plasmid target in living cells (*2-4*). These reports provided evidence that intracellular conditions will support a triple helix structure, that ODNs added to cells are capable of recognizing and binding to an intracellular DNA target sequence, and that triplex formation can lead to biological effects in the form of mutations. Subsequently, two reports of triplex formation in the intact chromatin of permeabilized cells have demonstrated the feasibility of intermolecular triplex formation in structured chromatin in cells. First, Giovannangeli and associates used digitonin permeabilized human cell lines with the HIV proviral sequence, and demonstrated triplex formation with an ODN-psoralen conjugate targeted to a polypurine tract (*5*). A second report by Belousov and co-workers demonstrated triple helix formation with a target sequence in the chemokine receptor gene, CCR5, in streptolysin O permeabilized mammalian cells (*6*).

We have also demonstrated triplex formation in chromatin by exposing isolated nuclei to TF-ODNs that target the *HER-2/neu* promoter (Figure 1). In these experiments, we made transcriptionally competent nuclei from HeLa (cervical cancer)

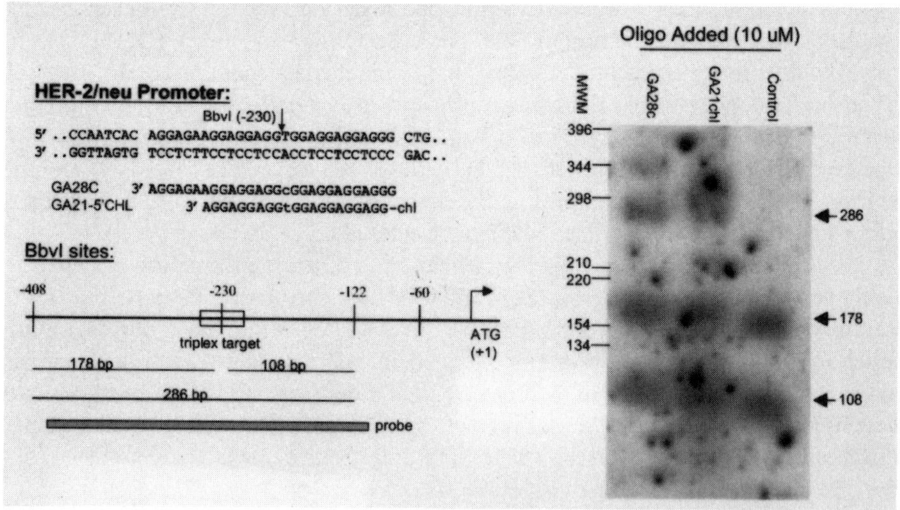

Figure 1. Demonstration of triplex formation in isolated SK-OV-3 nuclei by Southern blot. The triplex target is in the *HER-2/neu* promoter (*10*) and the TF-ODNs are shown. Chlorambucil (chl) is conjugated to the 5' terminus of the 21-mer TF-ODN (GA21-5'CHL).

or SK-OV-3 (ovarian cancer) cells by the methods used for nuclear run-on transcription assays. We suspended the nuclei in a solution containing TF-ODNs (an unmodified phosphodiester ODN or one containing an appended chlorambucil), and after triplex formation, the nuclei were extensively washed, naked genomic DNA was isolated, and the DNA was digested with *BbvI* followed by Southern blotting of the *HER-2/neu* promoter with a probe overlapping the triplex target region. Triplex formation was detected by protection from *BbvI* digestion in the triplex target sequence, visualized as a 286 bp restriction fragment in the presence of the TF-ODNs compared to bands of 178 bp and 108 bp in the presence of a control ODN (a representative Southern Blot from SK-OV-3 cells is shown in Figure 1). Based on band intensity in the Southern blot, we estimate that approximately 30% of the target sequences were accessible to the TF-ODN and protected from *BbvI* digestion. The 30% target site accessibility that we observed is very similar to the findings reported by both Giovannangeli *et al.* and Belousov *et al.* Several other interesting observations include: 1) the level of triplex formation by the unmodified ODN was essentially identical to the level of triplex formation by the chlorambucil conjugate, implying very stable triplex formation by the phosphodiester ODN; 2) similar levels of triplex formation were observed in both HeLa cells and SK-OV-3 cells, implying that the level of transcription of the *HER-2/neu* gene does not alter the accessibility of the triplex target site, since HER-2/neu is expressed at very low levels in HeLa cells while it is highly overexpressed (approximately 50-fold) in SK-OV-3 cells.

Taken together, these data provide support for the anti-gene strategy by demonstrating the ability of TF-ODNs to bind to intracellular and chromosomal targets. A cautionary note is provided by the relatively consistent ceiling of approximately 30% triplex formation in chromatin that is independent of cell type and the transcriptional activity of the target gene. The ability of triplex formation to occur in isolated nuclei or permeabilized cells emphasizes the need for an ODN delivery system which can lead to high concentrations of free ODN in the nucleus of cells, and suggests the feasibility of the anti-gene strategy if such a delivery system is found. Furthermore, techniques are now available to demonstrate the formation of a triple helix in chromatin, and future studies are needed that correlate the binding of the TF-ODN to the target gene with the effects on target gene expression.

IMMUNOHISTOCHEMICAL DETECTION OF TRIPLEX FORMING OLIGONUCLEOTIDES IN TISSUE CULTURE CELLS

In our laboratory, we have developed a novel and highly sensitive immunohistochemical method to analyze the cellular uptake, nuclear localization and intracellular stability of unmodified and chemically modified ODNs in various tumor cell lines and immortalized keratinocytes (9). Immunohistochemical detection of 5-bromodeoxyuridine (BrdU) modified ODNs in tissue culture cells has a number of advantages. The immunohistochemical assay has multiple amplification steps that increase the sensitivity of ODN detection compared to direct visualization of fluorescein labeled ODNs by fluorescence microscopy. Counterstaining of the nuclei with hematoxylin allows for the easy distinction of intranuclear localization of ODN from cytoplasmic peri- or para-nuclear aggregation of ODN. Finally, this technique is applicable to tissue sections processed by routine pathological methods, and it may be readily applied to the detection of ODNs in tissue after administration to animals or humans.

Triplex formation by BrdU substituted ODNs was evaluated by gel mobility shift analysis and DNase I footprinting, and these studies demonstrated that the binding of the TF-ODNs is not affected by BrdU substitution. However, we found that the anti-BrdU antibody does not recognize BrdU when the TF-ODN is bound to the major groove of the double-stranded DNA. We found that the anti-BrdU antibody is able to bind to triplex DNA resulting in a supershift when the TF-ODN was synthesized with a 5' BrdU "tail" consisting of eight BrdU nucleotides, due to the fact that the 5' BrdU tail remains single stranded to allow antibody binding.

The sensitivity of unmodified phosphodiester ODNs (PO-ODNs) to intracellular and serum nucleases represents an additional problem with the cellular bioavailability of TF-ODNs. The half-life of PO-ODNs microinjected into the cytoplasm of several eukaryotic cell lines can be measured in minutes (*11*). We compared the uptake, nuclear localization, and intracellular stability of unmodified and chemically modified ODNs in tumor cell lines and primary human epidermal keratinocytes (Figures 2). Human non-small cell lung carcinoma (Calu-1 and H460), pancreatic carcinoma (BxPC-3 and PANC-1), and cervical carcinoma (HeLa) cell lines were used in this study. BrdU substituted and fluorescein labeled ODNs were detected in cells by immunohistochemistry using monoclonal anti-BrdU and anti-fluorescein antibodies. The binding of these monoclonal antibodies to intracellular ODNs was demonstrated using a biotinylated secondary antibody, followed by an avidin-peroxidase conjugate and visualized using diaminobenzidine as a chromogenic substrate. Because BrdU is a nucleoside analog rather than a large, planar fluorophore such as fluorescein, the intracellular trafficking of a BrdU modified ODN should closely mimic the subcellular distribution of the native ODN. When we compared the cellular uptake of BrdU modified and fluorescein labeled phosphorothioate ODNs (PS-ODNs) in a number of cell lines, their rate of uptake and subcellular distribution were almost identical (Figure 2, A and B). BrdU monomers released during intracellular degradation of ODNs are efficiently incorporated into the genomic DNA, which can be detected by anti-BrdU antibodies only after denaturation of the chromatin. ODN degradation was observed by immunohistochemical evaluation of the denatured interphase chromatin and metaphase chromosomes from ODN treated cells (Figure 2, C and D). Immunohistochemical detection of BrdU modified ODNs in cells fixed without denaturation (in which nuclear staining represents intact ODN) and with denaturation of genomic DNA (in which nuclear positivity indicates BrdU in chromatin due to ODN degradation) is a novel method to study the intracellular stability of ODNs.

We observed that PS-ODNs are efficiently taken up by tumor cells and become localized predominantly in the cytoplasm. Interestingly, the cytoplasmic distribution of PS-ODNs in tumor cells follows two distinct patterns which appear to be cell type specific. In some tumor cells (e.g., H460, PANC-1), most of the internalized PS-ODNs become sequestrated as a single globular mass adjacent to the nuclei (paranuclear) or surrounding the nuclei (perinuclear) (Figure 2, A and B). These patterns can be easily misinterpreted as nuclear localization by fluorescence microscopy or in subcellular fractionation studies because of their close association with the nuclear surface. In other tumor cell lines (e.g., Calu-1, BxPC-3, HeLa), PS-ODNs distribute in a diffuse punctate pattern. As reported by Noonberg *et al.*, (*12*) we also found that the cytoplasmic entrapment of PS-ODNs appears to be specific for malignant tumor cell lines, whereas remarkable nuclear localization of PS-ODNs was noted in primary normal human epidermal keratinocytes (NHEK cells).

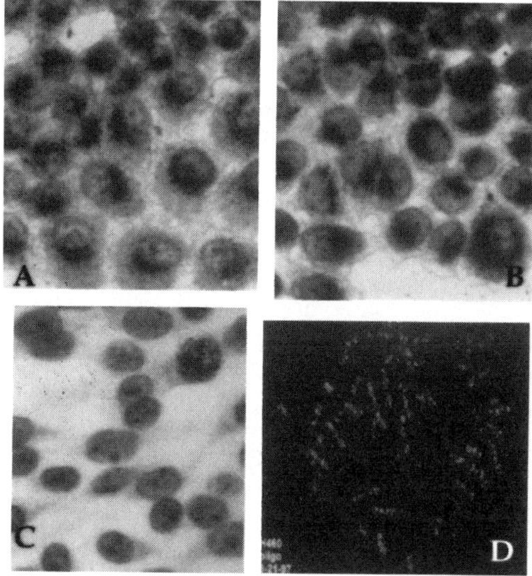

Figure 2. Immunohistochemical detection of BrdU substituted (**A**) and fluorescein labeled (**B**) phosphorothioate ODNs in H460 cells. Denaturation of the genomic DNA of H460 cells treated with a BrdU labeled phosphodiester ODN demonstrates positive immunostaining in chromatin (**C**) and metaphase chromosomes (**D**), suggesting that BrdU monomers are recycled into the genomic DNA following intracellular degradation of the ODN.

The use of nuclease resistant phosphorothioate backbones decreases the binding affinity for both DNA and RNA (*13,14*) and increases binding to serum and cellular proteins (*15*), the net effect being a decrease in the specificity of the ODN for target DNA or RNA versus non-target binding. Our data are consistent with a model wherein PS-ODNs become sequestered in different compartments of the cytoplasm, which is tumor cell type specific and may be due endosomal entrapment or cytoplasmic protein binding or both. We found that PS-ODNs bind strongly to cytoplasmic proteins, of various tumor cells, with molecular weights ranging from 42-67 kDa. In contrast, PO-ODNs, including alpha anomers, do not show any affinity for cytoplasmic protein as demonstrated by gel mobility shift and Southwestern analysis. This model of sequestration is further supported by the marked differences in the cellular uptake and subcellular distribution pattern of alpha anomeric PO-ODNs compared to PS-ODNs. Although uptake of alpha anomeric PO-ODNs by tumor cells appears to be significantly lower than PS-ODNs, the cells demonstrate diffuse non-punctate cytoplasmic immunostaining rather than the punctate or globular pattern observed with PS-ODNs, and they also demonstrate significant nuclear localization. Based on our studies, alpha anomeric PO-ODNs appear to be superior to PS-ODNs and 3'-amino capped PO-ODNs since they appear to have both intracellular stability and nuclear localization. Delivery strategies that enhance the intracellular bioavailability of alpha anomeric PO-ODNs will be useful in improving the potential efficacy of the triplex anti-gene strategy.

ADENOVIRUS-POLYLYSINE DELIVERY OF TRIPLEX FORMING ODNs TO TISSUE CULTURE CELLS

Established methods to accomplish gene transfer to human cells often employ viral vectors which take advantage of the natural ability of the virus to insert viral genes into the host cell. The adenovirus, like most other viruses, is taken into the cell by receptor-mediated endocytosis. Adenoviruses are potent endosomolytic agents, and since they are DNA viruses, the transfer of viral DNA to the nucleus is accomplished efficiently. Adenovirus-polylysine (AdpL)-DNA complexes have been used as highly effective vectors for the delivery of plasmid DNA to a wide variety of cells and tissues (16). AdpL-DNA complexes exploit the receptor-mediated endocytosis and endosomolysis of the adenovirus to deliver DNA that is carried on the viral capsid by a polylysine bridge. AdpL-DNA complexes make use of replication-defective viral particles that are not designed to express recombinant genes, and as double-stranded DNA viruses, adenoviruses do not integrate viral genes into the host genome. Thus, the viral particles in AdpL-DNA complexes act only as carriers for the DNA of interest, and this DNA can be produced by conventional methods at high efficiency and to the desired level of purity.

We have applied the AdpL vector to the delivery of TF-ODNs, and we demonstrated that adenovirus-polylysine-ODN (AdpL-ODN) complexes are capable of delivering high quantities of ODNs to the nucleus of a wide variety of tumor cells in culture (9). AdpL-ODN complexes were constructed from p259A, an E1A deleted, human type 5 adenovirus. Originally, the polylysine bridge was attached to the viral particle by a non-neutralizing antibody (MP301) against an epitope on a chimeric hexon protein of the p259A capsid, and the antibody was conjugated to polylysine (MP301pL). Subsequently, we have used AdpL complexes in which the polylysine was covalently linked to the viral capsid with a chemical cross-linking agent; this synthesis was shown not to diminish the ability of the virus to undergo endocytosis and endosomolysis (*17*). TF-ODNs were added to the AdpL complexes to achieve charge neutralization of the polyanionic phosphodiester backbone of the ODN with the polycationic amino groups of the polylysine. These complexes were then added to the media of a wide variety of tissue culture cells, and the uptake, persistence, and bioavailability of the ODNs were measured.

First, the uptake of radioactively labeled ODNs by tumor cells in culture were measured (Figure 3). In these experiments, the TF-ODN was internally radiolabeled and incorporated into the AdpL-ODN complexes. Various tumor cell lines that grow as monolayers were treated for twenty-four hours, the cells were washed, lysed, and

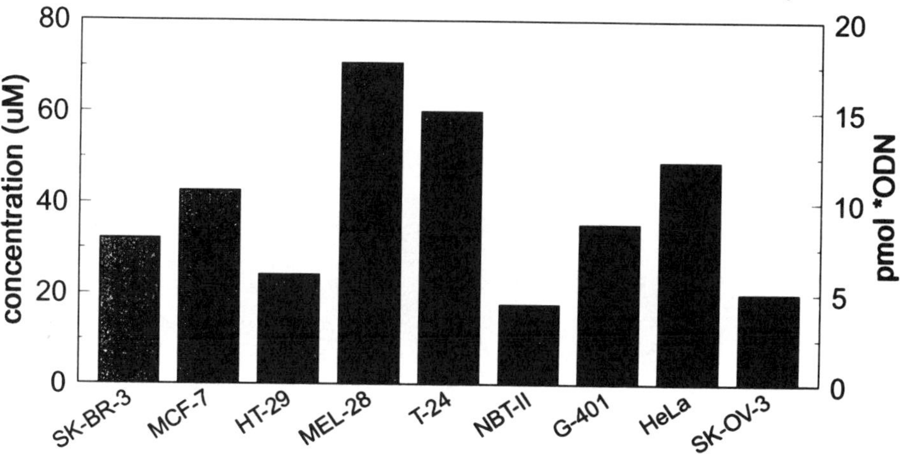

Figure 3. ODN delivery to several tumor cell lines by adenovirus-polylysine-ODN complexes. Radioactivity associated with the nuclear pellet after cell fractionation is reported as picomoles of labeled ODN (pmol *ODN, 75 pmol *ODN were added to each culture), based on specific activity.

the nuclei were isolated. Scintillation counting of the nuclear pellet, the supernatant (membrane/vesicle/cytoplasm fraction), and the media was performed, and the concentration of ODN in the nucleus was calculated based on estimated nuclear volumes. The resulting intranuclear ODN concentrations ranged from approximately 20-70 µM in all monolayer cell lines evaluated to date. These concentrations represent a 1000-fold increase compared with the concentration in the culture media, confirming the active uptake of AdpL-ODN complexes.

We used cell fractionation to compare the uptake and nuclear localization of internally labeled ODNs in AdpL-ODN complexes compared with cationic lipid-ODN complexes (Figure 4). For these experiments, we used the lipid preparation DOTAP:DOPE (1:1), and formed complexes with the TF-ODNs at a 4:1 lipid:ODN

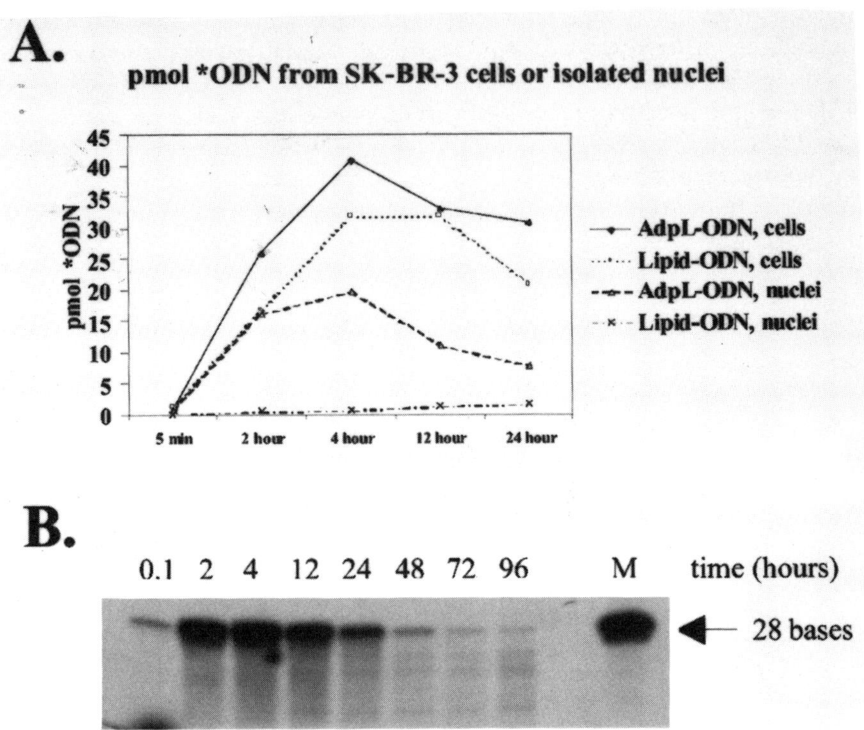

Figure 4. **(A)** Comparison of total cellular ODN uptake (cells) versus localization of ODN in the nucleus (nuclei) after treatment of SK-BR-3 cells with AdpL-ODN complexes or lipid-ODN complexes. **(B)** Electrophoresis of labeled ODN recovered from the nuclear fraction at various times after AdpL-ODN treatment of SK-BR-3 cells.

ratio. We treated SK-BR-3 breast cancer cells with AdpL-ODN, lipid-ODN, or free ODN and harvested the cells at various time points for nuclear preparations. The isolated nuclei were divided for scintillation counting and electrophoresis of the genomic DNA. The delivery of the ODN from the AdpL-ODN complex was rapid, peaking at 4 hours. Remarkably, the ODN persisted intact for at least 24 hours, and was detectable in the nuclear fraction for 4 days. Since the ODN was an unmodified phosphodiester, it is likely that the polylysine moiety protects the ODN from degradation, and that the ODN remains associated with the polylysine intracellularly. Furthermore, the uptake of lipid-ODN into the cell (as measured by the counts associated with washed, whole cells) was quantitatively similar to the uptake of AdpL-ODN complexes; however, nuclear localization of the lipid-ODN complexes was poor. These data are consistent with sequestration of the lipid-ODN complexes in the endosomal compartment, while the AdpL-ODN complexes are able to escape the endosome and localize in the nucleus.

In order to evaluate ODN uptake by an alternate methodology and visualize the percentage of the cell population that takes up the ODNs and their subcellular distribution, we used the immunohistochemical assay for BrdU substituted ODNs. With this assay, we demonstrated that the majority of cells were immunoreactive for BrdU after treatment with the AdpL-ODN complexes and confirmed the presence of the ODNs in the nucleus. Several other interesting features of this data include: 1) erratic uptake of free ODNs from the culture media; 2) degradation of free ODNs and incorporation of the BrdU into the chromatin visualized by a distinct immunohistochemical pattern; 3) uptake of both AdpL-ODN and lipid-ODN complexes in a globular pattern in a paranuclear location, suggesting endosomal vesicle mediated uptake; 4) nuclear localization of AdpL delivered ODNs.

Finally, the intracellular bioavailability of the ODNs after AdpL delivery was indirectly evaluated by the co-delivery of a reporter plasmid. AdpL complexes were constructed with a mixture of a luciferase reporter plasmid and the TF-ODN, and evidence for the delivery of functional DNA to the nucleus was obtained by luciferase expression from the reporter plasmid.

BIOLOGICAL ACTIVITY OF ADENOVIRUS-POLYLYSINE DELIVERY OF OLIGONUCLEOTIDES TO TISSUE CULTURE CELLS

Despite the high level of ODN delivery to several different cell lines, we have been unable to convincingly demonstrate an anti-gene effect with TF-ODNs designed to bind to the *HER-2/neu* promoter, delivered in either lipid-ODN complexes or AdpL-ODN complexes. There are several possible explanations for the lack of biological effects, and two general lines of reasoning apply: 1) *triplex formation does not occur in spite of efficient delivery*, due to inefficient release from the polylysine component of the AdpL-ODN complex, chromatin structure, intracellular cation composition, degradation of the unprotected phosphodiester by nucleases, or sequestration of the ODN in the nuclear matrix; or 2) *triplex formation does occur but is biologically irrelevant*, due to disruption of the triplex by DNA or RNA polymerase, due to a lack of importance of the triplex target for *HER-2/neu* transcription, or because only a limited amount of triplex is formed.

In contrast to the lack of efficacy that we have observed inhibiting HER-2/neu expression by TF-ODNs, there are limited data suggesting that triplex formation in the *HER-2/neu* promoter can reduce HER-2/neu expression in a cell line (MCF-7) with a single copy of the *HER-2/neu* gene when the TF-ODNs are delivered by a cationic lipid (*18*). Furthermore, p259A adenoviruses, used as endosomolytic agents, have enhanced the efficacy of antisense ODNs directed against herpes simplex virus RNA (*19*). AdpL-ODN complexes have now been used to deliver antisense ODNs against hepatitis B virus (HBV), strongly inhibiting HBV viral replication *in vitro* (*20*) as well as TF-ODNs directed against the rat aquaporin 5 gene which appear to be capable of reducing (chromosomal) gene expression in cell culture (*21*).

CONCLUSION

Methods are now available to detect the formation of intermolecular triplex DNA in chromatin. Triplex formation has been clearly demonstrated in permeabilized cells and isolated nuclei, situations in which the cell membrane and endosomal vesicle have been removed as barriers to ODN delivery. The delivery of TF-ODNs to the nucleus of tumor cells in culture can be efficiently accomplished, and it will be critical to re-evaluate adenovirus-polylysine complexes, cationic lipids and other ODN delivery methods to determine whether a sufficient concentration of free ODN is achieved in the nucleus of cultured cells for triplex formation with the target gene in structured chromatin. It will then be possible to correlate an observed biological anti-gene effect with the formation of a triple helix in the target gene.

ACKNOWLEDGMENTS

This research is supported by the National Cancer Institute (1K08-CA70262-01 and 1R29-CA76568-01 to S. W. E.) and the American Society of Clinical Oncology (to S. W. E.).

REFERENCES

1. Giovannangeli, C. and Hélène, C. (1997). Progress in the developments of triplex-based strategies. *Antisense and Nucleic Acid Drug Dev.* **7**, 413-421.
2. Svinarchuk, F., Debin, A., Bertrand, J. R. and Malvy, C. (1996). Investigation of the intracellular stability and formation of a triple helix formed with a short purine oligonucleotide targeted to the murine c-*pim-1* proto-oncogene promoter. *Nucleic Acids Res.* **24**, 295-302.
3. Wang, G., Seidman, M. M. and Glazer, P. M. (1996). Mutagenesis in mammalian cells induced by triple helix formation and transcription-coupled repair. *Science* **271**, 802-805.
4. Guieysse, A. L., Praseuth, D., Grigoriev, M., Harel-Bellan, A. and Hélène, C. (1996). Detection of covalent triplex within human cells. *Nucleic Acids Res.* **24**, 4210-4216.
5. Giovannangeli, C., Diviacco, S., Labrousse, V., Gryaznov, S., Charneau, P. and Hélène, C. (1997). Accessibility of nuclear DNA to triplex-forming oligonucleotides: the integrated HIV-1 provirus as a target. *Proc. Natl. Acad. Sci. USA* **94**, 79-84.
6. Belousov, E. S., Afonina, I. A., Kutyavin, I. V., Gall, A. A., Reed, M. W., Gamper, H. B., Wydro, R. M. and Meyer, R. B. (1998). Triplex targeting of a native gene in permeabilized intact cells: covalent modification of the gene for the chemokine receptor CCR5. *Nucleic Acids Res.* **26**, 1324-1328.
7. Stein, C. A. Controversies in the cellular pharmacology of oligodeoxynucleotides. *Ciba Found. Symp.* **209**, 79-89.
8. Temsamani, J., Kubert, M., Tang, J., Padmapriya, A. and Agrawal, S. (1994). Cellular uptake of oligodeoxynucleotide phosphorothioates and their analogs. *Antisense Res. Dev.* **4**, 35-42.

9. Ebbinghaus, S. W., Vigneswaran, N., Miller, R., Chee-Awai, R. A., Mayfield, C. A., Curiel, D. T. and Miller, D. M. (1996). Efficient delivery of triplex forming oligonucleotides to tumor cells by adenovirus-polylysine complexes. *Gene Ther.* **3**, 287-297.

10. Ebbinghaus, S. W., Gee, J. E., Rodu, B., Mayfield, C. A., Sanders, G. and Miller, D. M. (1993). Triplex formation inhibits *HER-2/neu* transcription *in vitro*. *J. Clin. Invest.* **92**, 2433-2439.

11. Fisher, T. L., Terhorst, T., Cao, X. and Wagner, R. W. (1993). Intracellular disposition and metabolism of fluorescently-labeled unmodified and modified oligonucleotides microinjected into mammalian cells. *Nucleic Acids Res.* **21**, 3857-3865.

12. Noonberg, S. B., Garovoy, M. R. and Hunt, C. A. (1993). Characteristics of oligonucleotide uptake in human keratinocyte cultures. *J. Invest. Dermatol.* **101**, 727-731.

13. Hacia, J. G., Wold, B. J. and Dervan, P. B. (1994). Phosphorothioate oligonucleotide-directed triple helix formation. *Biochemistry.* **33**, 5367-5369.

14. Kibler-Herzog, L., Zon, G., Uznanski, B., Whittier, G. and Wilson, W. D. (1991). Duplex stabilities of phosphorothioate, methylphosphonate, and RNA analogs of two DNA 14-mers. *Nucleic Acids Res.* **19**, 2979-2986.

15. Murakami, A., Nagahara, S., Uematsu, H., Otoi, K., Haruna, T., Ide, H. and Makino, K. (1992). Interaction of antisense DNA with nucleic acids/proteins. *Nucleic Acids Symp. Ser.* 123-124.

16. Curiel, D. T. (1994). High-efficiency gene transfer mediated by adenovirus-polylysine-DNA complexes. *Ann. N. Y. Acad. Sci.* **716**, 36-58.

17. Michael, S. I. and Curiel, D. T. (1994). Strategies to achieve targeted gene delivery *via* the receptor-mediated endocytosis pathway. *Gene Ther.* **1**, 223-232.

18. Porumb, H., Gousset, H., Letellier, R., Salle, V., Briane, D., Vassy, J., Amor-Gueret, M., Israel, L. and Taillandier, E. (1996). Temporary *ex vivo* inhibition of the expression of the human oncogene HER2 (NEU) by a triple helix-forming oligonucleotide. *Cancer Res.* **56**, 515-522.

19. Kulka, M. and Aurelian, L. (1995). Antiviral activity of an oligo(nucleoside methylphosphonate) that targets HSV-1 immediate-early pre-mRNA 4,5 is augmented by cotreatment with replication-defective adenovirus. *Antisense Res. Dev.* **5**, 243-249.

20. Madon, J. and Blum, H. E. (1996). Receptor-mediated delivery of hepatitis B virus DNA and antisense oligodeoxynucleotides to avian liver cells. *Hepatology.* **24**, 474-481.

21. Delporte, C., Panyutin, I. G., Sedelnikova, O. A., Lillibridge, C. D., O'Connell, B. C. and Baum, B. J. (1997). Triplex-forming oligonucleotides can modulate aquaporin-5 gene expression in epithelial cells. *Antisense and Nucleic Acid Drug Dev.* **7**, 523-529.

10 TRIPLE-STRANDED COMPLEXES AS ANTIGENE MOLECULES: TRANSCRIPTION INHIBITION *EX VIVO*

Carine Giovannangeli

INTRODUCTION

The design of molecules that recognize specific sequences on the DNA double helix would provide a rational basis for the development of antigene drugs. The double helix is a grooved cylindric molecule and the different facets of the three-dimensional shape of the helical sequence can be recognized by different classes of sequence-specific ligands.

Among the strongest and most specific ligands of the double-stranded DNA are:
- small molecules such as synthetic oligoamides which can bind the minor groove *(1, 2)*;
- triplex-forming molecules (TFMs): (i) oligodeoxyribonucleotides and their modified analogues (TFOs), and *in situ* generated RNA molecules which bind to the DNA major groove (for review: *3*) (ii) peptide nucleic acids (PNAs), which can locally unwind the double helix to form a triplex with one of its strands by strand invasion (for review: *4*).

Much progress has been made during the past decade to improve the sequence specificity and affinity of these double-stranded DNA ligands. This progress has been motivated in part by the hope of artificially modulating the various biological processes acting on DNA within a living organism, such as transcription, replication, repair (site-directed mutagenesis), and recombination. Consequently such specific DNA ligands are attractive candidates for gene-based therapeutics, and powerful tools for cellular and molecular biology.

This paper discusses the ability of triple-stranded complexes (composed of TFO, *in situ* generated RNA, and PNA) to interfere with transcription: it addresses both well-established and promising triplex-based developments and limitations in controlling the expression of specific genes *ex vivo* (in cultured cells) and *in vivo* (in animal models).

MECHANISM OF TRIPLEX-INDUCED
INHIBITION OF TRANSCRIPTION

Choice of the DNA target sequence

The use of TFMs is a strategy to inhibit gene expression at the transcriptional level. When the target is located within the control region (promoter or enhancer), the bound oligonucleotide may inhibit transcription factor binding or induce a conformational change in DNA precluding assembly of the transcription machinery. When the oligonucleotide binds downstream of the transcription start site, it may inhibit the elongation step of the transcription process, blocking RNA polymerase read-through.

The elongation process is much more difficult to inhibit because the stability of the triple-helical complex is usually not sufficient to arrest the transcription machinery once it is launched on its double-helical template. However, it was clearly demonstrated in a cell-free system (HeLa nuclear extract) that eukaryotic RNA polymerase II can be arrested, and transcription terminated, when a highly thermodynamically and kinetically stable triplex is encountered *(5, 6)*.

Consequently, TFMs are able to induce transcription inhibition independently of whether their oligopyrimidine•oligopurine targets are located in control regions or in transcribed parts of the gene. To achieve an efficient transcriptional arrest, the use of molecules that are tightly bound to DNA under physiological conditions is necessary.

The best triplex-forming molecules

A large effort has been devoted to modifying either the base, the sugar, or the backbone portion of oligonucleotides. The ideal TFM should bind strongly to its double-stranded target and resist degradation in a cellular environment. A description of all these modifications, treated elsewhere in this volume (see for example chapters 5, 6, 18, 19), is not included in this article. Here we briefly describe the TFMs that have been shown to be active *ex vivo* and thus seem to be the most promising.

(i) Some (G, A)-containing oligonucleotides form unusually stable short triplexes *(7)*. Competing self-structures (intermolecular homoduplex, or G-quartets) have been described, but in some cases solutions now exist to overcome this problem, e.g., 7-deazaguanine substitution, or use of an unmodified helper oligonucleotide *(8)*;

(ii) Modified oligonucleotides containing N3'→P5' phophoramidate linkage (pn) – the 3'-hydroxyl is substituted by a 3'-amino group on the sugar residue – form stable triplexes under conditions where the isosequential phosphodiester (po) oligonucleotides are unable to do so; this was observed for TFOs containing (T, C) or (T, G) and interacting via the Hoogsteen motif, but not for TFOs containing

(T, G) or (A, G) and interacting via the reverse Hoogsteen motif (*5, 9*; and see chapter 6, this volume);

(iii) Depending on their sequence, TFOs composed of RNA bind more tightly than the corresponding DNA oligonucleotides to a DNA double-helix (see chapter 17, this volume). Consequently it seems attractive to generate an RNA molecule containing the third-strand sequence *in situ*, using stable transfection of cells with a plasmid from which such an RNA molecule is transcribed *(10)*;

(iv) Peptide nucleic acid (PNA) is a DNA analogue in which the entire sugar-phosphate backbone is replaced by a polyamide (pseudopeptide) chain (see chapter 18, this volume). PNAs containing pyrimidines do bind tightly to double-stranded DNA by a mechanism involving strand displacement, which leads to the formation of 2:1 complexes with the oligopurine sequence, and to displacement of the second DNA strand (oligopyrimidine sequence), which remains single-stranded. Such a PNA/DNA triplex is sufficiently stable to arrest transcription elongation, but only if the PNA is targeted to the template strand. In contrast, when a triplex is formed with a TFO interacting by (reverse) Hoogsteen hydrogen bonds with the preexisting duplex, the RNA polymerase arrest is observed independently of whether the oligopurine- or the oligopyrimidine-containing strand is the coding strand *(11)*.

A rapidly growing body of work had described the ability of TFMs to inhibit transcription in cell-free systems (for review: *12*). Here we will focus only on triplex-induced effects in cultured cells.

TRANSIENT EXPRESSION ASSAYS USING EXOGENOUS PLASMID TARGETS : DEMONSTRATION OF TRIPLEX-MEDIATED EFFECTS

The available data describing results obtained within cells are rather rare. In addition, only a few studies provide evidence that the observed effect on gene expression is due to triplex formation; they are described in Table 1. In some cases the observed effect might be due to binding of the TFM to a transcription factor rather than to DNA. However, encouraging reports describe the possibility of (i) triplex formation and persistence intracellularly on exogenous plasmid targets transiently transfected into living cells, and (ii) triplex-mediated biological activities. In particular, four examples have investigated the binding of the TFO to its oligopyrimidine•oligo-purine targets in transfected cells (see Table 1).

Directed mutagenesis

Triplex formation is used successfully to induce mutations at the triplex recognition site (see chapter 13, this volume). The plasmid is transfected first, then the TFO (30-mer(G, A), or 13-mer(G, X = 7-deazaxanthine)) covalently linked to a psoralen moiety (Pso) is added to the culture medium, and irradiation is performed *(13, 14)*. In one study the effect of the TFO (30-mer(G, A)) was investigated in the absence of irradiation *(15)*.

DNA target	Triplex-forming molecule	Triplex formation	Triplex involvement	TFM-induced effect	Ref
Exogenous plasmid					
c-pim/luc (initiation)	13-mer(G, A)3'*	*In vitro* preformed triplex	(+) 60% DMS protection	30% luc protein inhibition	*(16)*
IL-2Rα/CAT (initiation)	15-mer(T, C^m)/Acr5'	*In vitro* preformed triplex	(+) No TFO effect with mutated target site Linear PCR	75% CAT protein inhibition	*(17)*
	15-mer(T, C^m)/Pso5'	TFO free in medium (20 μM)		20% modified DNA	*(18)*
PGK/PPT/*luc* (elongation)	15-mer(T, C, G)pn/Acr5'	TFO/cationic complex (1 μM)	(+) No TFO effect with mutated target site	50% luc protein inhibition	*(19)*
SV40/SupF (mutagenesis)	30-mer(G, A)3'* 30-mer(G, A)3'*/Pso5' 13-mer(G, X)3'*/Pso5' (X=7-deazaxanthine)	TFO free in medium (2 μM)	(+) TFO-induced mutations on triplex site	0.3% 2.1% 1.4% mutation frequency	*(15)* *(13)* *(14)*
Endogenous gene					
CCR-5 gene	12-mer(A, G°)3'*/ 5' nitrogen mustard (G° = modified guanine)	Streptolysin-O (20 μM)	(+) Ligation-mediated PCR	20% modified DNA	*(26)*
HIV-1/PPT/*pol* gene	15-mer(T, C, G)pn/Pso5'	Digitonin (5 μM)	(+) No TFO effect with mutated target site and REPA/competitive PCR	30% modified DNA	*(25)*
SupF gene (mutagenesis)	bisPNA (8-8 or 10-10 bases long)	Streptolysin-O	(+) PNA-induced mutations on triplex site	0.1% mutation frequency	*(31)*
c-myc gene (initiation)	27-mer(G, A)3'*	TFO free in medium (25 μM)	(+) DNase I protection (-) CNBP protein	50% *c-myc* RNA inhibition	*(37)* *(38)*
IGF-I gene (5'UTR) IGF-IR gene (3'UTR)	RNA containing 23- or 24-mer(G, A)	Vector transcription from MT or CMV promoter	(+/-) RNA containing 23- or 24-mer(T, C) or without triplex insert	100% IGF-I or IGF-IR RNA and protein inhibition	*(40)* *(10)*

Table 1. Examples of TFM-induced biological effects *ex vivo*; triplex involvement (+) or not (-) is discussed (see text and references for details, and Abbreviations section)

The location of induced mutations around the site where the TFO binds, demonstrates both the possibility of *forming* a triplex inside cells, and of generating site-specific mutations in mammalian cells with high efficiency, at least on a plasmid vector.

Inhibition of transcription initiation

The demonstration of triplex-induced inhibition of transcription initiation was reported using two systems.

Svinarchuk *et al. (16)* have detected directly the binding of a (G, A)-containing TFO (13-mer) on an oligopyrimidine•oligopurine sequence present in the murine c-*pim*-1 proto-oncogene promoter. Triplex formation was demonstrated using a dimethylsulfate (DMS) footprinting assay: the N-7 atoms of the guanines present on the oligopurine strand of the DNA target were protected from methylation by DMS when the TFO was bound. Triplex formation was detected only when the complex was preformed *in vitro* before transfection. Then DMS protection was observed up to 72 hours after transfection, demonstrating the intracellular *persistence* of the triplex during this period. In addition, under conditions where 60% of the target was protected from DMS methylation because of the presence of the triplex on the c-*pim* promoter, a partial inhibition of the expression of the reporter gene was reported.

However, no triplex formation is detected in cells which are first transfected with the plasmid and then further incubated with the (G, A)-containing TFO. The reasons for this lack of detection are not completely understood, but one of them is the low sensitivity of the detection method (see chapter 11, this volume).

Grigoriev *et al. (17)* have demonstrated a transcriptional inhibition induced by triplex formation on a 15 base pair long sequence present in the α subunit of the interleukin-2 receptor (IL-2Rα) promoter. The demonstration of triplex involvement in the observed inhibition makes use of a plasmid that has been mutated within the oligopyrimidine•oligopurine target sequence to provide the appropriate control system without changing the oligonucleotide sequence. This mutant sequence is unable to form a triplex with the anti-(IL-2Rα) TFO, but the location of the mutations in the target sequence is such that it inhibits neither transcription factor binding, nor transcriptional activation from the mutated promoter. Following an incubation with an acridine-oligonucleotide conjugate (15-mer(T, C^m)/Acr5') to preform the triplex *in vitro*, the plasmid bearing the wild-type or mutated promoter of IL-2Rα (driving expression of the CAT reporter gene) is transfected into mammalian cells. Transcription is inhibited (75% at 5 μM of TFO/Acr) from the plasmid with the wild-type sequence that permitted triplex formation, but not from the plasmid with the mutant sequence.

More recently *(18)*, it has been reported in the same system that a triplex can be detected even if the triplex is not preformed *in vitro*. Cells are first electroporated with the plasmid, then incubated with the TFO (15-mer(T, C^m)/Pso5'), and finally irradiated to irreversibly damage the DNA at the triplex site. The location and the amount of TFO-induced DNA modifications are evaluated by linear PCR: 20% of triplex-induced photoproducts (at 20 µM of TFO/Pso) are detected on the oligopyrimidine•oligopurine target sequence. This result demonstrates that the triplex can be *formed* inside cells on a plasmid target.

Inhibition of transcription elongation

Faria *et al. (19)* have demonstrated an inhibition of transcription at the elongation level induced by formation of a non-covalent triplex. A 16 base pair long oligopyrimidine•oligopurine sequence, wild type (HIV-1/PPT sequence: 5' AAAAGAAAAGGGGGGA 3') or mutated (5' AAAAGAAAAaGGaGGA 3' with 2 mismatches, in lower case) was cloned upstream from the *luciferase* gene, in the transcribed and untranslated 5'-part of the gene. The plasmid and the TFO (15-mer(T, C, G)pn/Acr5') were transfected separately to be sure that the triplex could only be formed in cultured cells and not *in vitro* before transfection. Inhibition of luciferase expression was observed for the wild-type plasmid (50% at 2 µM of TFO/Acr) and not for the plasmid with the mutated triplex sequence.

These results demonstrate both the possibility of *forming* a triplex inside cells, and the ability of such a complex to arrest transcription elongation with an oligophosphoramidate in the absence of crosslinking reactions.

Concluding remarks

The cellular model systems described above with exogenously transfected plasmid targets are useful to demonstrate the feasibility of the triplex-based strategy: they allow one to demonstrate that transcription inhibition is really induced by TFO binding to its complementary oligopyrimidine•oligopurine sequence. When the gene of interest is carried by a plasmid vector, it is possible to introduce mutations in the target sequence. This is a way to clearly prove the molecular mechanism of the TFO-induced expression inhibition: the same TFO is tested in two cellular systems that differ only by a few mutations in the targeted oligopyrimidine•oligopurine sequence; all the additional interactions that the TFO might have with cellular components other than the targeted DNA sequence are perfectly identical in the two systems.

Several examples of transcription inhibition by oligonucleotides that are not listed in Table 1 do exist (α1(I)collagen promoter *(20, 21)*; granulocyte-macrophage colony stimulating factor (GM-CSF) promoter *(22)*; immunoglobulin enhancer *(23)*; progesterone responsive element *(24)*), but they often lack the appropriate controls to conclusively demonstrate triplex involvement.

Nevertheless, the structure of the targeted sequence carried by the transfected plasmid is likely to be quite dissimilar to the one that exists in the chromosomal DNA.

ENDOGENOUS GENE AS A TARGET: PROOF-OF-CONCEPT

For the majority of the targeted genes, the TFMs that were found to be able to bind their complementary target under physiological conditions contained G residues. A large number of studies have described a biological response after administration of such molecules which is due to sequence-specific interactions with extra/intra-cellular proteins. Consequently, transcription inhibition induced by a molecule designed to form a triplex must be carefully analyzed; the only convincing demonstration that the TFM works through the postulated antigene mechanism is based on the proof of binding to the oligopyrimidine•oligopurine DNA targeted sequence, as described in the previous paragraph for exogenously transfected DNA targets.

Accessibility of chromosomal nuclear DNA

One of the main questions raised by the development of the antigene strategy *ex/in vivo* deals with the recognition of DNA sequences in the context of chromatin. Three recent examples describe the accessibility of a DNA sequence to TFMs within the chromatin (see Table 1).

Triplex as a covalent lesion on nuclear DNA: detection and quantification. Two examples are based on the use of TFOs covalently linked to a molecule that is able to react with DNA (psoralen *(25)*, or nitrogen mustard *(26)*). This is a powerful tool to trap the triple-helical complex formed in the cell nucleus. This approach exploits the reaction of these conjugates at the triplex site (photo-induced crosslink formation with the DNA strands for the psoralen moiety; alkylation of guanine on the oligopurine target by phenylacetate mustard), which converts a non-covalent triple-helical complex into a covalent one and results in a localized damage to genomic DNA.

Different assays were used to quantitatively analyze the TFO-induced covalent modification of genomic DNA (restriction enzyme protection assay: REPA; competitive PCR; ligation-mediated PCR: LMPCR). These procedures allow a direct examination of triplex formation within the cell nuclei and evaluation of both the specificity and the efficacy of the reaction. In both examples about 30% of the target sequence was shown to be covalently modified by the oligonucleotide conjugate following a cell permeabilization procedure (digitonin or streptolysin-O treatment of the cells) that increases penetration of the oligonucleotide and maintains nuclear integrity. In contrast, when naked deproteinized chromosomal DNA was used as a target *in vitro*, 95% could be reached.

The fraction of modified DNA reflects both the coverage of the double-stranded target and the yield of the covalent modification. Thus the amount of modified DNA is expected to depend on several factors such as reaction conditions, oligonucleotide concentration and penetration, and nucleosome positioning during the reaction time. In particular, DNA accessibility depends on the cell cycle, and especially on the transcriptional state of the targeted oligopyrimidine•oligopurine sequence in the gene

136

in question; the promoter and enhancer regions of actively transcribed genes are known to be accessible enough to allow binding of transcription factors to their DNA target sequences *(27)*. In a similar fashion, a TFM should be able to reach its DNA target, at least at protein-binding sites. The two targeted sequences presented above (CCR5 gene and HIV-1/*pol* gene) are located in the coding regions of the respective gene. However, for the HIV system, the 16 base pair long targeted PPT/HIV-1/*pol* sequence belongs to a 500 base pair fragment which has been shown to act as a transcription enhancer and to contain several transcription factor binding sites.

The accessibility of a DNA sequence cannot be predicted *a priori*, but it does not seem to be limited to control regions of genes. Therefore it is important to develop techniques - such as those described above and in *(25, 26)* - that may be used to identify and analyze target accessibility in cells. The ability to reconstitute nucleosomal structures *in vitro (28)* may soon provide a consistent way to predict accessibility of any DNA sequences. By analogy it can be noted that various experimental predictive methods do exist and seem very effective in answering the question of accessibility of an RNA molecule to heteroduplex formation *(29, 30)*, which is a crucial question for the antisense strategy. There is a good correlation between heteroduplex yield measured in *in vitro* predictive assays, and antisense potential *ex/in vivo*.

Triplex-directed mutagenesis. Using exogenously transfected oligonucleotide-psoralen conjugates or high affinity TFOs provokes site-specific mutagenesis via the induction of a repair-dependent pathway.

Faruqi *et al. (31)* have recently described the possibility of inducing mutations within a chromosomally integrated vector sequence: a single dimeric PNA molecule ((8-8) or (10-10) / bis-PNA) can form a triple-stranded complex with the complementary oligopurine targeted sequence by strand invasion, resulting in the displacement of one DNA strand and formation of a D-loop; the PNA molecule is introduced into the cells by streptolysin-O treatment, which allows efficient entry into the cell nuclei. Reporter gene analysis has revealed the induction of mutations located at or adjacent to the PNA binding site; this clearly indicates that the chromosomal target site is reached by the PNA molecule within living cells.

Around 10-fold greater induction – as compared to the background – of mutagenesis was observed at the targeted chromosomal site in mouse cells with an (8-8) or (10-10) / bis-PNA, as previously described with an extrachromosomal vector in monkey and human cells using non-covalent triplex formed with a (G, A)-containing TFO (see Table 1). The yield of mutations (around 0.1 - 5% depending on the TFM) does not reflect the occupancy of the oligopyrimidine•oligopurine target sequence, since it is related to the small fraction of triplex that is repaired (*32*; and chapter 14, this volume), and the fraction without the original sequence of DNA. However, even low levels of triplex can provoke TFM-induced mutations that are detectable: this is a very sensitive method, contrary to triplex-induced transcription inhibition.

These three examples clearly demonstrate the accessibility of chromosomal DNA to TFMs (TFO and PNA) in cultured cells, even if the examples described do not allow one to evaluate the amount of coverage of the oligopyrimidine•oligopurine target, and consequently to estimate the level of the expected inhibition of transcription.

Gene regulation induced by molecules designed to form a triplex on DNA

There have been several reports describing the inhibition of gene expression in eukaryotic cells using TFOs to target endogenous genes (granulocyte-macrophage colony stimulating factor (GM-CSF, *33*); HER2(NEU) *(34)*; tumor necrosis factor (TNF, *35*); aldehyde dehydrogenase (ALDH$_2$, *36*). However, no direct evidence is provided that this inhibition is due to triplex formation rather than to other oligonucleotide-directed mechanisms. Nevertheless, two examples will be further discussed here (see Table 1).

Mechanisms of action. Postel *et al. (37)* have described c-*myc* transcriptional inhibition using a G-rich TFO. An indication of triplex formation is also reported, based on inhibition of a DNase I hypersensitive site overlapping the triplex site in the presence of the TFO. However, it has been shown that the mechanism of gene expression inhibition more likely involves transcription factor binding to the TFO, which acts as a decoy *(38, 39)*. Even if triplex formation is involved in transcription inhibition observed in the presence of the TFO, its potential partial contribution remains to be clearly established.

Triplex-induced gene therapy. Ilan and co-workers (IGF-IR gene *(10)*; IGF-I gene *(40)*) have described an original triplex approach in the gene therapy field. The strategy is based on the ability of RNA to bind as a third strand to DNA. Cells were transfected with a vector allowing transcription of an RNA molecule containing a sequence that could potentially bind to double-helical DNA to form a triplex. The target sequences were either a 23 base pair oligopyrimidine•oligopurine sequence present in the IGF-I gene (in the 5'-untranslated and transcribed part of the gene), or a 24 base pair sequence present in the IGF-IR gene (in the 3'-untranslated and transcribed part of the gene). Inhibition of IGF-I and IGF-IR expression was observed in cultured cells, and tumor growth inhibition in an animal model was reported. Two of the data described will be further discussed.

Firstly, the results presented by Ilan and co-workers show that an RNA strand containing a 23 or 24 nucleotide long (G, A) sequence designed to form a triplex on an IGF-I or IGF-IR sequence induces transcription inhibition of the targeted gene. Two control RNA transcripts containing either the polylinker sequence, or the 23 or 24 nucleotide long (C, T) sequence (requiring pH dependent triplex formation in contrast to the (G, A) sequence) do not affect IGF-I or IGF-IR transcription. Other mechanisms cannot be excluded, such as binding and sequestering of a regulatory protein factor by the (G, A) sequence of the RNA transcript or by the oligopyrimidine•oligopurine sequence of the DNA vector; but triplex formation

138

remains the most plausible mechanism for the inhibition of IGF-I gene expression. Nevertheless, an apparent discrepancy exists between the *ex/in vivo* results and *in vitro* data. *In vitro* studies have shown that the use of an RNA (C, U)-containing third strand strongly enhances triplex stability compared to the DNA (C, T) third strand, but under conditions where a (G, A) DNA third strand binds to the target oligopyrimidine•oligopurine sequence, the isosequential RNA does not bind. The molecular basis of these observations is not known at present. However, the conditions within cells might favor the binding of the RNA molecule containing the (G, A) portion; and, also, flanking regions that are not present in *in vitro* assays may play a crucial role in the presentation of the (G, A) sequence to the DNA target. In addition, cellular proteins that recognizes such a complex may stabilize (G, A) RNA-directed triple helices *ex/in vivo*. Even if the mechanism of action of the (G, A) RNA molecule remains to be clarified, these results are encouraging because they provide a basis for a fresh approach to gene therapy for effecting inhibition of gene expression, by using a triplex-based strategy.

Secondly, it must be noted that the RNA targeted to the IGF-I gene had no effect on IGF-IR gene expression; in contrast, inhibition of IGF-IR led to an inhibition of IGF-I as expected from previous studies on the regulation of IGF-I by IGF-IR expression. In addition, simultaneous to the IGF-I decrease, an increased expression of protease nexin I is seen in this system. This observation highlights the interest of using methods (such as differential display) that allow an analysis and quantification of a large number of RNA species *(41, 42)*. A comparison of expression patterns obtained in the presence and in the absence of TFMs might be very informative; in particular in systems involving cross- or retro-regulations, information concerning the TFM-induced modifications in the level of various RNAs would be useful to choose the most appropriate gene to target.

PERSPECTIVES

Oligonucleotide-induced activity in cell nucleus

Even though it was demonstrated that endogenous genes can be targeted by TFMs in permeabilized cells, the demonstration that a TFM is able to inhibit transcription under more physiological conditions by the postulated mechanism of binding to a oligopyrimidine•oligopurine double-stranded target, and the determination of the level of such inhibition remains to be firmly established. It has been shown that an antisense oligonucleotide is active in the cell nucleus by binding to the pre-mRNA *(43)*. Consequently it seems that it is possible to deliver to the nucleus a concentration of oligonucleotide that allows it to reach its nucleic acid target and to induce the expected antisense effect. This results is promising for the antigene approach.

Triplex-induced gene cleavage

It is now well established that a triplex can be formed within cells, but the intracellular fate of such a complex is completely unknown. In particular, it would be very interesting to efficiently damage the DNA where the triplex is formed. This would increase the efficiency of the antigene approach and could be described as an equivalent to the RNase H, or the recently described double-stranded RNase *(44)* activities involved in the antisense approach. Different strategies can be studied.

(i) The existence of eukaryotic proteins that recognize a triplex structure has been reported *(18, 45)*. However they have not been identified yet; they might be associated with a pre-existing cleavage activity (DNase-like), or be able to be converted into a DNA cleaving agent by chemical modification (for example : *(46)*).

(ii) The TFM can be covalently linked to a substance that is able to recruit a cellular enzyme with a cleaving activity that would induce irreversible damage at the triplex site (for example : TFO-3'-camptothecin and topoisomerase I: *(47)*).

(iii) The TFM can be conjugated to a variety of reactive groups such as adduct inducers (psoralen, platin), alkylating agents (nitrogen mustard), and cleaving agents (Fe^{2+}-EDTA, Cu^+-OP) (see chapter 7, this volume).

The ability to obtain covalent modification at the TFO binding site has been largely documented by *in vitro* data. But only a few data describe their use in cell cultures (alkylating groups *(26)*; psoralen *(13, 25)*). In this kind of approach, the activateable moiety must be chosen for its ability to efficiently and specifically modify the triplex site and to resist inactivation by other cellular components before reacting with its oligopyrimidine•oligopurine target.

Conclusions

Strategies based on triple helix formation still suffer from constraints with respect to target sequence recognition and requirements for stabilizing triple-helical complexes. Recently, progress has been made toward alleviating these constraints and improving affinity. The demonstration of oligonucleotide binding to its chromosomal DNA target in cell nuclei is available at present, and in the near future the evaluation of its efficacy at inhibiting will emerge. TFMs provide useful tools for the dissection of mechanisms of regulation involved in gene transcription. Also, their use might potentially lead to novel therapeutics, as is the case for the older antisense molecules. Much more work to improve *in vivo* oligonucleotide binding, stability and delivery to the nucleus that contains the nucleic acid target will help move this triplex technology from molecular biology to clinical applications.

ACKNOWLEDGMENTS

C.G. thanks Claude Hélène for critical evaluation of the manuscript. C.G. acknowledges the French agency for AIDS (ANRS) for its support.

140

ABBREVIATIONS

TFO, triplex-forming oligonucleotide; TFM, triplex-forming molecule; 3'*, 3' modification to resist nuclease-mediated degradation by incorporation of various chemical moieties (such as: triethyleneglycol, propylamine group,...); Acr, acridine; Pso, psoralen; C^m, 5-methyl cytosine; pn, phosphoramidate N3'→P5' oligonucleotide analogues; PNA, peptide nucleic acid; ; CAT, chloramphenicol acetyltransferase; *luc*, *luciferase* gene; PPT, polypurine tract sequence; IL-2Rα, α subunit of the interleukin-2 receptor promoter; PGK, phosphoglycerate kinase promoter; MT, metallothionein I promoter; CMV, cytomegalovirus promoter; CNBP, cellular nucleic acid binding protein; DMS, dimethyl sulfate; REPA, restriction enzyme protection assay; PCR, polymerase chain reaction.

REFERENCES

1. Gottesfeld, J. M., Neely, L., Trauger, J. W., Baird, E. E. and Dervan, P. B. (1997). Regulation of gene expression by small molecules. *Nature* **387**, 202-205.
2. Bremer, R. E., Baird, E. E. and Dervan, P. B. (1998). Inhibition of major-groove-binding proteins by pyrrole-imidazole polyamides with an Arg-Pro-Arg positive patch. *Chemistry & Biology* **5**, 119-133.
3. Chan, P. P. and Glazer, P. M. (1997). Triplex DNA: fundamentals, advances, and potential applications for gene therapy. *J. Mol. Med.* **75**, 267-282.
4. Nielsen, P. E. (1995). DNA analogues with nonphosphodiester backbones. *Annu. Rev. Biophys. Struct.* **24**, 167-183.
5. Giovannangeli, C., Perrouault, L., Escudé, C., Gryaznov, S. and Hélène, C. (1996). Efficient inhibition of transcription elongation *in vitro* by oligonucleotides phosphoramidates targeted to proviral HIV DNA. *J. Mol. Biol.* **261**, 386-398.
6. Wang, Z. and Rana, T. M. (1997). DNA damage-dependent transcriptional arrest and termination of RNA polymerase II elongation complexes in DNA template containing HIV-1 promoter. *Proc. Natl. Acad. Sci. USA* **94**, 6688-6693.
7. Svinarchuk, F., Paoletti, J. and Malvy, C. (1995). An unusually stable purine (purine-pyrimidine) short triplex. *J. Biol. Chem.* **270**, 14068-14071.
8. Svinarchuk, F., Cherny, D., Debin, A., Delain, E. and Malvy, C. (1996). A new approach to overcome potassium-mediated inhibition of triplex formation. *Nucleic Acids Res.* **24**, 3858-3865.
9. Escudé, C., Giovannangeli, C., Sun, J. S., Gryaznov, S., Garestier, T. and Hélène, C. (1996). Stable triple helices formed by oligonucleotide N3'->P5' phosphoramidates inhibit transcription elongation. *Proc. Natl. Acad. USA* **93**, 4365-4369.
10. Rininsland, F., Johnson, T. R., Chernicky, C. L., Schulze, E., Burfeind, P., Ilan, J. and Ilan, J. (1997). Suppression of insulin-like growth factor type I receptor by a triple-helix strategy inhibits IGF-I transcription and tumorigenic potential of rat C6 glioblastoma cells. *Proc. Natl. Acad. Sci. USA* **94**, 5854-5859.
11. Giovannangeli, C., Perrouault, L., Escudé, C., Thuong, N. T. and Hélène, C. (1996). Specific inhibition of transcription elongation in vitro by triplex-forming oligonucleotide intercalator-conjugates on HIV proviral DNA. *Biochemistry* **35**, 10539-10548.
12. Maher, L. J., III (1996). Prospects for the therapeutic use of antigene oligonucleotides. *Cancer Investigation* **14**, 66-82.

13. Wang, G., Levy, D. D., Seidman, M. M. and Glazer, P. M. (1995). Targeted mutagenesis in mammalian cells mediated by intracellular triple helix formation. *Mol. Cell. Biol.* **15**, 1759-1768.

14. Faruqi, A. F., Krawczyk, S. H., Matteucci, M. D. and Glazer, P. M. (1997). Potassium-resistant triple helix formation and improved intracellular gene targeting by oligode-oxyribonucleotides containing 7-deazaxanthine. *Nucleic Acids Res.* **25**, 633-640.

15. Wang, G., Seidman, M. M. and Glazer, P. M. (1996). Mutagenesis in mammalian cells induced by triple helix formation and transcription-coupled repair. *Science* **271**, 802-805.

16. Svinarchuk, F., Debin, A., Bertrand, J. R. and Malvy, C. (1996). Investigation of the intracellular stability and formation of a triple helix formed with a short purine oligonucleotide targeted to the murine c-pim-1 proto-oncogene promoter. *Nucleic Acids Res.* **24**, 295-302.

17. Grigoriev, M., Praseuth, D., Guieysse, A. L., Robin, P., Thuong, N. T. and Hélène, C. (1993). Inhibition of interleukin-2 receptor alpha-subunit gene expression by oligonucleotide-directed triple helix formation. *C. R. Acad. Sci. Paris* **316**, 492-495.

18. Guieysse, A. L., Praseuth, D., Grigoriev, M., Harel-Bellan, A. and Hélène, C. (1996). Detection of covalent triplex within human cells. *Nucleic Acids Res.* **24**, 4210-4216.

19. Faria, M., Wood, C., White, M., Hélène, C. and Giovannangeli, C. (1998). *Ex vivo* inhibition of transcription elongation by phosphoramidate triplex-forming oligonucleotides (*in preparation*).

20. Kovacs, A., Kandala, J. C., Weber, K. T. and Guntaka, R. V. (1996). Triple helix forming oligonucleotide corresponding to the polypyrimidine sequence in the rat alpha1 (I) collagen promoter specifically inhibits factor binding and transcription. *J. Biol. Chem.* **271**, 1805-1812.

21. Joseph, J., Kandala, J. C., Veerapanane, D., Weber, K. T. and Guntaka, R. V. (1997). Antiparallel polypurine phosphorothioate oligonucleotides form stable triplexes with the rat α1(I) collagen gene promoter and inhibit transcription in cultured rat fibroblasts. *Nucleic Acids Res.* **25**, 2182-2188.

22. Kochetkova, M. and Shannon, M. F. (1996). DNA triplex formation selectively inhibits granulocyte-macrophage colony-stimulating factor gene expression in human T cells. *J. Biol. Chem.* **271**, 14438-14444.

23. Neurath, M. F., Max, E. E. and Strober, W. (1995). Pax (BSAP) regulates the murine immunoglobulin 3' α enhancer by suppressing binding of NF-αP, a protein that controls heavy chain transcription. *Proc. Natl. Acad. Sci. USA* **92**, 5336-5340.

24. Ing, N. H., Beekman, J. M., Kessler, D. J., Murphy, M., Jayaraman, K., Zendegui, J. G., Hogan, M. E., O'Malley, B. W. and Tsai, M. J. (1993). *In vivo* transcription of a progesterone-responsive gene is specifically inhibited by a triplex-forming oligonucleotide. *Nucleic Acids Res.* **21**, 2789-2796.

25. Giovannangeli, C., Diviacco, S., Labrousse, V., Gryaznov, S., Charneau, P. and Hélène, C. (1997). Accessibility of nuclear DNA to triplex-forming oligonucleotides: the integrated HIV-1 provirus as a target. *Proc. Natl. Acad. Sci. USA* **94**, 79-84.

26. Belousov, E. S., Afonina, I. A., Kutyavin, I. V., Gall, A. A., Reed, M. W., Gamper, H. B., Wydro, R. M. and Meyer, R. B. (1998). Triplex targeting of a native gene in permeabilized intact cells: covalent modification of the gene for the chemokine receptor CCR5. *Nucleic Acids Res.* **26**, 1324-1328.

27. Hurt, E. and Schütz, G. (1998). Nucleus and gene expression. *Current Op. Cell Biol.* **10**, 301-303.

28. Schultz, M. C., Hockman, D. J., Harkness, T. A. A., Garinther, W. I. and Altheim, B. A. (1997). Chromatin assembly in a yeast whole-cell extract. *Proc. Natl. Acad. Sci. USA* **94**, 9034-9039.

29. Matveeva, O., Felden, B., Audlin, S., Gesteland, R. F. and Atkins, J. F. (1997). A rapid *in vitro* method for obtaining RNA accessibility patterns for complementary DNA probes: correlation with an intracellular pattern and known RNA structures. *Nucleic Acids Res.* **25**, 5010-5016.

30. Southern, E. M., Milner, N. and Mir, K. U(1997). Discovering antisense reagents by hybridization of RNA to oligonucleotide arrays. In *Oligonucleotides as therapeutic agents*, Wiley, Chister (Ciba foundation symposium) pp.38-46.

31. Faruqi, A. F., Egholm, M. and Glazer, P. M. (1998). Peptide nucleic acid-targeted mutagenesis of a chromosomal gene in mouse cells. *Proc. Natl. Acad. Sci. USA* **95**, 1398-1403.

32. Sandor, Z. and Bredberg, A. (1995). Deficient DNA repair of triple helix-directed double psoralen damage in human cells. *FEBS Lett.* **374**, 287-291.

33. Kochetkova, M., Iversen, P. O., Lopez, A. F. and Shannon, M. F. (1997). Deoxyribonucleic acid triplex formation inhibits granulocyte macrophage colony-stimulating factor gene expression and suppresses growth in juvenile myelomonocytic leukemic cells. *J. Clin. Invest.* **99**, 3000-3008.

34. Porumb, H., Gousset, H., Letellier, R., Salle, V., Briane, D., Vassy, J., Amor-Gueret, M., Israel, L. and Taillandier, E. (1996). Temporary *ex vivo* inhibition of the expression of the human oncogene HER2 (NEU) by a triple helix forming oligonucleotide. *Cancer Res.* **56**, 515-522.

35. Aggarwal, B. B., Schwar, L., Hogan, M. E. and Rando, R. F. (1996). Triple helix-forming oligodeoxyribonucleotides targeted to the human tumor necrosis factor (TNF) gene inhibit TNF production and block the TNF-dependent growth of human glioblastoma tumor cells. *Cancer Res.* **56**, 5156-5164.

36. Tu, G. C., Cao, Q. N. and Israel, Y. (1995). Inhibition of gene expression by triple helix formation in hepatoma cells. *J. Biol. Chem.* **270**, 28402-28407.

37. Postel, E. H., Flint, S. J., Kessler, D. J. and Hogan, M. E. (1991). Evidence that a triplex forming oligodeoxyribonucleotide binds to the c-myc promoter in HeLa cells thereby reducing c-myc mRNA levels. *Proc. Natl. Acad. Sci. USA* **88**, 8227-8231.

38. Michelotti, E. F., Tomonaga, T., Krutzsch, H. and Levens, D. (1995). Cellular nucleic acid binding protein regulates the CT element of the human c-myc protooncogene. *J. BioL. Chem.* **271**, 9494-9499.

39. Simonsson, T., Pecinka, P. and Kubista, M. (1998). DNA tetraplex formation in the control region of c-*myc*. *Nucleic Acids Res.* **26**, 1167-1172.

40. Shevelev, A., Burfeind, P., Schulze, E., Rininsland, F., Johnson, T. R., Trojan, J., Chernicky, C. L., Hélène, C., Ilan, J. and Ilan, J. (1997). Potential triple helix-mediated inhibition of IGF-I gene expression significantly reduces tumorigenicity of glioblastoma in an animal model. *Cancer Gene Ther.* **4**, 105-12.

41. Kuang, W. W., Thompson, D. A., Hoch, R. V. and Weigel, R. J. (1998). Differential screening and suppression subtractive hybridization identified genes differentially expressed in an estrogen receptor-positive breast carcinoma cell line. *Nucleic Acids Res.* **26**, 1116-1123.

42. Meier-Ewert, S., Lange, J., Gerst, H., Herwig, R., Schmitt, A., Freund, J., Elge, T., Mott, R., Herrmann, B. and Lehrach, H. (1998). Comparative gene expression profiling by oligonucleotide fingerprinting. *Nucleic Acids Res.* **26**, 2216-2223.

43. Condon, T. P. and Bennett, C. F. (1996). Altered mRNA splicing and inhibition of human E-selectin expression by an antisense oligonucleotide in human umbilical vein endothelial cells. *J. Biol. Chem.* **271**, 30398-30403.

44. Wu, H., McLeod, A. R., Lima, W. F. and Crooke, S. T. (1998). Identification and partial purification of double-stranded RNase activity. *J. Biol. ChEm.* **273**, 2532-2542.

45. Musso, M., Wang, J. C. and Dyke, M. W. V. (1996). *In vivo* persistence of DNA triple helices containing psoralen-conjugated oligodeoxyribonucleotides. *Nucleic Acids Res.* **24**, 4924-4932.
46. Chen, C. B., Landgraf, R., Walts, A. D., Chan, L., Schlonk, P. M., Terwilliger, T. C. and Sigman, D. S. (1998). Scission of DNA at a preselected sequence using a single-strand-specific chemical nuclease. *Chemistry & Biology* , 283-292.
47. Matteucci, M., Lin, H. Y., Huang, T., Wagner, R., Sternbach, D. D., Mehrotra, M. and Besterman, J. M. (1997). Sequence-specific targeting of duplex DNA using a camptothecin-triple helix forming oligonucleotide conjugate and topoisomerase I. *J. Am. Chem. Soc.* **119**, 6939-6940.

11 GENE-TARGETING TRIPLE HELIX FORMING PURINE OLIGONUCLEOTIDES

Fedor Svinarchuk and Claude Malvy

INTRODUCTION

The recognition of DNA sequences by drugs opens up the possibility of regulating gene expression at the level of genes. Natural systems employ large polypeptides to identify DNA sequences. In the last 10 years smaller molecules endowed with sequence specificity have been discovered. Among them are the PNA of Nielsen (*1, 2*) and cyclic polyamides of Dervan (*3*). As the title indicates, we shall devote this review to applications at the cellular level of a particular class of compounds, the triple-helix forming purine oligonucleotides. We will not discuss the "clamp" approach, where *triplex forming oligonucleotides* (TFOs) have been used to inhibit DNA synthesis (*4, 5*).

A lot of *in vitro* results have demonstrated "positive" properties of TFOs for gene-targeted therapy. These are mainly (i) a high selectivity of binding, (ii) a high affinity constant, which in some cases is close to those of DNA-binding proteins, (iii) a composition made up of natural bases. This last point implies either a low or an absence of toxicity. Theoretically, only the necessity that the target sequence be a long purine/pyrimidine tract distinguishes them from the ideal gene-targeting drug. And advances in the chemistry of TFOs make one optimistic about overcoming this limitation (*6*). However, in view of results actually obtained to date, the potential biological significance of TFOs is still an open question.

In the last two years much progress has been made in the chemistry of TFOs, and the number of publications in this field has been growing very fast. In consequence, we have made the choice: 1) to compare recent developments in the chemistry of TFOs with the progress of their *in vivo* applications; 2) to limit our discussion to a particular class of TFO – the purine TFOs (puTFOs), including some data for TFOs composed of GT bases, since their properties and rules for triplex formation are often very similar to puTFOs. We hope that this kind of analysis will prove useful for successful application of TFOs for gene-targeted regulation.

146

What kind of properties should one expect from the compounds, which we would like to be gene-targeting drugs? Taking into account only the cellular level and not that of the organism, such compounds should: (i) penetrate into the nucleus; (ii) find their target sequence; (iii) either prevent polymerase movement through the target, or compete with regulatory protein(s) interacting with this region, or irreversibly damage the target. From this point of view, TFOs should have the following properties:

1. Be able to specifically recognize a single-copy gene in the human genome. This implies a factor of selectivity of around 100,000. According to the estimations of different authors, this means that the TFO has to recognize sequences ranging between 15 and 17 base pairs long.

2. Display a strong binding to their targets at physiological salt concentration, pH and temperature.

3. Their design should be easily accomplished starting from the knowledge of the target sequence. The possibility of regulating a gene's expression starting from its sequence is indeed the most attractive advantage of the antisense and antigene strategies.

These general criteria are necessary but not sufficient to predict the results of gene regulation by TFOs, because: 1) cellular DNA is under a chromatin structure, and the influence of the DNA conformation in such a structure as well as the influence of the chromatin proteins may be an important factor for *in vivo* triplex formation; 2) TFOs should really compete with regulatory proteins to find their target in the regulatory region of the gene, because even in an inactive state the regulatory sites are often covered by inactive forms of regulatory proteins or by repressor proteins, which are released only after chemical modifications by their active counterparts. As a consequence, the estimation of the potency of TFOs is indeed dependent to a large extent on the rate of triplex formation; 3) theoretically, some proteins could stabilize TFO interactions with the target, since proteins that recognize triplexes have now been identified (7).

The problems of TFOs delivery at the level of cells and, in future, at the level of the whole organism, are very similar for TFOs and for antisense oligonucleotides. In consequence, even TFOs which satisfy all of these conditions are far from assured of becoming gene-targeting drugs.

The physico-chemistry and *in vitro* application of TFOs have been well reviewed recently (8, 9). Nevertheless, we shall discuss some properties of triple helices, which from our point of view are important for *in vivo* applications, and, after that, the data available for purine TFOs *in vivo*.

In vitro STUDIES OF (R∗Y)∗R[1] TRIPLEXES

The correct rules for the formation of purine triplexes date from 1991 (*10, 11*). Purine-rich oligonucleotides bind in the major groove of the purine/pyrimidine tract

---

[1] R, purine; Y, pyrimidine

of double-helical DNA. They are antiparallel to the Watson-Crick purine strand. The GC base pair of the double-stranded DNA is recognized by a G in the TFO, and an AT pair is recognized either by an A or a T (*11*). Because of this particularity of recognition (AT pair by A or T), we will include some data for TFOs composed of GT bases in our discussion, even though they are not really puTFOs. A very important property of these triple helices is their non-dependence on pH, at least in a range from 6 to 8. Their formation depends on multivalent cations such as spermine or Mg^{+2} at concentrations which are estimated to be compatible with the composition of the intracellular medium. Target recognition by a TFO is very sensitive to a single mismatch (*12*). The prediction of the stability of these complexes is still very empirical; however, this stability can be extremely high for some sequences. For instance, the T_m of the three-strand complex with the 13-mer TFO 5'-GGGGAGGGGGGAGG-3' exceeds by 3°C that of the double-stranded target DNA: 64 and 67° C for the T_m of the triplex and duplex, respectively (*13*). The triplex Tm decreases to 30° C with two mismatches in the TFO (*14*). Another triplex has been described which was stable in a co-migration assay at 55° C (*15*). Equilibrium dissociation constants as low as 10^{-7} M to 10^{-10} M have been reported for several puTFOs (*11, 16*). This is quite comparable to the affinity of the peptide transcription factors regulating gene expression *in vivo*.

Despite these very interesting features, three problems have appeared for triplex formation with puTFOs.

Firstly, a growing number of publications report that the formation of triplexes involving guanine-rich oligonucleotides is inhibited at physiological ionic strength, particularly by the presence of K^+ cations (*17, 18, 19*). It can be assumed that this effect is caused by the formation of competing structures such as guanine quartets and parallel-stranded homoduplexes (same references).

Secondly, the rate of triplex formation is low. It does not exceed by much that of $(Y*R)*Y$ triplex formation and is usually near 10^3 mol^{-1} sec^{-1}, which is 10^3 times lower than the rate of duplex formation (*20*). The highest rate which has been reported for puTFOs is near 10^3 mol^{-1} sec^{-1} for a 13-mer TFO targeted to the *c-pim* protooncogene promoter (*21*), and for a 7-mer TFO coupled to the intercalating agent oxazolopyridocarbazole and targeted to the HIV-1 integration sequence (J.-F. Mouscadet, personal communication).

Another problem for the application of the puTFOs is the unpredictability of their binding affinity, which can vary from undetectable for some sequences (such as the sequence 5'-AGGAGGAGGAGA-3' described by Malkov *et al.* (*22*), where the authors did not find triplex formation in the presence of Mg^{+2} ions) to those which increase the stability of the targeted duplexes.

An interesting feature of purine TFOs is the existence of high stability triplexes. As emerges from the combinatorial approach of Hardenbol and Van Dyke (*23*), 13 bases is enough to form a triplex stable at 37° C. These authors have found equilibrium dissociation constants for the 13, 14 and 15-mers to be equal to $5-8\times10^{-8}$ M, whereas the constant for the sequence with 12 bases of homology is only 10^{-6} M. In our experiments we have also found that the stability was much higher for TFOs longer than 12-mer than for an 11-mer. The half time of dissociation at 37° C and 150 mM K$^+$ was, respectively, 30 min for an 11-mer and more than 4 days for the TFOs of lengths 14, 17, and 20 (24). This high stability of some sequences gives rise to an unexpected problem: a recognition by 13 base pairs, although endowed with a good specificity, is nevertheless not sufficient to provide single gene recognition by the TFO. As one can see, for some purine TFOs the high stability may trigger unexpected consequences.

During the last two years some new modifications, and some already known for pyrimidine TFOs have been applied in purine TFO chemistry in order to increase stability and to overcome potassium-mediated inhibition of triplex formation. Among them are an alkylating agent – chlorambucyl (25, 26) – or an N-bromacetyl electrophile (27), the intercalators acridine and psoralen (28, 29), and the specific triplex-stabilizing intercalator coraline (25, 26, 28). Purine TFOs with modified backbones were also tested : N,N-diethyl-ethylenediamine phosphoroamidates, bearing positively charged linkages (30); and phosphorothioates (31). Most of these modifications improve parameters of triplex formation and stability in *in vitro* studies. In the following paragraph we discuss whether there is evidence for triplex formation with purine TFOs inside the cells.

In vivo STUDIES OF (R∗Y)∗R TRIPLEXES

There are not so many papers describing gene regulation by puTFOs as one might expect from their *in vitro* properties. The main features of the 9 publications, where biological activity of TFOs, composed of A and G bases was studied, are summarized in Tables 1 and 2. Roughly the range of inhibitory activity of puTFOs can be described as going from around 100 % at 1 μM for the best to 50 % at 50 μM for the least efficient. This means that, up until now and despite the attractiveness of DNA as a target, their activity appears lower when compared to antisense oligonucleotides, which are able to inhibit 100% of the expression of a targeted gene at concentrations ranging from 1 to 100 nM. This comparison should nevertheless be put in perspective taking into account the huge number of publications about antisense oligonucleotides. From our point of view, the most interesting is to point out whether the observed biological effects are really due to the formation of triplexes inside the cells or not. Obviously, to prove TFO action through a triplex mechanism, one needs to demonstrate the formation of a triplex inside the cell on the endogenous gene in parallel with the biological activity of the corresponding TFO. In reality only a few publications are devoted to proving that the intracellular action mechanism relies on the properties of a triple helix.

As one can see (Tables 1 and 2), the biological activity of puTFOs has been tested on endogenous genes, on viruses, and on plasmid constructs bearing reporter genes controlled by promoters containing the target sequence for the TFOs. Most of the publications report a biological activity of the puTFOs and the ability of TFOs to inhibit in a sequence-specific manner the expression of the targeted genes. Most of

First author (Date)	Ref.	Targeted gene	Cell system	Delivery of oligo to cells	Cell target status	Arguments for TH in cells	TH: percent inhibition of gene activity	Effect of control oligo	TH: inhibition of control gene
POSTEL (1991)	32	C-MYC promoter	HeLa (human)	passive addition	endogen.	DNase I inhibition	50% at 50 µM	no effect at 125 µM	0% on β-actin
ROY (1993)	33	CAT under control of Interferon Response Element	HeLa (human)	cotransfec. with plasmid	transfec.	CAT activity	100% at 1 µM	50% at 1.5 µM	40% at 1 µM (control target)
PORUMB (1994)	34	polypurine tract of the Friend retrovirus	Dunni (mouse)	passive addition	infected	antiviral activity	25.8% at 10 µM	8.5% at 10 µM	
PORUMB (1996)	35	upstream regulatory region of HER2	MCF7 (human)	passive addition	endogen.	Northern Blot	40% at 2 µM	0% at 2 µM	0% (GPDH)
SVINARCHUK (1996)	14	luciferase under control of PIM promoter	G355-5 (cat)	preformed TH	transfect.	*in vivo* DMS footprint	20%	0%	

Table 1. Purine TFO gene regulation in cells (1991-1996)

Ref., reference number; cotransfect., cotransfected; endogen., endogenous; transfec., transfected.

First author (Date)	Ref.	Targeted gene	Cell system	Delivery of oligo to cells	Cell target status	Arguments for TH in cells	TH: percent inhibition of gene activity	Effect of control oligo	TH: inhibition of control gene
SVINARCHUK (1996)	14	luciferase under control of PIM promoter	G355-5 (cat)	electropor. independently of plasmid transfect.	transfect.	*in vivo* DMS footprint No TH	no effect	no effect	
ALUNNI-FABBRONI (1996)	15	murine c-Ki-RAS promoter	NIH 3T3 (mice)	preformed TH	transfect.	CAT activity	90%	40%	40% on transfect. β-galactosidase
DELPORTE (1997)	36	rat aquaporin 5	MDCK epithelial cells (dog)	adenovirus polylysine	recombi. adenovir.	Western Blot	61%	6%	0% luciferase
KIM (1998)	31	Cyclin D1 promoter	HeLa (human)	phosphoro. electropor.	stable transfect.	luciferase reporter gene	60% at 10 μM	no effect	no effect
KIM (1998)	37	human c-MYC promoter	HeLa (human)	phosphoro. cotransfect. with plasmid	transfect.	luciferase reporter gene	90%	low inhibition	

Table 2. Purine TFO gene regulation in cells (1996-1998)

Ref., reference number; electropor., electroporation; transfect., transfected; recombi., recombinant adenov., adenovirus; phosphoro., phosphorothioate; cotransfec., cotransfected.

the quoted works follow the same scheme: firstly, there is an *in vitro* demonstration of triplex formation between the TFO and the target sequence; and secondly, in cellular experiments, the activity of the TFOs and those of some control oligonucleotides (in some cases mutated target sequences have been used) are compared. In our opinion this type of experiment does not prove that the biological activity is linked to triplex formation inside the cells, since: (i) a biological activity of some antisense oligonucleotides has already been described which does not relate to the antisense mechanism (*38*) and, moreover, oligonucleotides composed of G and T bases display a biological activity greater than those composed of all four bases (*39*); (ii) modern methods of molecular biology such as *in vivo* footprinting (*40*) give the possibility of detecting triplex formation directly in cells. And, for some cases (Tables 1 and 2), one could imagine another mechanism to explain the biological effects. For example, when cotransfection of the reporter gene and TFOs has been used, we cannot exclude the possibility that triplex formation takes place outside the cell during the period of preincubation of the cells with the compounds. This point of view is supported by the data of the same author that the observed results are not changed by the presence or absence of the TFO in the medium after transfection (*33*). Moreover, in some publications *in vitro* conditions for triplex formation are not comparable to the physiological ones. For example, one has to be careful when triplex formation with phosphorothioate TFOs is tested in a buffer without K^+ (*31, 37*). In our hands, triplex formation with this type of TFO was much more sensitive to K^+ addition than was triplex formation with phosphodiester TFOs.

Did anyone try to show triplex formation with the target gene inside the cells in parallel with testing their biological activity? Of course, yes. The first work was reported in 1991. The authors monitored triplex formation inside the cells by the DNase I protection method in isolated nuclei (*32*). However, this experiment being performed with isolated nuclei, there is a step of nuclei purification. This opens up the possibility for triplex formation, not in live cells, but rather after cell lysis during nuclei preparation. The concentration of the TFOs used by the authors to treat the cell was very high: 125 μM. Absorption on the cell surface and cellular uptake of the TFOs could induce the binding of a rather high quantity of oligonucleotides to the cell. During the cell lysis the oligonucleotides are released from different cellular compartments and become accessible for triplex formation. As far as this possibility cannot be excluded in the experiment described, we consider that the formation of the triplex inside the cells remains to be proven.

In another more recent work the authors tried to follow triplex formation on an electroporated plasmid inside the cell by *in vivo* DMS footprint method (*14*). DMS can easily penetrate into living cells and modify the N7 position of guanines due to the formation of Hoogsteen base pairing (*41*) leading to phosphate backbone chain cleavage after treatment with piperidine – but not within a purine-purine-pyrimidine triplex. The short time of reaction (usually 2 -5 minutes) allows this reagent to make a flash picture of the DNA inside the cell. The scheme of the experiment is presented in Figure 1. An important feature in this experiment is that the triplex preformed with the puTFO is absolutely dependent on divalent cations: in the presence of EDTA the third strand immediately dissociates from the target DNA (*14*). So the addition of EDTA to the cell culture ensures that all triplexes registered by the footprint method are really inside live cells, since triplexes outside and inside dead cells are dissociated upon EDTA treatment. The following conclusions have been drawn from these experiments:

a) The triplex, preformed with the 13-mer TFO 5'-GGGGAGGGGGAGG-3' targeted to the *c-pim* protooncogene promoter region, is stable inside the cells for at least 24 hours after electroporation.

b) The preformed triplex inhibits the expression of a reporter gene under the control of the *c-pim* promoter.

c) Being electroporated first, the targeted DNA is no longer accessible for triplex formation.

Because of the unexpected last conclusion, other more efficient methods of oligonucleotide delivery were then tried: cationic lipids and electroporation. All methods used have shown no triplex formation inside the cells. Finally, coelectroporation of the targeted DNA with a 3000-fold molar excess of the TFO was tried. In this case, when the electroporation was carried out in DMEM medium, which contains 0.8 mM Mg^{+2}, it was possible to register 20 to 30% protection of the targeted guanines. But when the same experiment was performed in the absence of magnesium, in PBS buffer, no triplex formation was found inside the cells. These results show that special precautions need to be taken to avoid triplex formation outside the cells before the target DNA and TFO are transfected into the cells. At the same time these experiments indicate the existence of a very fast and efficient mechanism of TFOs separation from the target DNA inside the cells after electroporation: the methodology used made it possible to register triplex formation, which takes place during 10-15 minutes incubation in ice before electroporation. However, no triplex formation was found inside the cells at 37° C after 24 hours of culture. (*In vitro* experiments have shown, that the rate of triplex formation with this TFO is 80 times higher at 37° C than at 0° C (*21*).

Recently, DMS footprint experiments were performed with other cell types (a cat fibroblast cell line (G355-5) and two murine cell lines (NIH 3T3 and Dunni)) and other target sequences (polypurine sequences from the HIV-2 genome (5'-GGAGGGGGAGGAGGAGGAGG-3') and the Friend virus genome (5'-GAAGGGGGAAGAAGGAGGGGAGGAGG-3'), as well as the sequence 5'-GAGGGGGAGGGGGTGGTGGGGGGGGGAAGGA-3', which has been used by Wang *et al.* (*42*) in their experiments on TFO mutagenesis (these experiments are discussed below). The results of these experiments confirm the conclusions obtained in the work with the *c-pim* promotor target : a) preformed triplexes were stable inside the cells (the delay before the measurement in the case of the HIV-2 TFO was extended and no difference between the protection level after 24 and 72 hours after electroporation was found); b) no triplex formation occurs inside the cells. Taking into account the fact that accuracy of the *in vivo* DMS footprint method is between 5 and 10% (10% of protection can be easily registered by this method, but 5% is already below the threshold of sensitivity), the conclusions can be rephrased in the following way:

1) Preformed triplexes are stable inside the cells for at least 72 hours;

2) Not more than 10% of the electroporated DNA is accessible for triplex formation.

More recently, two groups tried to detect triplexes inside the cells using either ligation-mediated PCR (*43*) or Southern blot analysis and a competitive PCR based display (*44*). In the work of the first group the authors carefully characterized the parameters of triplex formation (estimated by modification of a guanine in the double-stranded DNA by an alkylating agent, chlorambucil) *in vitro,* and found that this modification can be very efficient in the presence of coraline, a selective triplex intercalator (*25, 28*). The same team then showed sequence-specific targeting and covalent modification of a DQ_β allele of the human MHC in genomic DNA (*26*). Finally, covalent modification of the gene for the chemokine receptor CCR5 by TFO in streptolysine O permeabilized cells was achieved. Modification of G by an isosteric guanine analog and utilisation of coraline were important for triplex formation (efficiency of site directed labelling was 5% and 24% at 5 μM and 20 μM TFO) (*43*). A second group has shown triple helix formation in digitonin-permeabilized cells on the integrated genomic copy of HIV-1 (*44*). Two important conclusions can be drawn from these experiments : 1) Triplex formation is possible in some cases on the native chromatin structures. 2) To achieve this triplex formation one needs to permeabilize cells. It looks quite possible that cell permeabilization allows TFOs to enter the cell and saturate all "non-specific" binding sites. After that, the real TFO concentration inside the cell becomes sufficient to see triplex formation on the target DNA.

Several explanations could be suggested to explain the failure to form detectable amounts of triplexes inside intact cells. We do not know which of them is in fact correct, but we intend to discuss those which, although reasonable, have to be eliminated. Firstly, it is not due to a special sequence of the TFO, as could have been thought from an initial publication (*14*): now the same result has been obtained with three other sequences. Secondly, it is not due to the "chromatinization" of the target DNA (see experiments with permeabilized cells). The hypothesis that a high intracellular concentration of potassium can prevent triplex formation was also tested. The negative effect of K^+ on triplex formation can be overcome by the prior formation of a short duplex either on the 3'- or 5'-end of the 20-mer TFO. These partial duplexes are referred to as "zipper" TFOs (*24*). It was demonstrated that a "zipper" TFO can form a triplex over the full length of the target, thus unzipping the short complementary strand in the presence of 150 mM K^+. But in cellular experiments the efficiency of these potassium-independent "zipper" TFOs was quite similar to unmodified ones. Unfortunately we have not found any data in the literature about *in vivo* experiments for such promising modifications of puTFOs as N,N-diethyl-ethylenediamine phosphoroamidates, bearing positively charged linkages (*30*), which have been shown to improve parameters of triplex formation *in vitro* in the presence of K^+.

154

Some more positive data about triplex formation with puTFOs inside nonpermeabilized cells have been published recently (*45, 46*), although it is difficult to know the yield of triplex formation in these experiments. To register intracellular triplex formation the authors have used TFOs linked to a psoralen derivative. In this approach site-specific triplex formation delivers the psoralen to the target site (transiently transfected supF reporter gene). Photoactivation of the psoralen yields adducts which can proceed into mutated sequences inside the eukaryotic cells. The recovered targeted DNA was then transfected into *E. coli* to allow the identification of mutants as white colonies among the wild-type blue ones. Speculatively, we can estimate the efficiency of triplex formation in these experiments. First, when the triplex was preformed *in vitro* followed by electroporation into eukaryotic cells, the percent of mutants was between 5 and 7%. One can suggest that in *in vitro* experiments, TFO binds nearly 100% of the target sequence (this is possibly an overestimation, since in this experiment the authors used a 10-mer TFO with a rather low equilibrium dissociation constant, between 10^{-6} and 10^{-7} M). *In vivo* experiments indicated a value of 1 to 2% for the mutants. So, one could estimate the efficiency of triplex formation inside the cells from these experiments as 20-30%. This value is substantially higher than those (less than 10%) described by (*14*) and closer to the data obtained on permeabilized cells (*43, 44*). Among the possible reasons for this discrepancy, we can suggest that either our method of calculation of the efficiency of triplex formation in the experiments of mutagenesis is very speculative and/or that the psoralen moiety probably influences the intracellular distribution of the TFOs. Moreover, in a similar experiment, these authors have found a mutagenic effect of unmodified TFOs (*47*). Unfortunately, the efficiency of triplex formation cannot be estimated from this experiment.

CONCLUSIONS AND PERSPECTIVES

The main conclusion which is emerging from the literature is that the mechanism of gene regulation by TFOs remains to be proven. Right now we cannot use TFOs as tools to investigate the influence of the targeted gene's expression on cellular metabolism. But the work by Glazer *et al.* makes one optimistic about the possibility of developing gene-targeting compounds based on TFOs. What could be done in this direction? In our opinion, the most important is to find the reason(s) underlying the failure of TFOs to interact quantitatively with the tested target sequences inside intact cells. We suggest that these cells have either an active defense mechanism to protect their genome from puTFOs, or that puTFOs react with undesired targets before reaching the intended target in DNA. In both cases important fundamental work is needed to find out the details of this mechanism. More generally, in the field of TFOs and sequence-targeting drugs, the particular case of puTFOs should stimulate scientists to improve delivery of compounds in the nucleus, to enhance the stability of the complex with the target, and to look for correlations inside cells between gene inhibition and target binding.

Of course a clinical therapeutic agent could be developed on the basis of some TFOs which have already been tested, but knowledge of the mechanism of action is important for development of specific gene-targeting compounds; and all steps of TFO interaction with the cells should be clarified. This information will help to develop new TFO modifications which could make them more efficient gene-targeting drugs.

REFERENCES

1. Peffer, N. J., Hanvey, J. C., Bisi, J. E., Thomson, S. A., Hassman, C. F., Noble, S. A. and Babiss, L. E. (1993). Strand-invasion of duplex DNA by peptide nucleic acid oligomers. *Proc. Natl. Acad. Sci. USA* **90,** 10648-10652.
2. Betts, L., Josey, J. A., Veal, J. M. and Jordan, S. R. (1995). A nucleic acid triple helix formed by a peptide nucleic acid-DNA complex. *Science* **270,** 1838-1841.
3. Cho, J., Parks, M. E. and Dervan, P. B. (1995). Cyclic polyamides for recognition in the minor groove of DNA. *Proc. Natl. Acad. Sci. USA* **92,** 10389-10392.
4. Volkmann, S., Jendis, J., Frauendorf, A. and Moelling, K. (1995). Inhibition of HIV-1 reverse transcription by triple-helix forming oligonucleotides with viral RNA. *Nucleic Acids Res.* **23,** 1204-1212.
5. Giovannangeli, C., Thuong, N. T. and Helene, C. (1993). Oligonucleotide clamps arrest DNA synthesis on a single-stranded DNA target. *Proc. Natl. Acad. Sci. USA* **90,** 10013-10017.
6. Kessler, D. J., Pettitt, B. M., Cheng, Y.-K., Smith, S. R., Jayaraman, K., Vu, H. M. and Hogan, M. E. (1993). Triple helix formation at distant sites: hybrid oligonucleotides containing a polymeric linker. *Nucleic Acids Res.* **21,** 4810-4815.
7. Musso, M., Nelson, L. D. and Van Dyke, M. W. (1998). Characterization of purine-motif triplex DNA-binding proteins in HeLa extracts. Biochemistry **37,** 3086-3095.
8. Maher, L. J., III (1996). Prospects for the therapeutic use of antigene oligonucleotides. *Cancer Invest.* **14,** 66-82.
9. Frank-Kamenetskii, M. D. and Mirkin, S. M. (1995). Triplex DNA structures. *Annu. Rev. Biochem.* **64,** 65-95.
10. Beal, P. A. and Dervan, P. B. (1991). Second structural motif for recognition of DNA by oligonucleotide-directed triple-helix formation. *Science* **251,** 1360-1363.
11. Durland, R. H., Kessler, D. J., Gunnell, S., Duvic, M., Pettitt, B. M. and Hogan, M. E. (1991). Binding of triple helix forming oligonucleotides to sites in gene promoters. *Biochemistry* **30,** 9246-9255.
12. Beal, P. A. and Dervan, P. B. (1992). The influence of single base triplet changes on the stability of a pur.pur.pyr triple helix determined by affinity cleaving. *Nucleic Acids Res.* **20,** 2773-2776.
13. Svinarchuk, F., Paoletti, J. and Malvy, C. (1995). An unusually stable purine(purine-pyrimidine) short triplex. The third strand stabilizes double-stranded DNA. *J. Biol. Chem.* **270,** 14068-14071.
14. Svinarchuk, F., Debin, A., Bertrand, J. R. and Malvy, C. (1996). Investigation of the intracellular stability and formation of a triple helix formed with a short purine oligonucleotide targeted to the murine *c-pim-1* proto-oncogene promotor. *Nucleic Acids Res.* **24,** 295-302.

156

15. Alunni-Fabbroni, M., Pirull, D., Manzini, G. and Xodo, L. E. (1996). (A,G)-oligonucleotides form extraordinary stable triple helices with a critical R.Y sequence of the murine c-Ki-ras promoter and inhibit transcription in transfected NIH 3T3 cells. *Biochemistry* **35**, 16361-16369.

16. Vasquez, K. M., Wensel, T. G., Hogan, M. E. and Wilson, J. H. (1995). High-affinity triple helix formation by synthetic oligonucleotides at a site within a selectable mammalian gene. *Biochemistry* **34**, 7243-7251.

17. Olivas, W. M. and Maher, L. J. (1995). Competitive triplex/quadruplex equilibria involving guanine-rich oligonucleotides. *Biochemistry* **34**, 278-284.

18. Suda, T., Mishima, Y., Asakura, H. and Kominami, R. (1995). Formation of a parallel-stranded DNA homoduplex by d(GGA) repeat oligonucleotides. *Nucleic Acids Res.* **23**, 3771-3777.

19. Noonberg, S. B., Francois, J. C., Garestier, T. and Helene, C. (1995). Effect of competing self-structure on triplex formation with purine-rich oligodeoxynucleotides containing GA repeats. *Nucleic Acids Res.* **23**, 1956-1963.

20. Xodo, L. E., Pirulli, D. and Quadrifoglio, F. A. (1997). A kinetic study of triple-helix formation at a critical R x Y sequence of the murine c-Ki-ras promoter by (A,G)- and (G,T) oligonucleotides. *Eur. J. Biochem.* **248**, 424-432

21. Svinarchuk, F., Bertrand, J. R. and Malvy, C. (1994). A short purine oligonucleotide forms a highly stable triple helix with the promoter of the murine c-pim-1 proto-oncogene. *Nucleic Acids Res.* **22**, 3742-3747.

22. Malkov, V. A., Voloshin, O. N., Soyfer, V. N. and Frank-Kamenetskii, M. D. (1993). Cation and sequence effects on stability of intermolecular pyrimidine-purine-purine triplex. *Nucleic Acids Res.* **21**, 585-91.

23. Hardenbol, P., and Van Dyke, M. W. (1996). Sequence specificity of triplex DNA formation: Analysis by a combinatorial approach, restriction endonuclease protection selection and amplification. *Proc. Natl. Acad. Sci. USA* **93**, 2811-2816.

24. Svinarchuk, F., Cherny, D., Debin, A., Delain, E. and Malvy, C. (1996). A new approach to overcome potassium-mediated inhibition of triplex formation. *Nucleic Acids Res.* **24**, 3858-3865.

25. Lampe, J. N., Kutyavin, I. V., Rhinehart, R., Reed, M. W., Meyer, R. B. and Gamper, H. B. (1997). Factors influencing the extent and selectivity of alkylation within triplexes by reactive G/A motif oligonucleotides. *Nucleic Acids Res.* **25**, 4123-4131.

26. Belousov, E. S., Afonina, I. A., Podyminogin, M. A., Gamper, H. B., Reed, M. W., Wydro, R. M. and Meyer, R. B. (1997). Sequence-specific targeting and covalent modification of human genomic DNA. *Nucleic Acids Res.* **25**, 3440-3444.

27. Grant, K. B., Dervan, P. B. (1996). Sequence-specific alkylation and cleavage of DNA mediated by purine motif triple helix formation. *Biochemistry* **35**, 12313-12319.

28. Gamper, H. B., Jr., Kutyavin, I. V., Rhinehart, R. L., Lokhov, S. G., Reed, M. W. and Meyer, R. B. (1997). Modulation of Cm/T, G/A, and G/T triplex stability by conjugate groups in the presence and absence of KCl. *Biochemistry* **36**, 14816-14826.

29. Orson, F. M., Klysik, J., Glass, G. A. and Kinsey, B. M. (1996). Comparison of triple helix formation by polypurine versus polypyrimidine oligodeoxynucleotides when conjugated to a DNA intercalator. *J. Exp. Ther. Oncol.* **1**, 177-185.

30. Dagle, J. M. and Weeks, D. L. (1996). Positively charged oligonucleotides overcome potassium-mediated inhibition of triplex DNA formation. *Nucleic Acids Res.* **24**, 2143-2149.

31. Kim, H. G. and Miller, D. M. (1998). A novel triplex-forming oligonucleotide targeted to human cyclin D1 (bcl-1, proto-oncogene) promoter inhibits transcription in HeLa cells. *Biochemistry* **37**, 2666-2672.

32. Postel, E. H., Flint, S. J., Kessler, D. J. and Hogan, M. E. (1991). Evidence that a triplex-forming oligodeoxyribonucleotide binds to the c-myc promoter in HeLa cells, thereby reducing c-myc mRNA levels. *Proc. Natl. Acad. Sci. USA* **88**, 8227-8231.

33. Roy, C. (1993). Inhibition of gene transcription by purine rich triplex forming oligodeoxyribonucleotides. *Nucleic Acids Res.* **21**, 2845-2852.

34. Porumb, H., Dagneaux, C., Letellier, R., Malvy, C. and Taillandier, E. (1994). Triple-helices targeted to the polypurine tract of a murine retrovirus. *Gene* **149**, 101-107.

35. Porumb, H., Gousset, H., Letellier, R., Salle, V., Briane, D., Vassy, J., Amor-Gueret, M., Israel, L. and Taillandier, E. (1996) Temporary ex vivo inhibition of the expression of the human oncogene HER2 (NEU) by a triple helix-forming oligonucleotide. *Cancer Res.* **56**, 515-522.

36. Delporte, C., Panyutin, I. G., Sedelnikova, O. A., Lillibridge, C. D., O'Connell, B. C. and Baum, B. J. (1997). Triplex-forming oligonucleotides can modulate aquaporin-5 gene expression in epithelial cells. *Antisense Nucleic Acid Drug Dev.* **7**, 523-529.

37. Kim, H. G, Reddoch, J. F, Mayfield, C., Ebbinghaus, S., Vigneswaran, N., Thomas, S., Jones, D. E., Jr. and Miller, D. M. (1998). Inhibition of transcription of the human c-myc protooncogene by intermolecular triplex. *Biochemistry* **37**, 2299-2304.

38. Gura, T. (1995) Antisense has growing pains. *Science* **270**, 575-577.

39. Fennewald, S. M., Mustain, S., Ojwang, J. and Rando, R. F. (1995). Inhibition of herpes simplex virus in culture by oligonucleotides composed entirely of deoxyguanosine and thymidine. *Antiviral Res.* **26**, 37-54.

40. Garrity, P. A. and Wold, B. J. (1992). Effects of different DNA polymerases in ligation-mediated PCR: enhanced genomic sequencing and in vivo footprinting. *Proc. Natl. Acad. Sci. USA* **89**, 1021-1025.

41. Voloshin, O. N., Mirkin, S. M., Lyamichev, V. I., Belotserkovskii, B. P. and Frank-Kamenetskii, M. D. (1988). Chemical probing of homopurine-homopyrimidine mirror repeats in supercoiled DNA. *Nature* **333**, 475-476.

42. Debin, A., Malvy, C. and Svinarchuk, F. (1997). Investigation of the formation and intracellular stability of purine (purine/pyrimidine) triplexes. *Nucleic Acids Res.* **25**, 1965-1974.

43. Belousov, E. S., Afonina, I. A., Kutyavin, I. V., Gall, A. A., Reed, M. W., Gamper, H. B., Wydro, R. M. and Meyer, R. B. (1998). Triplex targeting of a native gene in permeabilized intact cells: covalent modification of the gene for the chemokine receptor CCR5. *Nucleic Acids Res.* **26**, 1324-1328.

44. Giovannangeli, C., Diviacco, S., Labrousse, V., Gryaznov, S., Charneau, P. and Helene, C. (1997). Accessibility of nuclear DNA to triplex-forming oligonucleotides: the integrated HIV-1 provirus as a target. *Proc. Natl. Acad. Sci. USA* **94**, 79-84.

45. Havre, P. A. and Glazer, P. M. (1993). Targeted mutagenesis of simian virus 40 DNA mediated by a triple helix-forming oligonucleotide. *J. Virol.* **67**, 7324-7331.

46. Havre, P. A., Gunther, E. J., Gasparro, F. P. and Glazer, P. M. (1993). Targeted mutagenesis of DNA using triple helix-forming oligonucleotides linked to psoralen. *Proc. Natl. Acad. Sci. USA* **90**, 7879-7883.

47. Wang, A., Seidman, M. M. and Glazer, P. M. (1996). Mutagenesis in mammalian cells induced by triple helix formation and transcription-coupled repair. *Science* **271**, 802-805.

12 DNA TRIPLE HELIX AS A TOOL TO REGULATE CYTOKINE GENE EXPRESSION

Marina Kochetkova and Mary Frances Shannon

INTRODUCTION

The recognition of promoter DNA by proteins plays a central role in the regulation of gene expression (*1*). Methods or agents that can selectively alter the interaction of a protein at a certain site on DNA could potentially modulate the transcription of a gene. Changes in the rate of transcription are an important mechanism of regulation of gene expression. In cases of undesirable overexpression of proteins implicated in disease, transcriptional repressors represent potential therapeutic agents.

A decade ago, Hogan and co-workers first reported that in the presence of Mg^{2+}, guanosine-rich natural oligodeoxynucleotides bound to a specific purine/pyrimidine site upstream of the human c-Myc gene and inhibited transcription of c-Myc mRNA *in vitro* (*2*). These findings initiated investigations directed to search for DNA triplex-based transcription inhibitors which led to the development of triplex forming oligonucleotides (TFOs) capable of modulating gene expression in living cells. The advances of "antigene" oligonucleotide-based techniques, structure of TFOs, kinetics, thermodynamics and other aspects of the DNA triple helix formation have been extensively studied and comprehensively discussed in numerous reviews (for example *3-5*) and elsewhere in this book. This chapter will illustrate the importance of triplex target choice in the context of a complex gene promoter and the advantages and limitations of this approach. As a prototypic example we will discuss the inhibition of Granulocyte-Macrophage Colony-Stimulating Factor (GM-CSF) gene expression utilizing DNA triplex technology.

CYTOKINES – POTENTIAL THERAPEUTIC TARGETS

Cytokines are a rapidly growing family of potent and multifunctional proteins or glycoproteins that act as local and/or systemic intercellular regulatory factors. Their varied physiological roles are slowly beginning to emerge and range from influencing "hemostasis" to "emergency actions" in response to infection or injury and in inflammatory responses. Recent data also show that members of the cytokine family of proteins are involved in a large number of pathological conditions including parasitic and viral infections, septic shock, asthma and other allergies, rheumatoid arthritis and cancer (6). It also should be noted that knockout mouse experiments demonstrated a high degree of redundancy within cytokines subfamilies where some but not all functions of the missing protein can be subserved by other related cytokines. One such example is GM-CSF, where inactivation of the GM-CSF gene does not alter "steady state" hemopoiesis but does reduce inflammatory responses to certain stimuli (7). These properties make hemopoietic cytokines such as GM-CSF good targets for the new potential therapeutics based on triplex technology.

One of the limitations in the *in vivo* application of triplex formation to inhibit gene expression may be the accessibility of the target sequence in context of tightly packaged chromatin. Tight packaging of DNA sequences in nucleosomes is likely to prevent effective recognition and binding of the third DNA strand. Cytokines are generally inducible proteins, i.e., they are produced in response to a specific combination of stimuli regulating transcription factor activity. It has been shown that, after activation, transcriptional regulatory regions frequently adopt altered chromatin structures which can be detected within nuclear DNA on the basis of their enhanced accessibility to DNase I. Most likely, these regions (termed DNase I hypersensitive sites or DH sites) represent nucleosome-free regions where specific DNA binding proteins associate with DNA and exclude nucleosomes. DH sites have been detected following induction upstream of many cytokine and cytokine receptor genes, including IL-2, IL-3, GM-CSF (8). Gene promoters or enhancers that are devoid of nucleosomes may provide more easily accessible targets for triplex formation. Thus, possible chromatin remodeling upon induction of cytokine genes might provide an additional advantage in attempts to repress gene transcription *in vivo* using TFOs.

INHIBITION OF GM-CSF EXPRESSION BY TRIPLE HELIX FORMATION ON THE GENE PROMOTER

GM-CSF is produced by many cells in the body including T cells, monocytes, and endothelial cells, in response to cellular activation by different immune (other cytokines, such as TNF or IL-1) or inflammatory stimuli (pathogens). Disregulated activation of GM-CSF can also occur as illustrated by the autocrine production of this protein in certain leukemias. Thus GM-CSF is abnormally expressed in leukemic cells from patients with Juvenile Myelomonocytic Leukemia, JMML (9) and in some cases of Acute Myeloid Leukemia, AML (10), and appears to be important for the growth and survival of these cells. In addition, constitutive production of GM-CSF has been associated with chronic inflammatory conditions, such as rheumatoid arthritis, psoriasis and asthma (11). There is little information available on mechanisms leading to abnormal GM-CSF production, but it is most likely due to defects in the normal signaling pathways for GM-CSF gene activation. Thus aberrant constitutive nuclear induction of NF-κB transcription factor complexes has been detected in primary JMML cells (*12*). Since the GM-CSF gene promoter has two well-defined NF-κB sites (Figure 1), it is likely that constitutive NF-κB in the nucleus will activate gene transcription.

We attempted to modulate GM-CSF production and function in stimulated human Jurkat T-cells and primary JMML cells by DNA triplex formation within regulatory elements of the gene. The GM-CSF gene contains a well characterized proximal promoter as well as an enhancer located 3 kb upstream of the transcription "start" site. Both appeared to be tightly regulated and contain elements required in the response of GM-CSF to most activators (reviewed in *13*).

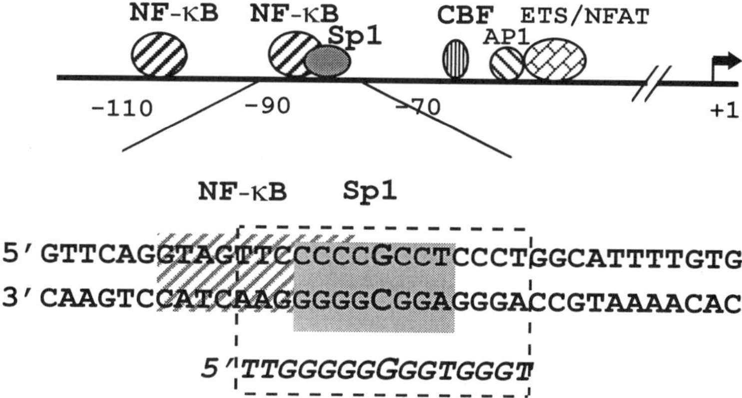

Figure 1. The GM-CSF gene proximal promoter with functionally important elements - transcription factor binding sites marked. An arrow indicates the transcription initiation point. The sequence of the promoter from -90 to -60 is shown as well as the 15 bp triplex target site (dashed rectangle) overlapping the κB element and Sp1 site (shaded rectangles). The TFO is aligned to its double-stranded target within the GM-CSF promoter and is shown in *ITALIC CAPITALS*. A point of "mismatch" in the duplex target and a G placed opposite it in the TFO is indicated in BIG CAPITALS.

The transcription factors that bind to the enhancer and promoter elements have been identified, and the role of various transcription factor binding sites has been studied in detail. Figure 1 schematically represents the structure of the GM-CSF proximal promoter. It encompasses multiple transcription factor binding sites including those for NF-κB, AP1, NFAT and Sp1, all of which appear to contribute to promoter function. However, extensive reporter construct studies have shown that the κB element proximal to the transcription "start" site plays a central role in GM-CSF promoter activation by different stimuli.

162

A 15 bp polypurine stretch overlapping the proximal κB and Sp1 elements was targeted for triplex formation by a GT-containing natural oligonucleotide antiparallel to the purine strand of the underlying duplex (Figure 1). This TFO selectively blocked NF-κB and Sp1 transcription factor binding to their respective sites (*12, 14*). When applied to the cell medium, a 3'-aminopropyl derivatized (to inhibit degradation by cellular exonucleases (*15*)) TFO of the same sequence specifically blocked transcription from a GM-CSF promoter luciferase reporter construct, in Jurkat T-cells activated by the HTLV-1 transactivator Tax, by approximately 65%. More importantly, when cells expressing Tax or cells stimulated with PMA/Ca^{2+} ionophore to mimic T-cell receptor activation were treated with the same 3'-aminopropyl modified TFO, GM-CSF protein concentrations in cell supernatants were specifically reduced in parallel with GM-CSF mRNA levels by approximately 60-70% (*14*). Thus triplex formation on the GM-CSF gene promoter allowed significant inhibition not only of a reporter construct but also of the endogenous gene expression in living cells.

It is also important that the use of the DNA triplex approach has confirmed that the κB/Sp1 element plays an important role in the activation of the GM-CSF gene. This suggests that DNA triple helix could represent a useful tool for investigating the function of individual elements in endogenous gene promoters. This in itself is an important advance, as it is a potentially simpler and more accessible approach for studying complex gene promoters *in vivo* than the time consuming and labor-intensive alternative of generating transgenic mice with altered promoter sequences. However, at present, sequence requirements for triplex formation restrict this approach to regulatory elements that contain polypurine/polypyrimidine stretches.

Inhibition of GM-CSF function by the TFO targeted to the κB/Sp1 region GM-CSF gene promoter has also been studied in primary JMML cells which produce GM-CSF in an autocrine manner and depend on this protein for cell growth and survival (*12*). Cells were treated with 3'-aminomodified oligonucleotide in the cell medium every 12 hours for 2 weeks. JMML cell colony growth in methylcellulose was specifically reduced by TFO treatment by approximately 60%, in parallel with a drop in secreted GM-CSF levels. The toxicity of the TFO was excluded by control experiments with normal bone marrow cells, suggesting inhibition of GM-CSF expression as the underlying mechanism of the observed growth suppression.

The 15-bp duplex DNA stretch targeted for triplex formation on the GM-CSF promoter contains one "mismatch" in an otherwise purine/pyrimidine tract, and the observed K_D value for the TFO binding to its duplex target was relatively high (>3 μM) (*14*). The original TFO used in the experiments described above (Figure 1) contained a G opposite the interrupting C:G pair. Aiming to increase the affinity of the TFO for its target on the GM-CSF promoter, we substituted a T base for the G base opposite the mismatch in the targeted duplex, since it has been demonstrated that a T, when placed opposite a C:G interruption, did not disrupt triplex formation and also showed substantial binding to the C:G base pair (*16*).

The affinity of the T-substituted TFO for the target was increased at least 10 fold with an observed apparent K_D <0.3 μM (*17*); and the T-substituted TFO was also more effective in specific blocking of the recombinant NF-κB Rel A binding to the κB element within the GM-CSF gene promoter fragment. However the increase in binding affinity of the TFO for its target site on the GM-CSF promoter failed to further potentiate its inhibitory effect on GM-CSF gene transcription in Jurkat T cells transiently transfected with a luciferase reporter construct driven by the GM-CSF promoter in comparison with the TFO initially used. This result could possibly be due to the already exhausted potential decrease in transcription activity that can be exerted on the GM-CSF promoter through the κB regulatory element; this interpretation is supported by our previous finding that mutation of the κB element within GM-CSF promoter yielded approximately a 70% reduction in luciferase reporter activity in Jurkat T cells stimulated with PMA/Ca^{2+} ionophore (*13*). Other factors such as different cellular uptake, intracellular stability or distribution of the two TFOs could also have contributed to the observed result.

DNA TRIPLEX FORMATION ON THE GM-CSF GENE ENHANCER

A powerful inducible enhancer has been identified 3 kb upstream of the GM-CSF gene transcription "start" site using a screen for DNase I hypersensitive sites (*18*). Detailed studies on the functional elements in this region have led to the identification of NFAT as a complex with AP1 as the major transcription factor driving GM-CSF enhancer function in PMA/Ca^{2+} ionophore activated T-cells (Figure 2). Binding sites for other transcription factors, such as CBF, have also been described (*13*).

To attempt to further inhibit GM-CSF gene expression above that seen with promoter-directed TFOs, a 28 bp polypurine/polypyrimidine stretch in the gene enhancer spanning the NFAT/AP1 element termed GM550 was targeted for triplex formation by a GT-containing natural oligonucleotide with the orientation antiparallel to the purine strand of the underlying duplex (Figure 2). Gel mobility shift assays and DNase I footprinting showed that this TFO bound to its target with high affinity. However, attempts to reduce transcription activity from a luciferase reporter construct containing the enhancer in front of the GM-CSF promoter, by treating transiently transfected and activated Jurkat T-cells in the culture medium with the 3'-modified enhancer TFO, did not yield any significant results. In addition, no effect was seen on endogenous GM-CSF mRNA levels in similar experiments with non-transfected cells. Treatment of primary JMML cells with a combination of enhancer and promoter targeted TFOs did not produce any further reduction in cell colony numbers compared to the promoter TFO alone, suggesting that the levels of GM-CSF protein remained unchanged. Four binding sites, including GM550, for NFAT/AP1 transcription factor complexes have been identified on the human GM-CSF gene enhancer using gel electrophoretic mobility shift assays (*19*). Each site functions independently when multimerized upstream of the basal promoter and reporter gene. Stepwise deletions were made into the enhancer from both directions, and resultant fragments were assayed in reporter constructs, to ascertain the relative contributions that different elements make to the inducible enhancer activity. Experiments with transiently transfected Jurkat T-cells stimulated with PMA/Ca^{2+} ionophore demonstrated that only a marginal decrease in reporter activity driven by the GM-CSF gene enhancer occurred upon deletion of the GM550 region from the enhancer. It is interesting to note that these experiments also showed that, despite the presence of four NFAT sites on the enhancer, almost all of its inducible activity was conferred by the enhancer core between nt 350 and 500, indicating a possible high degree of redundancy in the utilization of activation sites within the GM-CSF gene enhancer.

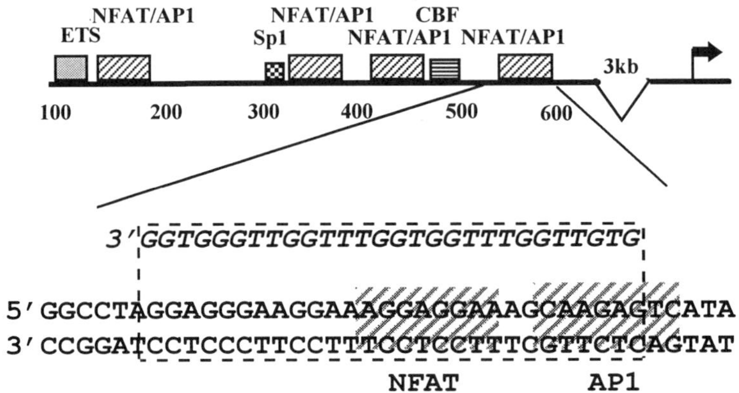

Figure 2. Schematic representation of the human GM-CSF enhancer with known NFAT/AP1 elements and other potential transcription factor binding sites. An arrow indicates the transcription "start" site located 3 kb downstream. Shown is a fragment of the enhancer sequence encompassing the GM550 element with NFAT and AP1 sites in shadowed boxes. The 28 bp region selected for triplex formation is indicated by the dashed rectangle with aligned sequence of the TFO targeted to the GM-CSF gene enhancer, which is shown in *ITALIC CAPITALS*.

The inability of the TFO targeted to the GM550 element of the GM-CSF enhancer to reduce transcriptional activity from reporter or endogenous genes supports the notion that this element does not play an important role in GM-CSF gene activation in T cells and in primary cells from JMML patients. However, it should be noted that the function of the entire GM-CSF enhancer has not yet been assessed in JMML cells.

The results from triplex formation experiments on the GM-CSF enhancer site also suggest that redundancy within regulatory sequences of complex and tightly controlled upstream elements of the target genes may pose an additional problem for the application of the "antigene" approach to modulate gene expression. On the other hand, in such cases DNA triplex formation may prove to be a very useful tool to assay the role of gene transcription regulatory sequences and transcription factor binding sites, especially if the sequence limitations of the DNA triplex approach can be overcome.

CONCLUDING REMARKS

Oligonucleotides are increasingly being viewed as tools for inhibition of the expression of disease-promoting proteins in *in vitro* and *in vivo* models of various diseases. Findings described in this chapter show that TFOs can act as selective gene repressors and can specifically inhibit the aberrant function of the targeted protein (growth promotion in JMML cells). However, extensive preliminary research is necessary to evaluate the input of the DNA element targeted for triplex formation into the transcriptional activity of the gene of interest.

A few other problems such as intracellular delivery and stability of the TFOs, possible *in vivo* toxicity and extension of recognition sites to more complex sequences still stand in the way of clinical applications of TFOs. In view of recent advances in the field, these limitations will ultimately be eliminated, and it seems most likely that utilization of synthetic oligonucleotide analogs and derivatives will allow existing obstacles in developing "antigene" therapeutics to be effectively overcome.

REFERENCES

1. Dynan, W. S. and Tjian, R. (1985). Control of eukaryotic messenger RNA synthesis by protein specific DNA binding proteins. *Nature* **316**, 774-777.
2. Cooney, M., Chernuszewicz, C., Postel, E. H., Flint S. J. and Hogan, M. E. (1988). Site-specific oligonucleotide binding represses transcription of the human *c-myc* gene *in vitro*. *Science (Wash. DC)* **241**, 456-459.
3. Helene, C., Thuong, N. T. and Harel-Bellan, A. (1992). Control of gene expression by triple helix-forming oligonucleotides. The antigene strategy. *Ann. NY Acad. Sci.* **660**, 27-36.
4. Stull, R. A. and Szoka F. C. (1995). Antigene, ribozyme and aptamer nucleic acid drugs: progress and prospects. *Pharm. Res. (NY)*. **12**, 465-483.
5. Maher, L. J., III (1996). Prospects for the therapeutic use of antigene oligonucleotides. *Cancer Invest.* **14**, 66-82.
6. Henderson, B. and Blake, S. (1992). Therapeutic potential of cytokine manipulation. *Trends Pharmacol. Sci.* **13**, 145-152.
7. Stanley, E., Lieschke, G. J., Drail, D., Metcalf, D., Hodgson, G., Gall, J. A. M., Maher, D.W., Cebon, J., Sinickas, V. and Dunn, A. R. (1994). Granulocyte-macrophage colony-stimulating factor-deficient mice show no major perturbation of haematopoiesis but develop a characteristic pulmonary pathology. *Proc. Natl. Acad. Sci.USA* **91**, 5592-5598.

8. Frolova, E. I., Dolganov, G. M., Mazo, I. A., Copeland, P., Stewart, C., O'Brien , S. J. and Dean, M. (1991). Linkage mapping of the human CSF2 and IL3 genes. *Proc. Natl. Acad. Sci.USA* **88**, 4821-4824.

9. Gualtieri, R. G., Emanuel, P. D., Zuckerman, K. S., Martin, G., Clark, S. C., Shadduck, R. K., Dracker, R. A., Akabutu, R., Nitschke, R., Hetherington, M. L., et al. (1989). Granulocyte-macrophage colony-stimulating factor is an endogenous regulator of cell proliferation in juvenile chronic myelogenous leukemia. *Blood* **74**, 925-929.

10. Young, D. C., Wagner, K. and Griffin, J. D. (1987). Constitutive expression of the granulocyte-macrophage colony-stimulating factor gene in acute myeloblastic leukemia. *J. Clin. Invest.* **79**, 100-104.

11. Gasson, J. C. (1991). Molecular physiology of granulocyte-macrophage colony-stimulating factor. *Blood* **77**, 1131-1145.

12. Kochetkova, M., Iverson, P. O., Lopez, A. F. and Shannon M. F. (1997). DNA triplex formation inhibits granulocyte-macrophage colony-stimulating factor gene expression and suppresses growth in Juvenile myelomonocytic leukemic cells. *J. Clin. Invest.* **99**, 3000-3008.

13. Shannon, M. F., Coles, L. S., Vadas, M. A. and Cockerill, P. N. (1997). Signal for activation of the granulocyte-macrophage colony-stimulating factor gene promoter and enhancer in T cells. *Critical Reviews Immunol.* **17**, 301-323.

14. Kochetkova, M. and Shannon, M. F. (1996). DNA triplex formation selectively inhibits granulocyte-macrophage colony-stimulating factor gene expression in T cells. *J. Biol. Chem.* **271**, 14438-14444.

15. Orson, F. M., Thomas, D. W., McShan, W. M., Kesler, D. J. and Hogan, M. E. (1991). Oligonucleotide inhibition of IL2Rα mRNA transcription by promoter region collinear triplex formation in lymphocytes. *Nucleic Acids Res.* **19**, 3435-3441.

16. Kochetkova, M. and Shannon, M. F. (1998). DNA triplex formation on the granulocyte-macrophage colony-stimulating factor gene proximal promoter. *Nucleosides, Nucleotides.* In press.

17. Durland, R. H., Rao, T. S., Revankar, G. R., Tinsley, J. H., Myric, A. M., Seth, D. M., Rayford, J., Singh, P. and Jayaraman, K. (1994). Binding of T and T analogs to CG base pairs in antiparallel triplexes. *Nucleic Acids Res.* **22**, 3233-3240.

18. Cockerill, P. N., Shannon, M. F., Bert, A. G., Ryan, G. R. and Vadas, M. A. (1993). The granulocyte-macrophage colony-stimulating factor / interleukin 3 locus in regulated by an inducible cyclosporin A sensitive enhancer. *Proc. Natl. Acad. Sci. USA* **90**, 2466-2470.

19. Cockerill, P. N., Bert, A. G., Jenkins, F., Ryan, G. R., Shannon, M. F. and Vadas, M. A. (1995). Human granulocyte-macrophage colony-stimulating factor enhancer function is associated with cooperative interactions between AP1 and NFATp/c. *Mol. Cell. Biol.* **15**, 2071-2080.

13 GENOME MODIFICATION BY TRIPLEX-FORMING OLIGONUCLEOTIDES

Karen M. Vasquez and Peter M. Glazer

INTRODUCTION

Gene expression is a process that is first regulated at the level of DNA. By introducing modifications in specific DNA sequences of living cells, it is possible to permanently alter the expression of genes relevant to disease. Triplex technology offers an approach to site-specific genome modification in mammalian cells such that it is possible to direct damage to specific sites in the DNA *via* triplex-forming oligonucleotides (TFOs) and thereby induce mutations or sensitize a site for gene replacement *via* homologous recombination [also reviewed in *(1, 2)*]. The work reported in this chapter describes the initial steps toward the development of a triplex-based strategy for site-specific genome modification *via* targeted mutagenesis and recombination.

Intermolecular triplexes

Triplex formation involves the specific recognition of duplex DNA by a single strand TFO. The TFO binds in the major groove of duplex DNA by forming specific Hoogsteen or reverse Hoogsteen hydrogen bonds between the bases in the TFO and the bases in the purine-rich strand of the underlying duplex *(3)* . It is this interaction that allows specific recognition of a segment of duplex DNA by the TFO, thus affording a variety of potential applications to modify a mammalian genome. There are several classes of DNA-DNA triplexes, including those formed by purine-rich TFOs which bind in an anti-parallel fashion through reverse Hoogsteen hydrogen bonds in the major groove *(4, 5)* . These TFOs are capable of forming stable triple-helical structures at physiological pH and are therefore applicable to cellular studies.

Mutagenesis

Many human diseases, including some types of cancer, are caused by mutations in the DNA. If disease-related mutations could be corrected or inactivated in cells, a significant step could be made toward the treatment of these genetic diseases. One strategy for correcting the defective gene is to attempt to change the mutated sequence to the wild-type sequence *via* triplex-induced site-specific DNA damage. For example, psoralen-modified TFO-directed DNA damage has been shown to induce T·A to A·T transversions at the targeted site in a reporter plasmid, both *in vitro* and *in vivo* (*6-8*). This transversion event is thought to be mediated by error-prone repair or replication at the triplex site. Incorporation of a site-specific change in the DNA sequence should lead to permanent, heritable corrections in gene expression.

Recombination

Homologous recombination involves the exchange of genetic information between homologous DNA sequences (*9*). Through the use of traditional gene targeting strategies *via* homologous recombination it is possible to replace or delete genetic information on chromosomes (*10*). The potential of gene targeting *via* targeted homologous recombination is very promising, but currently is limited. One of the major limitations of the current technology is the low frequency (10^{-8} to 10^{-5} per cell) at which homologous recombination occurs in mammalian cells (*11*). In addition, random integration of transfected DNA is nearly 1000-fold higher than targeted recombination (*12*). Triplex technology offers an alternative approach to targeting specific genes for manipulation of a mammalian genome.

It has been demonstrated that DNA damage (e.g., dsDNA breaks) can stimulate the frequency of recombination. In yeast it has been shown that dsDNA breaks can enhance the frequency of homologous recombination up to 1000-fold (*13*). In mammalian cells, dsDNA breaks have been shown to stimulate homologous recombination 10 to 300-fold (*14-17*).

Another class of DNA damage that may stimulate recombination is psoralen-induced photocrosslinking of DNA. These lesions must be repaired for cell survival, and recombination has been implicated in the repair of psoralen crosslinks. In yeast, the RAD (radiation sensitivity) genes are involved in responses to DNA damage. RAD52 is required for the repair of psoralen damage [reviewed in (*18*)], recombination, and the repair of dsDNA breaks. Another yeast RAD gene, RAD10, is required for recombination and its mammalian homologue, ERCC1, when overexpressed in Chinese hamster ovary (CHO) cells, increases sensitivity to psoralen crosslinks (*19*). This result implies that recombination may be involved in the repair of psoralen damage in mammalian cells.

ASSAYS TO DETECT MUTAGENESIS AND RECOMBINATION

Experimental systems have been constructed using both plasmid DNA and chromosomal DNA triplex targets in mammalian cells to study both the mutagenic and recombinagenic potentials of triplex-forming oligonucleotides.

Mutagenesis assays

The *supF* gene as a reporter for extrachromosomal mutational analysis: The *supF* gene encodes a tRNA that suppresses amber mutations; in the bacterial strain ClacZ125(am) the *supF* gene produces blue colonies on plates containing X-Gal and

IPTG, while mutant versions produce white colonies. The plasmid, pSupFG1, is an SV40-based shuttle vector that can replicate in both COS cells, due to the SV40 origin and T antigen gene, and in bacterial cells by virtue of the pBR327 origin. The general strategy for triplex-targeted mutagenesis of the *supF* gene on an extrachromosomal plasmid in mammalian cells is shown in Figure 1. For *in vitro* studies, the TFO is first incubated with the plasmid to allow for triplex formation, then the oligonucleotide-plasmid complex is UVA irradiated to form crosslinks at the targeted base pair (bp 167). Subsequently, the TFO-plasmid complex is transfected into COS cells and allowed to replicate for at least 48 hours prior to analysis. To assay for intracellular triplex formation (i.e., *in vivo* studies) the plasmid is first transfected into COS cells, and the TFO is added directly to the cells and incubated prior to UVA irradiation. The plasmid DNA can then be isolated for blue/white screening and the mutation frequency calculated as the number of mutant (white) colonies divided by the number of total (blue + white) colonies. This reporter system allows for facile investigation of the potential *in vivo* applications of triplex formation in an extrachromosomal context.

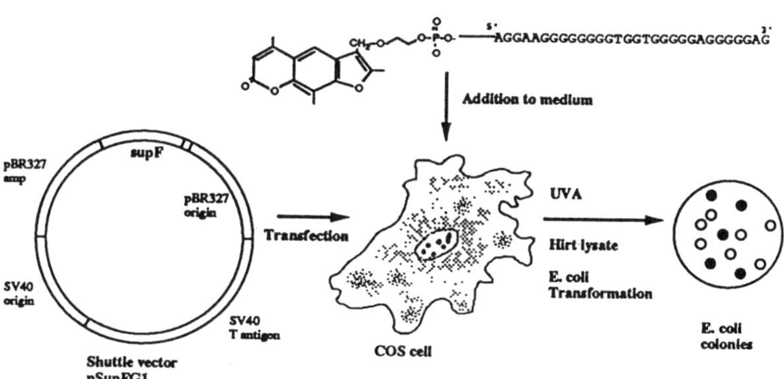

Figure 1. Experimental strategy for detecting triplex-directed extrachromosomal mutagenesis. The shuttle vector, pSupFG1, is transfected into COS cells *via* electroporation with the psoralen-TFO covalently crosslinked (*in vitro* protocol) or added separately to the growth medium after plasmid transfection (*in vivo* protocol). Cells are incubated to allow for TFO uptake and binding; then cells are subjected to UVA irradiation to generate psoralen adducts. Following replication, the vector molecules are isolated for transformation into *E. coli* for genetic selection of the *supF* gene.

The *supFG1* gene as a reporter for triplex-induced mutagenesis on a chromosomal locus in mouse cells: In order to study mutagenesis in mammalian cells in a chromosomal context, a mouse fibroblast cell line (3340) was constructed containing ~15 copies of λsupFG1 (a lambda vector containing the *supFG1* mutation reporter gene (*20*). The *supFG1* gene contains a high-affinity 30-bp TFO binding site and encodes a tRNA that suppresses amber mutations (*20*). Psoralen-modified TFOs designed to bind the purine-rich region in the *supFG1* gene can be added directly to the cells (alternatively, various delivery systems can be used; e.g., cationic lipids, streptolysin, electroporation) to allow for oligo uptake and binding to the target site. The cells are then UVA irradiated to photoactivate psoralen, and allowed to recover for 48 hours so that replication and/or repair can occur. The vector DNA can then be isolated from the mouse genomic DNA and packaged into phage particles for mutational analysis using lambda packaging extracts (*21, 22*) as shown in Figure 2. If a mutation has occurred in the *supFG1* gene, the amber mutation in the β-galactosidase gene in a lacA(am) strain of *E. coli* will not be suppressed, and the resulting plaque will be white in the presence of IPTG and X-Gal. The wild-type *supF* gene will suppress the amber mutation in the β-galactosidase gene and will produce blue colonies in the presence of IPTG and X-Gal. This detection system allows for analysis of mutations occurring at a reporter gene in a chromosomal locus in mouse cells. For more detailed analysis of the mutants, DNA sequencing can be performed.

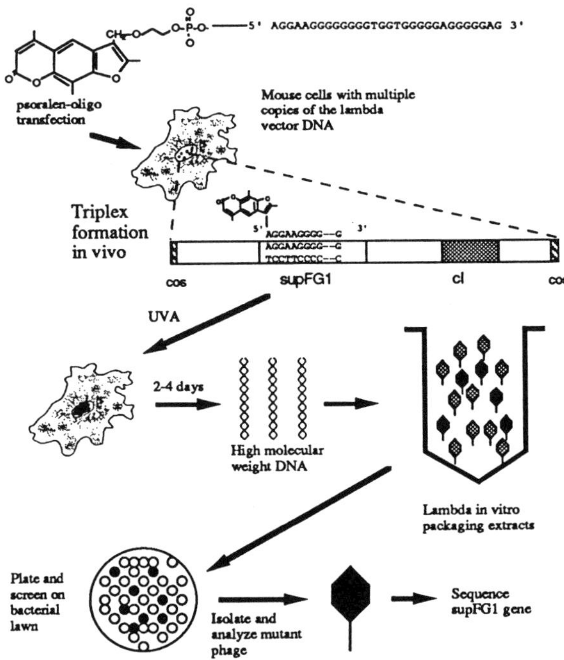

Figure 2. Experimental system for detecting triplex-directed chromosomal mutations. Psoralen-modified TFOs are added to mouse cells (3340) containing copies of the *supFG1* gene integrated on the chromosome. Cells are UVA irradiated to activate the psoralen and then incubated to allow for mutations to form. Genomic DNA is isolated, and the vector DNA is excised and packaged into phage particles for analysis in bacteria to detect mutations that occurred in the mouse cells.

Recombination assays

The *supF* gene as a reporter assay for triplex-induced extrachromosomal recombination: The vector, pSupFAR, was constructed to assess the recombinagenic potential of triplex-directed DNA damage between tandem *supF* genes flanking the triplex target site (*23*) . By incorporating 2 mutant copies of the *supF* gene, intrachromosomal recombination can be measured by screening for a reconstructed wild-type *supF* gene. Using this system, recombinant colonies will be blue in the presence of X-Gal and IPTG; parental colonies will be light blue, whereas vectors that have undergone mutations or gene conversion events will be white (see Figure 3).

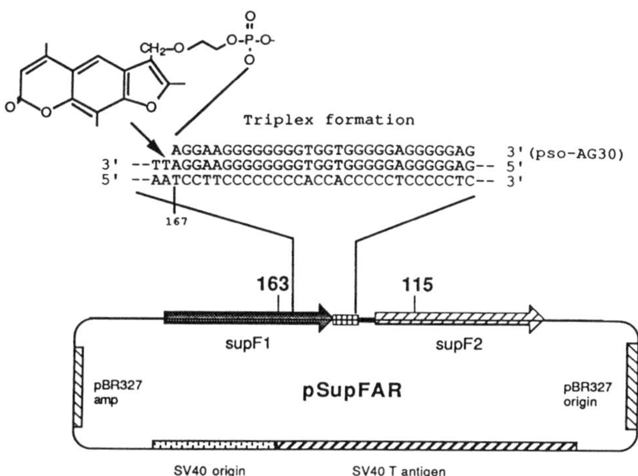

Figure 3. Schematic representation of shuttle vector, pSupFAR, used to detect triplex-induced extrachromosomal recombination. The vector contains two mutant copies of the *supF* gene in a tandem array. The triplex-binding site is located at the 3' end of *supF1*.

The APRT gene as a chromosomal test system for recombination and mutation: The adenine phosphoribosyltransferase (APRT) gene offers many advantages for recombination and mutation studies of a native chromosomal gene in mammalian cells. The hamster APRT gene is a small (2.3 kb), well-characterized gene with hemizygous cell lines available (*24*), making it ideal as a genetic marker for mammalian cell studies. Drug selections for both gain and loss of function are available, and many modified cell lines are available to facilitate recombination and mutation studies on the chromosome (*24-27*).

The APRT gene assay for triplex-induced recombination and mutagenesis in a native gene in mammalian cells: In order to test the effects of site-specific DNA damage on recombination and mutation, a sensitive genetic selection assay must be employed. Cell lines were constructed with a tandem duplication of the APRT gene, which can serve as a substrate for measuring intrachromosomal recombination (*26*). The general strategy for detecting chromosomal recombination in the APRT gene in

CHO cells is shown in Figure 4. The upstream copy of the gene has a large 3' deletion, rendering the gene inactive (APRT-), while the downstream copy is functional (APRT+). The herpes virus thymidine kinase (TK) gene and the bacterial guanine phosphoribosyltransferase (GPT) gene are located between the two APRT genes as additional genetic markers. By selecting for cells that have lost both APRT and GPT functions following triplex-mediated DNA damage, the effects on both recombination and mutation can be assessed at a level of detection near 1 event in 10^5 to 10^6 cells (Vasquez, K. M., Ph.D. thesis, 1996).

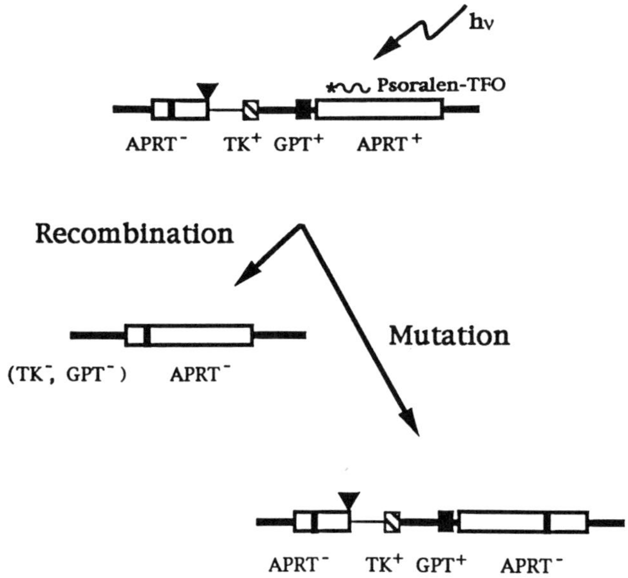

Figure 4. APRT assay to detect triplex-induced recombination and mutation on the chromosome in CHO cells. The chromosomal substrate contains a tandem duplication of the APRT gene. The upstream copy contains a 3' deletion (indicated by the triangle) and a mutation in exon 2 (indicated by the bar) rendering the gene inactive (APRT⁻). The downstream copy is wild-type (APRT⁺). The genotype associated with a recombination event can be selected for by using drug selections for APRT⁻, TK⁻, and GPT⁻. A mutation event can be selected by APRT⁻, TK⁺, GPT⁺.

BIOLOGICAL EFFECTS OF TFOs IN CELLS

Effects of TFOs on mutagenesis

Triplex-induced mutagenesis of lambda DNA using psoralen-modified TFOs: Triplex-induced mutagenesis *in vitro* was demonstrated in 1993 when a psoralen-conjugated 10-base TFO targeted to a site in the *supF* gene was shown to induce site-specific mutations in a λ phage genome (7). When the λ DNA was incubated with TFOs and then subjected to UVA irradiation, the frequency of mutations in the targeted gene was stimulated 100-fold. Sequence analysis of the mutants showed that 96% were in the region of the triplex target site. This result demonstrated the feasibility of using TFOs to target mutations to sites in duplex DNA, leading to studies to determine the potential application of TFOs in cells.

Triplex-induced mutagenesis of SV40 DNA in COS cells using psoralen-modified TFOs: The demonstration of TFO-targeted mutagenesis in the *supF* gene in a lambda phage genome led to further studies to explore the possibility of targeting mutations to a viral genome in mammalian cells (*6*). The general strategy used to investigate this application is shown in Figure 1. First, the SV40 vector DNA was incubated with a psoralen-conjugated 10-mer TFO and irradiated with UVA (1.8 J/cm^2). The TFO-plasmid complex was then transfected into monkey COS cells and incubated for 48 hours to allow replication and repair processes. Upon isolation and screening of the SV40 vector DNA, it was found that over 6% of the viral genomes were mutated at the TFO-target site (*6*). Sequence analysis showed that all mutations were in the targeted region with 55% being T·A to A·T transversions at the targeted psoralen crosslinking site. The next step was to extend this *in vitro* application to target mutations to DNA *in vivo*.

Triplex-induced mutagenesis on the SV40 shuttle vector in mammalian cells using psoralen-modified TFOs: Triplex-directed mutagenesis as an approach for gene therapy must be demonstrated *in vivo*. Toward this goal, triplex-directed mutations in cells have been induced with a psoralen-conjugated TFO plus UVA irradiation on plasmid DNA (*8*). Mutations were generated on an SV40 shuttle vector transfected into COS cells, directed by intracellular triplex formation (see Figure 1). When the psoralen-modified TFO was added to COS cells, *supF* mutations were stimulated approximately 100-fold, and nearly 70% of the total mutations were T·A to A·T transversions at the triplex/duplex junction (*8*). This result provides evidence for the occurrence of intracellular triplex formation on a plasmid in mammalian cells.

Triplex-induced mutagenesis on the SV40 shuttle vector in mammalian cells using unmodified TFOs: The ability of psoralen-modified TFOs to induce mutagenesis both *in vitro* and *in vivo* is well documented (*6-8*), but there may be disadvantages to this approach for genetic manipulation in intact animals. For example, psoralen is a very active mutagen, and nonspecific reactivity upon UVA irradiation may be detrimental. Additionally, the effective administration of UVA irradiation, required for psoralen activation, may prove challenging in intact animals. It would therefore be advantageous if unmodified TFOs could induce mutagenesis in the absence of a tethered mutagen. Toward this goal, intracellular triplex-induced mutagenesis has been demonstrated on plasmids in COS cells using unmodified TFOs (*28*). COS cells were transfected with the plasmid, pSupFG1, and incubated for 12 hours to allow chromatinization of the vector DNA. TFOs were then added to the medium and, 48 hours later, the vector DNA was isolated for mutational analysis. The 30-mer TFO designed to bind the polypurine target site in the *supFG1* gene was able to induce mutations at a frequency greater than 10-fold above the background. Sequence analysis of the mutants revealed mostly scattered point mutations with a few deletions around the triplex binding site. Although the frequency of triplex-induced mutagenesis was lower than that induced by psoralen-modified TFOs (10-fold induction vs. 100-fold induction, respectively), the results are promising in that the use of unmodified TFOs abrogates the need for both UVA irradiation and a tethered mutagen. This should prove advantageous for therapeutic applications.

Repair of triplex damage on the SV40 shuttle vector in mammalian cells: To elucidate the mechanism of triplex-induced mutagenesis in the absence of psoralen damage, similar studies were performed in repair-deficient cells. Human xeroderma pigmentosum group A (XPA) cells, which are deficient in nucleotide excision repair, were used to provide insight into the involvement of DNA repair in triplex-induced mutagenesis. When the pSupFG1 plasmid was transfected into XPA cells and subsequently treated with TFOs, no triplex-targeted mutagenesis was detected (*28*). This result suggests that nucleotide excision repair is required in the process of triplex-induced mutagenesis. In addition, triplex-induced mutagenesis was not detected in Cockayne's syndrome group B (CSB) cells which are defective in transcription-coupled repair (*28*), indicating a requirement for transcription-coupled repair as well as excision repair in the triplex-induced mutagenic pathway.

To further investigate the involvement of transcription-coupled repair in triplex-induced mutagenesis, the induction of DNA repair synthesis of the TFO-bound plasmid was measured in HeLa extracts by incorporation of α-[^{32}P]-dCTP. The specific 30-mer TFO stimulated labeling of pSupFG1, but not of a control plasmid (pUC19). Therefore, repair synthesis was induced by triplex formation on the pSupFG1 target vector (*28*). Next, triplex-mediated transcription inhibition was measured using the pSupFG1 vector as substrate for *in vitro* transcription assays. When the experiments were performed in the presence of the TFO, a truncated transcript was detected, indicating triplex-mediated inhibition of transcription elongation (*28*).

Taken together, these results demonstrated that the TFOs that were capable of inducing mutagenesis and stimulating repair were also specifically inhibiting transcription of the *supFG1* gene. The proposed mechanism by which these events occur is one of "gratuitous repair" (*29*), in which the TFO blocks transcription at the target site leading to gratuitous, and potentially error-prone repair.

Triplex-induced mutagenesis in the chromosomally integrated supFG1 reporter gene in mouse cells: Although targeted extrachromosomal mutagenesis in cells *via* intracellular triplex formation is encouraging, a similar effect on a chromosomal target is required for gene therapeutic applications. To address this issue, studies were performed using a mouse fibroblast cell line (3340) containing multiple copies of the λsupFG1 shuttle vector in a chromosomal locus. The cells were treated with psoralen-conjugated TFOs in the presence of UVA irradiation. The results demonstrated up to a 10-fold induction in mutagenesis in cells treated with a specific TFO as compared to a control TFO or to untreated cells (*30*). Sequencing data showed that the majority of the mutations were single or double base pair insertions or deletions within the triplex-binding site. This result provides evidence for the utility of TFOs to enhance the frequency of mutagenesis at a specific site in a mammalian chromosome, and lends additional support to the potential therapeutic applications of TFOs.

When the psoralen-modified TFOs were added to the 3340 cells, but not subjected to UVA irradiation, induction of mutagenesis was still detected near 10-fold above the background frequency. This result suggests a triplex-mediated, photoproduct-independent mechanism of mutagenesis. Sequencing data were consistent with this finding, as no T·A to A·T transversions were detected at the psoralen crosslinking site (*30*). These results are consistent with the extrachromosomal reports mentioned above.

Repair of triplex-directed psoralen damage in mammalian cells: The nucleotide excision repair pathway is responsible for removal of covalent DNA lesions and is the only known system for removing bulky DNA adducts in human cells. The damaged DNA is removed by incision of the phosphodiester bonds on either side of the damage by excinuclease complexes (22 nucleotides 5' and 6 nucleotides 3' from the site of damage) resulting in the release of an oligonucleotide containing the damage, followed by repair synthesis and ligation (*31*). Repair of psoralen crosslinks is thought to occur by nucleotide excision repair and may also involve recombinational repair pathways (*32, 33*).

There is evidence that triplex-directed psoralen adducts are repaired in mammalian cells. An example is provided by the ability of a plasmid covalently crosslinked *in vitro* by a TFO to replicate in mammalian cells (*6, 34*). In the work reported by Sandor and Bredberg, an SV40 shuttle vector containing the mutational *supF* reporter gene was treated with a psoralen-linked 22-mer purine-rich TFO and UV irradiated to form a covalent triplex in the target gene. The plasmid-TFO complex was then passaged in human Jurkat cells and the recovered vector was analyzed for mutants. The presence of triplex-mediated monoadducts and crosslinks in the target sequence did not reduce the yield of replicated vector molecules, indicating that repair of the triplex-associated DNA damage had occurred. Interestingly, psoralen-coupled TFOs designed to introduce lesions at both ends of the *supF* triplex target site in the were not repaired efficiently in the SV40 vector passaged in human Raji cells (*35*).

In another study, a plasmid containing a triplex binding site in the β-galactosidase reporter gene was transfected into human cells following incubation with a 15-mer psoralen-conjugated TFO (*36*). Using a PCR-based assay, the removal of the triplex-directed psoralen crosslink was shown to occur within a few hours. This result lends additional support to the idea that triplex-directed psoralen lesions are recognized and repaired by the cellular repair machinery.

It is not yet clear how psoralen-TFO-induced lesions are repaired, although the length of the TFO appears to play a role. Previous work revealed that different patterns of mutations were induced by psoralen-conjugated TFOs of 10 or 30 bases that were targeted to the same site (*6, 8, 34*) suggesting that the length of the TFO may be a critical factor in the repair of these lesions leading to the altered mutational spectra. The psoralen-modified TFO 10 nucleotides long induced mutations similar to that of psoralen alone, while the psoralen-modified 30-mer TFO produced a different spectrum of mutations, suggesting an altered repair process. Since the excision repair patch generated in nucleotide excision repair is generally between 28-30 bases in mammalian cells, the 30-mer TFO may have been capable of blocking the incision step at the site 5' to the psoralen damage. To assess the whether the triplex-directed psoralen adducts were excised, the TFOs were 3' radiolabelled and the psoralen-TFO-damaged plasmids were incubated in HeLa cell extracts. When the products were analyzed by denaturing polyacrylamide gel electrophoresis, it was shown that the 30-mer TFO inhibits the excision of the damaged DNA, while the damage induced by the 10-mer TFO was excised (*37*).

These results suggest that triplex structures can influence DNA repair pathways, and may inhibit nucleotide excision repair by blocking an excinuclease incision site.

It has been shown that the length of the triplex itself can effect DNA metabolic processes in the absence of psoralen damage. For example, TFOs were capable of blocking the progression of primer extension by DNA polymerase (*38*). Interestingly, when a series of TFOs of varying lengths were tested for their ability to inhibit the polymerase, the longer TFOs were found to be more effective than the shorter TFOs (*38*). These results suggest that the length of the third strand may be important in it's removal from the duplex DNA.

Effects of TFOs on recombination

Since the frequency of homologous recombination is low compared to random integration (i.e., non-homologous events), it has been a goal to enhance homologous recombination at specific sites in the genome for therapeutic applications. One approach to stimulating recombination is through introducing double strand breaks or other types of DNA damage. By covalent modification of a TFO with a DNA damaging agent, this approach can be tested at specific sites recognized by the TFO. There have been several examples of triplex-induced recombination reported in the literature.

Triplex-induced recombination in an SV40 shuttle vector in human cells: Triplex-mediated psoralen DNA damage can induce a low frequency of intermolecular homologous recombination in an SV40 shuttle vector transfected into human cells (*39*). An SV40 shuttle vector containing a mutated *supF* gene was treated with a psoralen-conjugated TFO and cotransfected into human cells with another plasmid containing the wild-type *supF* gene. By selecting for recombinants with a reactivated reporter gene, a low frequency (near 3-fold above background) of triplex-induced recombinants was detected. The low level of induction suggests that psoralen-modified TFOs are capable of stimulating recombination. Further studies to optimize this effect are warranted.

Triplex-induced recombination in an SV40 shuttle vector in repair proficient and repair deficient mammalian cells: To further investigate the ability of TFOs to stimulate recombination and to discern the mechanism of repair of triplex-directed psoralen lesions, the plasmid, pSupF2, containing a tandem duplication of the *supF* gene (described previously), was used to measure the effects of triplex-directed psoralen lesions on the recombination frequency. In these experiments psoralen conjugated TFOs were added to cells previously transfected with the pSupF2 plasmid and then UVA irradiated. Vector DNA was isolated and analyzed 48 hours after treatment to allow time for repair of the triplex-induced psoralen lesions. Blue colonies (indicating a recombinant) were generated at a frequency 25-fold above the background when treated with the specific, psoralen-conjugated TFO. In this system the frequency of recombination was clearly enhanced by the specific psoralen-conjugated TFO. When the recombinants were analyzed, a variety of products were detected, including those produced by nonconservative recombination, gene conversion, and mutation. Interestingly, gene conversion events were not detected when the TFO was removed after induction of the psoralen adducts (*23*). This result implies that the TFO itself has an influence on DNA metabolic processes. When similar experiments were performed in cell lines deficient in mismatch repair (MSH2 or MLH1 cells), the triplex-induced stimulation in extrachromosomal recombination was not significantly different from that of wild-type cells treated with the same TFOs (Faruqi, A. F., personal communication). However, when experiments were performed in cells deficient in nucleotide excision repair (XPA cells), the increase in the frequency of recombination was reduced nearly 2-fold compared to the wild-type cells (corrected by the expression of an integrated cDNA of the XPA gene). This finding implicates nucleotide excision repair in the process of triplex-induced recombination. Collectively, these results provide some insight into the mechanism(s) of repair of triplex-induced DNA damage and the potential interactions between cellular recombinational repair, nucleotide excision repair and mismatch repair pathways.

Triplex-induced recombination in the native chromosomal APRT gene in CHO cells: In an attempt to stimulate homologous recombination at a chromosomal locus, the effects of a psoralen-modified TFO on the frequencies of recombination and mutation in CHO cells were studied at a well characterized, high affinity $(K_d = 10^{-10} M)$ triplex-forming site in intron 1 of the hamster APRT gene *(40, 41)*. CHO cells containing a tandem duplication of the APRT gene were incubated with increasing concentrations of either a specific TFO or a scrambled sequence control TFO and then subjected to UVA irradiation (366 nm, 1.2 J/cm^2). Cells were then plated in selective media to assay for the occurrence of recombination or mutation events. The results demonstrated a slight enhancement (approximately 3-fold) in recombination and/or mutation in a dose-dependent fashion that was specific to the TFO designed to bind the target site duplex (Vasquez, K. M., Ph.D. thesis, 1996). Thus, TFOs may be capable of stimulating recombination of a native gene in mammalian cells.

FUTURE PROSPECTS AND CHALLENGES

Considerable progress has been made in the development of triplex technology as an approach to modify a mammalian genome. However, further advances must be made in several areas including: 1) extending the "triplex code" so that TFOs can be designed and targeted to any DNA sequence of interest; 2) improving the uptake and stability of oligonucleotides in cells; 3) increasing the binding affinities of TFOs to their target site duplexes; 4) enhancing site-specific DNA damage on the chromosome; and 5) providing direct evidence for the formation of triplexes on the chromosome. Efforts are currently ongoing in many laboratories to address these and other issues to improve the practical utility of TFOs as genome-modifying agents.

REFERENCES

1. Chan, P. P. and Glazer, P. M. (1997). Triplex DNA: fundamentals, advances, and potential applications for gene therapy. *J. Mol. Med.* **75**, 267-282.
2. Vasquez, K. M. and Wilson, J. H. (1998). Triplex-directed modification of genes and gene activity. *Trends Biochem. Sci.* **23**, 4-9.

3. Frank-Kamenetskii, M. D. and Mirkin, S. M. (1995). Triplex DNA structures. *Ann. Rev. Biochem..* **64**, 65-95 .
4. Postel, E. H., Flint, S. J., Kessler, D. J. and Hogan, M. E. (1991). Evidence that a triplex-forming oligodeoxyribonucleotide binds to the c-myc promoter in HeLa cells, thereby reducing c-myc mRNA levels. *Proc. Natl. Acad. Sci. USA* **88**, 8227-8231.
5. Beal, P. A. and Dervan, P. B. (1991). Second structural motif for recognition of DNA by oligonucleotide-directed triple-helix formation. *Science* **251**, 1360-1363.
6. Havre, P. A. and Glazer, P. M. (1993). Targeted mutagenesis of simian virus 40 DNA mediated by a triple helix-forming oligonucleotide. *J Virol* **67**, 7324-7331.
7. Havre, P. A., Gunther, E. J., Gasparro, F. P. and Glazer, P. M. (1993). Targeted mutagenesis of DNA using triple helix-forming oligonucleotides linked to psoralen. *Proc. Natl. Acad. Sci. USA* **90**, 7879-7883.
8. Wang, G., Levy, D. D., Seidman, M. M. and Glazer, P. M. (1995). Targeted mutagenesis in mammalian cells mediated by intracellular triple helix formation. *Mol. Cell. Biol.* **15**, 1759-1768.
9. Thaler, D. S. and Stahl, F. W. (1988). DNA double-chain breaks in recombination of phage lambda and of yeast. *Ann. Rev. Genet.* **22**, 169-197.
10. Capecchi, M. R. (1989). Altering the genome by homologous recombination. *Science* **244**, 1288-1292.
11. Bollag, R. J., Waldman, A. S. and Liskay, R. M. (1989). Homologous recombination in mammalian cells. *Ann. Rev. Genet.* **23**, 199-225.
12. Roth, D. B. and Wilson J H. (1986). Non-homologous recombination in mammalian cells: role for short sequence homologies in the joining reaction. *Mol. Cell Biol.* **6**, 4295-4304.
13. Averbeck, D. (1985). Relationship between lesions photoinduced by mono- and bi-functional furocoumarins in DNA and genotoxic effects in diploid yeast. *Mutat. Res.* **151**, 217-233.
14. Kucherlapati, R. S., Eves, E. M., Song, K. Y., Morse, B. S. and Smithies, O. (1984). Homologous recombination between plasmids in mammalian cells can be enhanced by treatment of input DNA. *Proc. Natl. Acad. Sci. USA* **81**, 3153-3157.
15. Brenner, D. A., Smigocki, A. C. and Camerini-Otero, R. D. (1985). Effect of insertions, deletions and double-strand breaks on homologous recombination in mouse L cells. *Mol. Cell Biol.* **5**, 684-691.
16. Wake, C. T., Vernaleone, F. and Wilson, J. H. (1985). Topological requirements for homologous recombination among DNA molecules transfected into mammalian cells. *Mol. Cell Biol.* **5**, 2080-2089.
17. Sargent, R. G., Brenneman, M. A. and Wilson, J. H. (1997). Repair of site-specific double-strand breaks in a mammalian chromosome by homologous and illegitimate recombination. *Mol. Cell Biol.* **17**, 267-277.
18. Friedberg, E. C., Walker, G. C. and Siede, W. (1995). DNA repair and mutagenesis. American Society for Microbiology, Washington, DC.
19. Bramson, J. and Panasci, L. C. (1993). Effect of ERCC-1 overexpression on sensitivity of Chinese hamster ovary cells to DNA damaging agents. *Cancer Res.* **53**, 3237-3240.
20. Narayanan, L., Fritzell, J. A., Baker, S. M., Liskay, R. M. and Glazer, P. M. (1997). Elevated levels of mutation in multiple tissues of mice deficient in the DNA mismatch repair gene, *Pms2*. *Proc. Natl. Acad. Sci. USA* **94**, 3122-3127.
21. Glazer, P. M., Sarkar, S. N. and Summers, W. C. (1986). Detection and analysis of UV-induced mutations in mammalian cell DNA using a lambda phage shuttle vector. *Proc. Natl. Acad. Sci. USA* **83**, 1041-1044.
22. Gunther, E. J., Murray, N. E. and Glazer, P. M. (1993). High efficiency, restriction-deficient in vitro packaging extracts for bacteriophage lambda DNA using a new *E. coli* lysogen. *Nucleic Acids Res.* **21**, 3903-3904.

23. Faruqi, A. F., Seidman, M. M., Segal, D. J., Carroll, D. and Glazer, P. M. (1996). Recombination induced by triple helix-targeted DNA damage in mammalian cells. *Mol. Cell. Biol.* **16**, 6820-6828.
24. Adair, G. M., Nairn, R. S., Wilson, J. H., Seidman, M. M., Brotherman, K. A., MacKinnon, C. and Scheerer J. B. (1989). Targeted homologous recombination at the endogenous adenine phosphoribosyltransferase locus in Chinese hamster cells. *Proc. Natl. Acad. Sci. USA* **86**, 4574-4578.
25. Pennington, S. L. and Wilson, J. H. (1991). Gene targeting in Chinese hamster ovary cells is conservative. *Proc. Natl. Acad. Sci. USA* **88**, 9498-9502.
26. Sargent, R. G., Merrihew, R. V., Nairn, R., Adair, G., Meuth, M. and Wilson, J. H. (1996). The influence of a (GT)29 microsatellite sequence on homologous recombination in the hamster adenine phosphoribosyltransferase gene. *Nucleic Acids Res.* **24**, 746-753.
27. Merrihew, R. V., Sargent, R. G. and Wilson, J. H. (1995). Efficient modification of the APRT gene by FLP/FRT site-specific targeting. *Somat. Cell Mol. Genet.* **21**, 299-307.
28. Wang, G., Seidman, M. M. and Glazer P. M. (1996). Mutagenesis in mammalian cells induced by triple helix formation and transcription-coupled repair. *Science* **271**, 802-805.
29. Hanawalt, P. C. (1994). Transcription-coupled repair and human disease. *Science* **266**, 1957-1958.
30. Vasquez, K. M., Wang, G., Havre, P. A. and Glazer P. M. (1998). Chromosomal mutations induced by triplex-forming oligonucleotides in mammalian cells. *Submitted.*
31. Sancar, A. (1994). Mechanisms of DNA excision repair. *Science* **266**, 1954-1956.
32. Sancar, A. and Tang, M. S. (1993). Nucleotide excision repair. *Photochem.. Photobiol.* **57**, 905-921.
33. Sladek, F. M., Munn, M. M., Rupp, W. D. and Howard-Flanders, P. (1989). *In vitro* repair of psoralen-DNA cross-links by RecA, UvrABC, and the 5'-exonuclease of DNA polymerase I. *J. Biol. Chem.* **264**, 6755-6765.
34. Sandor, Z. and Bredberg, A.. (1994). Repair of triple helix directed psoralen adducts in human cells. *Nucleic Acids Res.* **22**, 2051-2056.
35. Sandor, Z. and Bredberg, A. (1995). Deficient DNA repair of triple helix-directed double psoralen damage in human cells. *FEBS Lett.* **374**, 287-291.
36. Degols, G., Clarenc, J. P., Lebleu, B. and Leonetti, J. P. (1994). Reversible inhibition of gene expression by a psoralen functionalized triple helix forming oligonucleotide in intact cells. *J. Biol. Chem.* **269**, 16933-16937.
37. Wang, G. and Glazer, P. M. (1995). Altered repair of targeted psoralen photoadducts in the context of an oligonucleotide-mediated triple helix. *J. Biol. Chem.* **270**, 22595-22601.
38. Hacia, J. G., Dervan, P. B. and Wold, B. J. (1994). Inhibition of Klenow fragment DNA polymerase on double-helical templates by oligonucleotide-directed triple-helix formation. *Biochemistry* **33**, 6192-6200.
39. Sandor, Z. and Bredberg, A. (1995). Triple helix directed psoralen adducts induce a low frequency of recombination in an SV-40 shuttle vector. *Biochim. Biophys. Acta* **1263**, 235-240.
40. Vasquez, K. M., Wensel, T. G., Hogan, M. E. and Wilson, J. H. (1995). High-affinity triple helix formation by synthetic oligonucleotides at a site within a selectible mammalian gene. *Biochemistry* **34**, 7243-7251.
41. Vasquez, K. M., Wensel, T. G., Hogan, M. E. and Wilson, J. H. (1996). High-efficiency triple-helix-mediated photo-cross-linking at a targeted site within a selectable mammalian gene. *Biochemistry* **35**, 10712-10719.

14 PSORALEN-COUPLED OLIGONUCLEOTIDES: *IN VIVO* BINDING AND REPAIR

François-Xavier Barre, Linda L. Pritchard
and Annick Harel-Bellan

INTRODUCTION

Psoralen-coupled oligonucleotides are a subclass of chemically modified oligonucleotides that have aroused considerable interest. Part of this interest comes from the double rationale behind the design of those compounds, which increase the efficiency of TFOs and improve the use of psoralens. When coupled to an oligonucleotide, psoralens improve its binding to the target DNA sequence as many other DNA intercalating agents do, albeit with a lower efficiency (*1*). But most importantly, psoralens form covalent bonds with pyrimidine bases when photoactivated with near-UV light, offering the possibility of covalently linking the oligonucleotide to its target and rendering its action, at least in theory, irreversible. Coupling of psoralens to TFOs offers the possibility of limiting their genotoxic, recombinagenic and mutagenic action to specific DNA sequences. Thus psoralen-coupled TFOs (Pso-TFOs) could represent efficient tools for gene modification.

Another reason for the interest in psoralen-coupled oligonucleotides arises from their rapid success in achieving both these ends. Moreover, the efficiency with which Pso-TFOs can be crosslinked to their target DNA sequences and the high mutagenicity of the targeted lesions are valuable tools to demonstrate triple helix formation inside live cells, a question of considerable interest in the field of triple helix research (*2*). Finally, the strong biological effects of these compounds have stimulated a new field of investigation on the repair of Pso-TFO targeted lesions.

COUPLING OF PSORALENS TO TRIPLE HELIX FORMING OLIGONUCLEOTIDES

Psoralens are naturally-occurring compounds, known for more than 3000 years for their deleterious effects when combined with phototreatment (*3*). They have a strong affinity for DNA, where they stack between bases. Upon UVA irradiation, they can form covalent bonds with pyrimidines (mainly thymines) (*3*).

The two psoralen molecules that have been used in the literature for coupling to TFOs, 5-MOP and HMT (Figure 1A), belong to the subfamily of bifunctional psoralens (*4*) which can induce interstrand crosslinks (ICs) provided that the molecule is intercalated between opposite pyrimidines (Figure 1B). ICs are mainly responsible for the genotoxic, recombinagenic and mutagenic effects of these compounds. In a first step, the 4', 5' double bond of the furan ring (F) reacts with an adjacent pyrimidine. The 3, 4 double bond of the pyrone ring (P) of the mono-adducted psoralen can then react with the opposite pyrimidine. Crosslinkable sites have been shown to be restricted mainly to 5'-TpA-3' sites and, to a much smaller extent, to 5'-ApT-3' sites (*5*). When a psoralen molecule only reacts with one thymine, the lesion is referred to as a mono-adduct (MA).

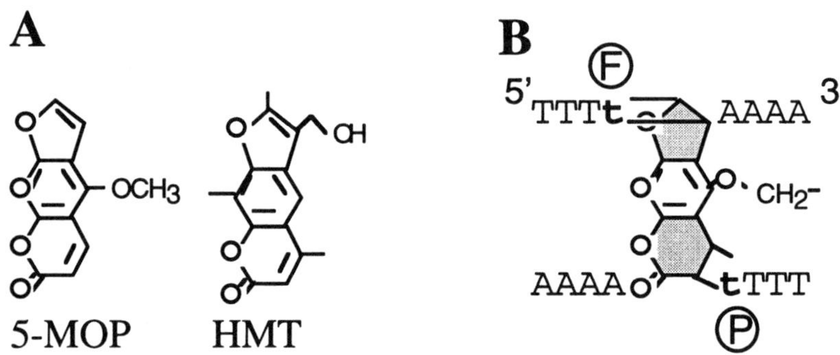

Figure 1. Psoralens used for coupling to TFOs **(A)** and representation of an interstrand crosslink **(B)**. The adducted thymines are shown in lower case.

Coupling to TFOs improves psoralens' photocrosslinking efficiency and restricts their action to the immediate vicinity of the triplex binding site

There have been several reports describing the use of psoralen-modified oligonucleotides targeted against single stranded nucleic acid sequences (*6*). In the case of Pso-TFOs, there was, however, a risk that triple helix formation would impede the intercalation of the psoralen between bases and its efficient photocrosslinking. But, in fact, following a first success using 5-MOP and a 6-carbon linker arm (*7*), TFOs could be crosslinked to a target using various psoralens such as HMT and linker arms from 2 to 86 carbons long (*8-10*). In addition, ICs were observed on TpA sites and ApT sites. Thus, coupling of psoralens to TFOs does not impede their ability to form ICs. On the contrary, IC levels obtained for TpA targets (85% (*7*) to 90% (*11*)) and ApT targets (25% (*9*) to 65%

(*12*)) are above expected IC levels for free psoralens. The fact that linker arms of more than 6 carbons give more freedom to the psoralen moiety but decrease crosslinking efficiency (*10*) further suggests that triple helix formation increases crosslinking efficiency by stabilizing the psoralen in an optimal position. Indeed, when targeted by TFOs, ICs adopt essentially one conformation, where the furan ring of the psoralen is linked to the thymine of the polypurine-rich strand of the TFO target, and the pyrone ring is linked to the thymine of the polypyrimidine-rich strand (*7, 12*).

In addition, it was also demonstrated that while increasing the efficiency of psoralen photocrosslinking at the TpA or ApT triple helix targeted sites, coupling to a TFO prevented nonspecific reactions of the psoralen with other sites (*7, 8*).

Coupling to psoralens improves triple helix formation

Triple helix formation has a strong positive effect on psoralen crosslinking efficiency, but psoralens also have a strong stabilizing effect on triplex formation. Indeed, a ten-degree increase in melting temperature for the triplex structure was observed using a 15mer pyrimidine-rich oligonucleotide coupled to 5-MOP through a 6-linker arm and targeted against the polypurine track from HIV-1, a target sequence with an adjoining TpA site (*13*).

Biological effects

TFOs can be used as competitive inhibitors for cellular sequence-specific DNA binding ligands such as transcription factors (*14*). Such a TFO linked to a psoralen molecule strongly inhibited gene transcription, at least when assayed using an ectopic transiently transfected promoter (*15*).

ICs are genotoxic, recombinagenic and mutagenic. All three actions could be rendered sequence-specific using TFOs. Indeed, using Pso-TFOs targeted against a polypurine rich region of an SV40 shuttle vector bordered by an ApT and a TpA site, Sandor and Bredberg (*16*) observed that, after repair in human cells, the amount of plasmid that was recoverable in bacteria was reduced to 50% when using a Pso-TFO targeted against the TpA site, which showed good crosslinking yields, whereas it was not affected when using Pso-TFOs targeted against the ApT site, which yielded few ICs. Using an oligonucleotide coupled to two psoralen molecules, one at each end, and able to react with both the ApT and TpA site of the target DNA sequence, they observed still lower recoveries. In our laboratory, an assay was designed to quantify plasmid survival directly inside eukaryotic *S. cerevisiae* cells after treatment with Pso-TFOs. We chose to quantify transformation efficiency of vectors containing a TpA triple helix target sequence. Using this assay, we found that the genotoxicity of Pso-TFO targeted adducts correlated with IC yield, whereas Pso-TFO MAs showed no genotoxicity (*11*).

The recombinagenic potential of Pso-TFO targeted adducts was also investigated. An assay designed for detection of intermolecular homologous recombination in human cells showed that psoralen-adduct-induced intermolecular recombination could occur (*17*), but at very low levels (0.08%). In contrast, a high level of intramolecular gene conversion (0.49%) and a low level of single strand annealing homologous recombination (0.09%) were observed (*18*).

Finally, mutagenicity of Pso-TFO targeted adducts was analyzed in a variety of systems. Using a Pso-TFO targeted against a bacterial amber suppresser tRNA gene, SupF, it was possible to detect targeted mutagenesis in the λ phage genome at the level of 0.23% (*8*). More interestingly, a number of studies used an SV40 *E. coli* shuttle vector carrying the SupF reporter gene, treated in vitro with Pso-TFOs and transiently transfected in COS or human cells, which was subsequently recovered and screened in bacteria; up to 10% of induced mutagenesis could be detected (*9, 19, 20*). In our laboratory, we developed a system in which the level of targeted mutagenesis is directly quantified in the cell where repair and replication occur. A Pso-TFO target DNA sequence with a TpA intercalation site was introduced into the coding frame of *URA3*, a yeast auxotrophic marker. Plasmids carrying the reporter gene were used to transform yeast cells (after *in vitro* treatment with Pso-TFOs) and screening of mutants was performed by direct plating on selective medium. Using this assay, we obtained levels of mutagenesis similar to those obtained with mammalian cells (*11*).

IN VIVO BINDING OF TFOs TO THEIR TARGET SEQUENCE

One of the most challenging questions when dealing with TFOs concerns their ability to bind to their target within the genome of live cells. A first difficulty that may limit the use of TFOs is the complexity of mammalian genomes. It was not at all evident that a TFO could find its target within an entire genome. Another important point was to verify that triple helix formation could occur under physiological conditions inside live cells. In addition, the higher order structure of DNA in live cells could also block triplex formation. Finally, many biological processes, such as transcription and replication, may limit triplex formation inside live cells. The efficiency of Pso-TFO targeted crosslinking and the high level of Pso-TFO site-specifically targeted mutagenesis have been used in attempts to answer this question.

TFOs can find their target within the entire genome of mammalian cells

When a restriction enzyme site overlaps a TFO target DNA sequence, covalently linked triplex inhibits cleavage by the enzyme at this particular site irreversibly and specifically. Using such an assay, it was demonstrated that Pso-TFOs could find their target sequence *in vitro* even within the entire genome of a transgenic mouse (*21*).

TFOs can bind to their target in live cells

Wang *et al.* (*22*) found that, after transient transfection of mammalian cells with an SV40 vector carrying the SupF reporter gene, incubation of these cells with Pso-TFOs targeted against SupF followed by UVA irradiation resulted in site-specific targeted mutagenesis, thus demonstrating that triplex formation could occur within cells. However, this was observed only with a 30mer Pso-TFO; a 10mer had no detectable effects.

MAs and ICs are known to block DNA synthesis (*23*). This property was used by Guieysse *et al.* (*24*) to develop a primer extension assay for the detection of adducts. Using this assay, the authors were able to directly demonstrate triplex formation on a plasmid within intact cells after incubation of the transiently transfected cells with the oligonucleotide followed by UVA irradiation.

Triple helix formation can occur on chromatin folded DNA

In 1997, Giovannangeli *et al.* (*25*) reported triplex formation and psoralen photocrosslinking within the genome of permeabilized human cells. This was shown using a restriction-enzyme-based assay and a competitive PCR-based assay. This important finding demonstrates that TFOs can recognize and bind specifically to their target DNA sequence within the chromatin structure inside the cell nuclei. However, phosphoramidate Pso-TFOs were necessary for this demonstration.

A demonstration of TFO binding on a chromosomal target within live cells is still lacking

To our knowledge, no clear demonstration of triple helix formation within the genome of intact live cells has been reported. It is of interest that: 1) for the demonstration of triplex formation inside cells, an increase in affinity was necessary as attested by the need for a 30mer TFO (*26*); and 2) for detection of triplex binding within DNA in chromatin, a further increase in affinity was required as shown by the need for a phosphoramidate TFO (*25*). This strongly suggests that the affinity of oligonucleotides for their target is still the limiting parameter.

In our laboratory, we are developing an assay in view of comparing different Pso-TFOs for their ability to bind to their target within the genome of live eukaryotic cells. The assay is based on two yeast strains, one of them bearing a *URA3* target reporter gene and the other a *URA3* reporter gene with a control sequence that is not recognized by the oligonucleotide. An electroporation procedure has been developed that enables quantitative introduction of oligonucleotides into intact yeast cells with as few as 1 µg of oligonucleotide (*27*). The interaction between the TFO and its target sequence is monitored through the detection of mutations introduced following IC formation (by UVA irradiation) and repair. Any increase in the frequency of mutants in the strain bearing the *URA3* target reporter gene (but not in the strain bearing the *URA3* control gene) can be assumed to be the "signature" of triplex formation.

REPAIR OF Pso-TFO TARGETED DAMAGE

When Pso-TFOs are used as competitive inhibitors of cellular sequence-specific DNA binding ligands, the psoralen moiety enables covalent linking of the oligonucleotide to its target, rendering its action, at least in theory, irreversible. However, a number of mechanisms ensure the integrity of the cell genome, and the repair of Pso-TFO targeted adducts would lead to reversal of the TFO-mediated inhibition. In addition, biological effects of Pso-TFOs that derive from the genotoxic, recombinagenic and mutagenic effects of psoralen adducts are also dependent on the cell repair machinery. Consequently, a number of studies have been devoted to understanding the repair of Pso-TFO mediated DNA damage.

Psoralen-adduct repair

As is often the case for basic cellular processes, psoralen-adduct repair has best been characterized in bacteria, essentially *E. coli* (*28*). Despite the important genotoxic effect of ICs in bacteria, it was found that some of the lesions were repaired. From genetic data, a model was proposed where repair of psoralen ICs is mediated by the successive action of *N*ucleotide *E*xcision *R*epair (NER), the Pol I 5'→3' exonuclease activity, and *R*ecombinational *R*epair (RR). This model was later confirmed by *in vitro* reconstitution of its different steps. In *E. coli*, this pathway is responsible for most of DNA recovery. However, the mutagenicity of psoralen ICs and MAs, which is dependent on SOS function, has led to the proposal of another pathway where MAs can be bypassed by DNA polymerases, and ICs are processed through excision of one of the adducts and bypass of the other. In addition, it has been shown that another minor pathway can lead to DNA recovery in the absence of NER. This pathway involves the RecA protein, implicated in post-replicational repair.

In *S. cerevisiae*, genetic studies (*29*) also implicated Nucleotide Excision Repair genes from the *rad3* epistasis group, Recombinational Repair genes from the *rad52* epistasis group and Error-Prone Repair (EPR) genes from the *rad6* epistasis group. However, some findings suggest significant differences between eukaryotic and bacterial repair of ICs. Whereas NER action leads only to single strand breaks in *E. coli* cells, double strand breaks are formed in *S. cerevisiae*. This has implications for the mode of initiation of Recombinational Repair. In addition, levels of induced mutagenesis are far above those obtained in bacteria, and Error-Prone Repair, which seemed to be a minor pathway in bacteria, can be of great importance for overall survival, depending on the cell-cycle state (G1 or S).

Few data are available concerning psoralen-adduct repair in mammalian cells. However, it has been demonstrated that MAs can be bypassed and that ICs are excised (*30*). Mutation spectra induced by free psoralens have been obtained.

Correlation between mutagenesis and TpA sites, which are hotspots for ICs suggest that, as is the case in lower organisms, ICs are more mutagenic than MAs (*5*). In addition, there is a strand bias for the induction of mutation toward the non-transcribed strand, which suggests that mutation induction derives from some repair process coupled to transcription, such as NER (*31*). Indeed, recently, an *in vitro* reconstituted mammalian NER assay could be used to excise MAs (*32*).

Pso-TFO targeted adduct repair efficiency

After transfection of a Pso-TFO adducted plasmid, the amount of a PCR amplification fragment encompassing the IC site was measured as a function of time in a normal cell line and in a repair-deficient Xeroderma Pigmentosum (XP) cell line (*33*). The amount of PCR fragment increased as a function of time in normal cells but not in XP cells, demonstrating the action of repair processes on ICs. However, from such a PCR-based assay, it is difficult to ascertain quantitatively the efficiency of repair.

Pso-TFO IC repair efficiency was also investigated using the SupF SV40 shuttle vector assay that was designed for the study of mutagenesis. In a first article, no DNA recovery difference was observed between Pso-TFO treated plasmids and untreated control DNA (*9*). However, the amount of ICs was low (25%) and the need to recover plasmids from human cells into bacteria to determine the amount of repair is somehow a limit to quantification. Indeed, in another paper, the same authors detected a 50% decrease in plasmid recovery when using a distinct Pso-TFO that yielded more ICs (*16*).

We designed an assay, in *S. cerevisiae*, that enables direct quantification of recoverable plasmids. Using this assay, we were able to demonstrate that Pso-TFO IC repair efficiency was very low (on the order of 10%). In contrast, MAs did not impair plasmid replication inside the cells (*11*). These results are in good agreement with the known genotoxic effects of free psoralens.

Pathways for the repair of Pso-TFO targeted adducts

The study of the repair pathways involved in the processing of Pso-TFO targeted adducts has been rendered difficult by the low resolution of the techniques employed. In our own experiments in yeast, the amount of recoverable plasmids in a Pso-TFO treated sample was close to the contaminating amount of uncrosslinked plasmids in the sample. As a consequence, differences between wild-type strains and repair-defective mutants were rendered marginal.

However, mutagenesis values and recombinagenic potential can be used to compare wild type cells and repair-defective mutants. In the particular case of psoralen ICs, some knowledge of the repair processes can be gained from such studies, as it has been demonstrated that, for free psoralens, mutants arose not through passive damage tolerance but by an active process implicating the repair machinery. Indeed, the type and number of mutants that were obtained, using the

SupF shuttle vector assay, in repair-defective human cell lines differed from that of normal cells. Fanconi Anemia cells were shown to yield a strong proportion of deletion mutants (*20, 34*). XP cells of group A (XPA) yielded essentially the same types of mutants as normal cells but with a quantitative decrease, implicating the XPA gene product in a repair pathway that could be error prone (*20*). More interestingly, XPV (group V) cells showed an increase in the number of mutants (correlated with a decrease in survival) and a change in the type of mutations induced (*20*). Thus the XPV complementation group, believed to be implicated in bypass replication, was established as a major component of the repair process for Pso-TFO targeted ICs.

The low level of Pso-TFO induced intermolecular recombination in human cells (*17*), the absence of Pso-TFO induced double-strand break formation in *Xenopus* oocytes (*35*) and the low level of single-strand annealing recombination in COS cells (*18*) suggest that the recombinagenic pathway may not be very effective in the removal of Pso-TFO adducts. However, a high level of intramolecular genetic conversion was obtained in COS cells (*18*). Thus, Pso-TFO targeted adducts may be processed through Recombinational Repair, but the success of such a process is likely to be highly dependent on the sequence context and/or the presence of a non-adducted homologous sequence in the vicinity.

Using yeast strains defective in each of the three major repair epistasis groups, we studied Pso-TFO induced mutagenesis (*11*). We found that ICs were highly mutagenic in contrast to MAs. Two classes of mutations were obtained in normal cells, substitutions and single base pair insertions. In a *rad51* strain, defective in Recombinational Repair, both types of mutations were increased, suggesting that this pathway is relatively error free and competes with the Error-Prone pathway for Pso-TFO adducts. In a *rad18* strain, defective in damage avoidance and Error-Prone Repair, all types of mutations were abolished, clearly demonstrating that mutation induction is an active process. In a *rad1* strain, defective in Nucleotide Excision Repair, we found that insertions were increased whereas substitutions were decreased. Thus, Pso-TFO ICs can be processed through EPR without any involvement of NER, leading mainly to insertions. However, following the action of NER, Pso-TFO ICs can also be subsequently treated through an EPR process, yielding mainly substitutions.

Taken together these results suggest that, in the case of eukaryotic cells, repair of Pso-TFO ICs on plasmid vectors is mainly achieved though the combined action of the NER and the EPR pathway. The Recombinational Repair pathway can also process these lesions but to a lesser extent.

Influence of the oligonucleotide tail on psoralen-adduct repair

TFOs prevent the binding of proteins at the sites neighboring triplexes. It has been reported that the triple helix itself could be recognized as a lesion by repair processes, yielding an increase in site-specific mutagenesis (*36, 37*). The oligonucleotide tail could therefore interfere with the repair of Pso-TFO targeted adducts.

Using the SupF mutagenesis assay, a difference was found in the type of mutations induced by a 10mer and a 30mer Pso-TFO following UVA irradiation (*26*). The 10mer oligonucleotide yielded mainly substitutions on the pyrimidine rich strand of the target, whereas the 30mer yielded substitutions on both strands as well as an increase in the number of deletion events. A psoralen molecule delivered to the target by a detachable 10mer TFO (coupled through a reducing bond to the psoralen, Pso-ss-TFO, photoactivated by UVA and detached from the oligonucleotide tail after reduction) yielded the same kind of mutations as the 10mer Pso-TFO. These results were taken as evidence that the differential repair observed was a consequence of the 30mer oligonucleotide tail blocking NER action on one of the two strands. However, they could also be the consequence of differences in IC yield or orientation between 10mer and 30mer Pso-TFOs, the two triple helixes having different stabilities. Indeed, a pattern of mutation identical to that observed for the 10mer was later found using the 30mer oligonucleotide (*10, 18*).

A difference in induced recombination was also found between a 30mer Pso-TFO treated vector and a vector treated with a 13mer Pso-ss-TFO that could be detached from the psoralen moiety by a reducing agent (*18*). Again, variations in IC yield or orientation could be at the origin of the difference.

Using our yeast assay, we compared the effects of Pso-ss-TFO treatment of plasmids to Pso-TFO treatment. Both oligonucleotides were 15mers, and we compared IC yield and orientation. The Pso-ss-TFO gave somewhat fewer ICs but they were mainly in the same orientation as the Pso-TFO ICs. We did not observe any quantitative difference between the two molecules in genotoxicity (as observed through loss of plasmid transformation efficiency) nor in mutagenicity (as observed through differences in mutation frequency). At the molecular level, the distribution of substitutions on the furan-adducted strand of the target *versus* the pyrone-adducted strand was the same for the two TFOs. Thus, we can conclude that, in our system, a 15mer oligonucleotide tail has no influence on the repair process.

Pso-TFO adduct repair was also tested *in vitro*, and differences between excision efficiency of a 10mer oligonucleotide and a 30mer oligonucleotide were observed (*10*). An inability to repair Pso-TFO ICs as compared to free psoralen ICs (targeted with a Pso-ss-TFO) has been demonstrated recently (Anne-Laure Guieysse, personal communication). These results are in disagreement with our *in vivo* results on mutagenicity and genotoxicity, and this discrepancy raises the question of the half-life of the oligonucleotide tail inside cells: nuclease or exonuclease degradation of the oligonucleotide tail inside live cells could reduce the problem of Pso-TFO adduct repair to the more simple problem of psoralen adduct repair.

CONCLUSION

Coupling psoralens to TFOs has offered one of the most valuable tools to study triplex formation and crosslink repair inside cells. It helped to demonstrate triplex formation under physiological conditions, inside live cells or on DNA in chromatin inside cell nuclei. Repair of psoralen adducts was also investigated. Repair results support the idea that ICs are very toxic and that their repair, in eukaryotic cells, depends more on NER and EPR action than on RR action, successful RR action being limited by the need for a non-adducted homologous sequence in the immediate vicinity of the lesion. Finally, the oligonucleotide tail seems to block some steps of repair *in vitro*; but *in vivo*, repair of Pso-TFO ICs seems not to differ much from free psoralen IC repair.

Coupling of TFOs to psoralens will remain a valuable tool for *in vivo* comparison of binding efficiencies of TFOs. In addition, Pso-TFOs enable easy psoralen delivery to specific TpA or ApT sites, high levels of ICs, and strand orientation of these ICs. They will therefore continue to be valuable tools for the study of psoralen-adduct repair.

ACKNOWLEDGMENTS

This work was supported by funds from the Agence Nationale pour la Recherche contre le SIDA and from Rhône Poulenc Rorer. F.-X. Barre, a student at the Institut de Formation Supérieure Biomédicale, is a recipient of a fellowship from the Agence Nationale pour la Recherche contre le SIDA.

REFERENCES

1. Sun, J. S., François, J. C., Montenay-Garestier, T., Saison-Behmoaras, T., Roig, V., Thuong, N. T. and Hélène, C. (1989). Sequence-specific intercalating agents: intercalation at specific sequences on duplex DNA via major groove recognition by oligonucleotide- intercalator conjugates. *Proc. Natl. Acad. Sci. USA* **86**, 9198-9202.
2. Praseuth, D., Guieysse, A. L., Itkes, A. V. and Hélène, C. (1993). Unexpected effect of an anti-human immunodeficiency virus intermolecular triplex-forming oligonucleotide in an in vitro transcription system due to RNase H-induced cleavage of the RNA transcript. *Antisense Res. Dev.* **3**, 33-44.
3. Scott, B. R., Pathak, M. A. and Mohn, G. R. (1976). Molecular and genetic basis of furocoumarin reactions. *Mutat. Res.* **39**, 29-74.
4. Brendel, M. and Ruhland, A. (1984). Relationships between functionality and genetic toxicology of selected DNA-damaging agents. *Mutat. Res.* **133**, 51-85.
5. Sage, E. and Bredberg, A. (1991). Damage distribution and mutation spectrum: the case of 8- methoxypsoralen plus UVA in mammalian cells. *Mutat. Res.* **263**, 217-222.
6. Cheng, S., Van Houten, B., Gamper, H. B., Sancar, A. and Hearst, J. E. (1988). Use of psoralen-modified oligonucleotides to trap three-stranded RecA- DNA complexes and repair of these cross-linked complexes by ABC excinuclease. *J. Biol. Chem.* **263**, 15110-15117.
7. Takasugi, M., Guendouz, A., Chassignol, M., Decout, J. L., Lhomme, J., Thuong, N. T. and Hélène, C. (1991). Sequence-specific photo-induced cross-linking of the

two strands of double-helical DNA by a psoralen covalently linked to a triple helix-forming oligonucleotide. *Proc. Natl. Acad. Sci. USA* **88**, 5602-5606.

8. Havre, P. A., Gunther, E. J., Gasparro, F. P. and Glazer, P. M. (1993). Targeted mutagenesis of DNA using triple helix-forming oligonucleotides linked to psoralen. *Proc. Natl. Acad. Sci. USA* **90**, 7879-7883.

9. Sandor, Z. and Bredberg, A. (1994). Repair of triple helix directed psoralen adducts in human cells. *Nucleic Acids Res.* **22**, 2051-2056.

10. Raha, M., Lacroix, L. and Glazer, P. M. (1998). Mutagenesis mediated by triple helix-forming oligonucleotides conjugated to psoralen: effects of linker arm length and sequence context. *Photochem. Photobiol.* **67**, 289-294.

11. Barre, F.-X. and Harel-Bellan, A. (*unpublished results*).

12. Gasparro, F. P., Havre, P. A., Olack, G. A., Gunther, E. J. and Glazer, P. M. (1994). Site-specific targeting of psoralen photoadducts with a triple helix- forming oligonucleotide: characterization of psoralen monoadduct and crosslink formation. *Nucleic Acids Res.* **22**, 2845-2852.

13. Giovannangeli, C., Thuong, N. T. and Hélène, C. (1992). Oligodeoxynucleotide-directed photo-induced cross-linking of HIV proviral DNA via triple-helix formation. *Nucleic Acids Res.* **20**, 4275-4281.

14. Grigoriev, M., Praseuth, D., Robin, P., Hemar, A., Saison-Behmoaras, T., Dautry-Varsat, A., Thuong, N. T., Hélène, C. and Harel-Bellan, A. (1992). A triple helix-forming oligonucleotide-intercalator conjugate acts as a transcriptional repressor via inhibition of NF kappa B binding to interleukin-2 receptor alpha-regulatory sequence. *J. Biol. Chem.* **267**, 3389-3395.

15. Grigoriev, M., Praseuth, D., Guieysse, A. L., Robin, P., Thuong, N. T., Hélène, C. and Harel-Bellan, A. (1993). Inhibition of gene expression by triple helix-directed DNA cross- linking at specific sites. *Proc. Natl. Acad. Sci. USA* **90**, 3501-3505.

16. Sandor, Z. and Bredberg, A. (1995). Deficient DNA repair of triple helix-directed double psoralen damage in human cells. *FEBS Lett.* **374**, 287-291.

17. Sandor, Z. and Bredberg, A. (1995). Triple helix directed psoralen adducts induce a low frequency of recombination in an SV40 shuttle vector. *Biochim Biophys Acta* **1263**, 235-240.

18. Faruqi, A. F., Seidman, M. M., Segal, D. J., Carroll, D. and Glazer, P. M. (1996). Recombination induced by triple-helix-targeted DNA damage in mammalian cells. *Mol. Cell. Biol.* **16**, 6820-6828.

19. Havre, P. A. and Glazer, P. M. (1993). Targeted mutagenesis of simian virus 40 DNA mediated by a triple helix-forming oligonucleotide. *J. Virol.* **67**, 7324-7331.

20. Raha, M., Wang, G., Seidman, M. M. and Glazer, P. M. (1996). Mutagenesis by third-strand-directed psoralen adducts in repair- deficient human cells: high frequency and altered spectrum in a xeroderma pigmentosum variant. *Proc. Natl. Acad. Sci. USA* **93**, 2941-2946.

21. Gunther, E. J., Havre, P. A., Gasparro, F. P. and Glazer, P. M. (1996). Triplex-mediated, in vitro targeting of psoralen photoadducts within the genome of a transgenic mouse. *Photochem. Photobiol.* **63**, 207-212.

22. Wang, G., Levy, D. D., Seidman, M. M. and Glazer, P. M. (1995). Targeted mutagenesis in mammalian cells mediated by intracellular triple helix formation. *Mol. Cell. Biol.* **15**, 1759-1768.

23. Giovannangeli, C., Thuong, N. T. and Hélène, C. (1993). Oligonucleotide clamps arrest DNA synthesis on a single-stranded DNA target. *Proc. Natl. Acad. Sci. USA* **90**, 10013-10017.
24. Guieysse, A. L., Praseuth, D., Grigoriev, M., Harel-Bellan, A. and Hélène, C. (1996). Detection of covalent triplex within human cells. *Nucleic Acids Res.* **24**, 4210-4216.
25. Giovannangeli, C., Diviacco, S., Labrousse, V., Gryaznov, S., Charneau, P. and Hélène, C. (1997). Accessibility of nuclear DNA to triplex-forming oligonucleotides: the integrated HIV-1 provirus as a target. *Proc. Natl. Acad. Sci. USA* **94**, 79-84.
26. Wang, G. and Glazer, P. M. (1995). Altered repair of targeted psoralen photoadducts in the context of an oligonucleotide-mediated triple helix. *J. Biol. Chem.* **270**, 22595-22601.
27. Barre, F.-X., Mir, L. M., Lécluse, Y. and Harel-Bellan, A. (1998). Highly efficient oligonucleotide transfer into intact yeast cells using square wave pulse electroporation. *Biotechniques* **25**, 294-296.
28. Van Houten, B. (1990). Nucleotide Excision Repair in *Escherichia coli. Microbio. Review* **54**, 18-51.
29. Friedberg, E. C. (1988). Deoxyribonucleic Acid Repair in the Yeast *Saccharomyces cerevisiae. Microbio. Review* **52**, 70-102.
30. Vos, J. M. and Hanawalt, P. C. (1987). Processing of psoralen adducts in an active human gene: repair and replication of DNA containing monoadducts and interstrand cross-links. *Cell* **50**, 789-799.
31. Sage, E., Drobetsky, E. A. and Moustacchi, E. (1993). 8-Methoxypsoralen induced mutations are highly targeted at crosslinkable sites of photoaddition on the non-transcribed strand of a mammalian chromosomal gene. *EMBO J.* **12**, 397-402.
32. Bessho, T., Mu, D. and Sancar, A. (1997). Initiation of DNA interstrand cross-link repair in humans: the nucleotide excision repair system makes dual incisions 5' to the cross- linked base and removes a 22- to 28-nucleotide-long damage-free strand. *Mol. Cell. Biol.* **17**, 6822-6830.
33. Degols, G., Clarenc, J. P., Lebleu, B. and Leonetti, J. P. (1994). Reversible inhibition of gene expression by a psoralen functionalized triple helix forming oligonucleotide in intact cells. *J. Biol. Chem.* **269**, 16933-16937.
34. Bredberg, A., Sandor, Z. and Brant, M. (1995). Mutational response of Fanconi anaemia cells to shuttle vector site- specific psoralen cross-links. *Carcinogenesis* **16**, 555-561.
35. Segal, D. J., Faruqi, A. F., Glazer, P. M. and Carroll, D. (1997). Processing of targeted psoralen cross-links in Xenopus oocytes. *Mol. Cell. Biol.* **17**, 6645-6652.
36. Wang, G., Seidman, M. M. and Glazer, P. M. (1996). Mutagenesis in mammalian cells induced by triple helix formation and transcription-coupled repair. *Science* **271**, 802-805.
37. Faruqi, A. F., Egholm, M. and Glazer, P. M. (1998). Peptide nucleic acid-targeted mutagenesis of a chromosomal gene in mouse cells. *Proc. Natl. Acad. Sci. USA* **95**, 1398-1403.

15 STRUCTURE AND BIOLOGY OF H DNA

Sergei M. Mirkin

HISTORICAL OVERVIEW

The term H DNA was coined in 1985 by Victor Lyamichev, Maxim Frank-Kamenetskii and myself (*1*). Our work was inspired by numerous studies showing that many regulatory regions of eukaryotic genes, both in chromatin and in the cloned state, exhibited hypersensitivity to nucleases and chemicals specific towards single-stranded DNA. Mapping revealed that these hypersensitive sites are commonly located within homopurine-homopyrimidine stretches of varying lengths and base composition (reviewed in *2*). Except for being homopurine-homopyrimidine, those stretches lacked apparent similarities. Several structures were proposed to explain the hypersensitivity of different homopurine-homopyrimidine motifs (*3-11*), but none of them could be satisfactorily applied to all these sequences.

In our study (*1*), we analyzed one such hypersensitive element, a $d(G-A)_{16} \cdot d(T-C)_{16}$ repeat from a spacer between the histone genes of sea urchin, using two-dimensional gel electrophoresis of DNA topoisomers. We showed that this repeat underwent a supercoil-dependent structural transition, and that the resultant structure was responsible for the S1 hypersensitivity. Two observations were particularly important for unraveling the novel structure: (i) the transition was pH-dependent preferably occurring at acidic pH and (ii) the drop in topoisomer mobility accompanying the transition indicated that the novel structure was topologically equivalent to unwound DNA. In order to explain these two features, we originally suggested that the homopyrimidine strand of our repeat forms a hairpin stabilized by protonated C*C base pairs, while the homopurine strand remains unstructured.

Using a similar approach, Pulleyblank and his colleagues studied another homopurine-homopyrimidine repeat (*12*) which also exhibited a pH-dependent structural transition. However, their data indicated that the novel structure was topologically equivalent to only partially unwound DNA, and it interacted with anti-Z antibodies (these two observations were not subsequently confirmed). They suggested that a homopurine-homopyrimidine repeat adopts a peculiar double-helical conformation built of alternating B- and Z-DNA segments.

None of the proposed structures could satisfactorily explain the fine pattern of nuclease sensitivity (*13*). This led us to reconsider our original model and, in 1986, we proposed a different model which turned out to be correct (*14*). The main element of the H DNA structure (Figure 1) is an intramolecular triple helix formed by the entire pyrimidine strand and one half of the purine strand, while the other half of the purine strand remains single-stranded. As one can see from Figure 1, the two complementary DNA strands are not linked in this structure. The triple helix is stabilized by CG*C$^+$ and TA*T base triads (Figure 2A). These triads are isomorphous, allowing good stacking in a triplex. As is obvious from the structure of the triads, all purines must be located in the same DNA strand. The exigency for cytosine protonation leads to the observed pH-dependence of the transition. Finally, the unpaired portion of the homopurine strand primarily accounts for the nuclease and chemical hypersensitivity.

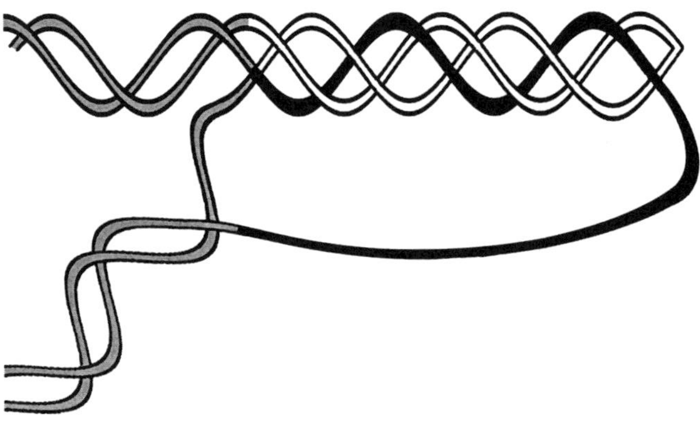

Figure 1. H DNA model. Black ribbon, homopurine strand; white ribbon, homopyrimidine strand; gray ribbons, adjacent double-helical DNA.

It took us another year to realize that the structural isomorphism of CG*C$^+$ and TA*T triads leads to a non-trivial requirement for homopurine-homopyrimidine sequences to form H DNA. Cytosines and thymines from one half of this sequence must interact with CG and TA base pairs, respectively, in its other half. Thus, to form H DNA, a sequence must be a homopurine-homopyrimidine mirror repeat (H palindrome). This hypothesis was proven in 1987 (*15*) using an elegant mutational approach illustrated in Figure 3. We found that the mirror symmetry of a homopurine-homopyrimidine sequence and, consequently, its H-forming ability was destroyed by a single A-to-G point substitution in either half of this sequence. However, a combination of two such substitutions in both halves of the sequence, restored the mirror symmetry and H-DNA formation. This study finally explained the clandestine similarity between natural S1 hypersensitive sequences: they were either perfect mirror repeats or contained extensive palindromic portions. The mutational approach used in this study also proved invaluable in subsequent studies of the biology of H DNA.

By 1988 the accuracy of the H DNA model was proven by several groups, including ours, using DNA probing with chemicals specific towards particular conformations of DNA bases (16-21). Chemicals specific to single-stranded DNA bases, such as osmium tetroxide, chloroacetaldehyde, diethyl pyrocarbonate and potassium permanganate, interacted precisely with those parts of H form which were predicted to be single-stranded, i.e., with the central part of the pyrimidine-rich strand

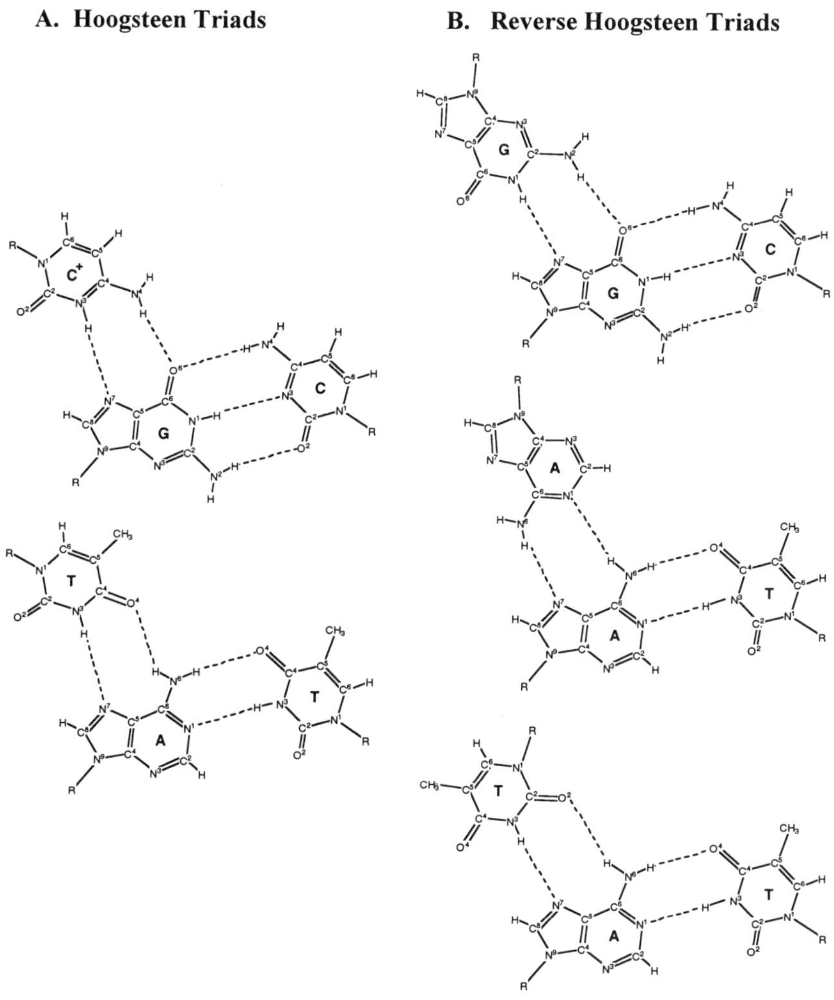

A. Hoogsteen Triads

B. Reverse Hoogsteen Triads

Figure 2. Triplex base triads.

and half of the purine-rich strand. At the same time, the N7-positions of guanines that were involved in the triplex were protected from methylation by dimethyl sulfate. In subsequent years, these chemicals became popular tools for detecting H DNA *in vitro* and *in vivo*.

H DNA became a prototype multistranded DNA structure formed by natural DNA. Since 1988, extensive studies of the structure and biology of H DNA have been conducted all over the world. These efforts revealed substantial structural polymorphism within what soon clearly became a family of similar structures. The effects of H DNA on polymerization processes such as replication and transcription *in vitro* were well understood. Finally, data and ideas about the biological functions of H DNA started to accumulate. In this chapter, I will concentrate on these topics with emphasis on recent developments. Readers interested in a more detailed description of the early data should consult (*22-24*).

STRUCTURAL POLYMORPHISM

H DNA results from the formation of an intramolecular triplex within otherwise double-stranded DNA. The chemical composition of a triplex, and the relative orientation of DNA strands within it, as well as triplex topology, can differ substantially. These differences depend on DNA sequence and ambient conditions. As of today, the H DNA family consists of four members which will be described below.

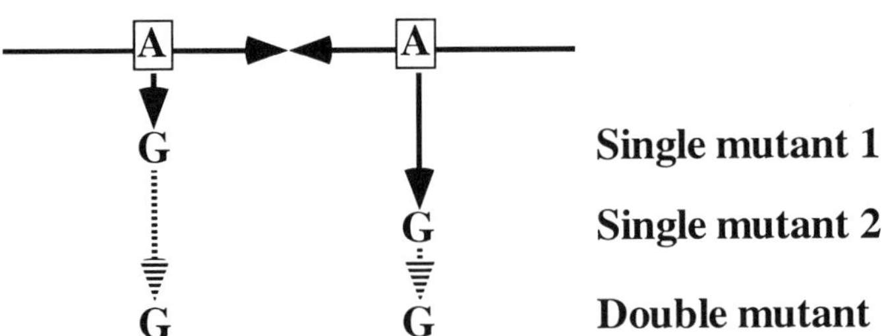

Figure 3. Mutational analysis of mirror symmetry required for H DNA. Horizontal arrows correspond to halves of an H palindrome.

H-y form

H-y structure (Figure 4A) is formed when the homopyrimidine strand of the homopurine-homopyrimidine repeat folds back to form a triplex. The major building blocks of this structure are CG*C+ and TA*T triads (Figure 2A), in which thymines and protonated cytosines form Hoogsteen hydrogen bonds (*25*) with AT and GC base pairs, respectively.

The need for cytosine protonation makes H-y formation dependent on pH (reviewed in *22*). Obviously, this pH dependence differs for different H palindromes depending on their GC content, so that it is the most extreme for $d(G)_n \cdot d(C)_n$ (*26*) and non-existent for the $d(A)_n \cdot d(T)_n$ repeats (*27*). However, only very long $d(A)_n \cdot d(T)_n$

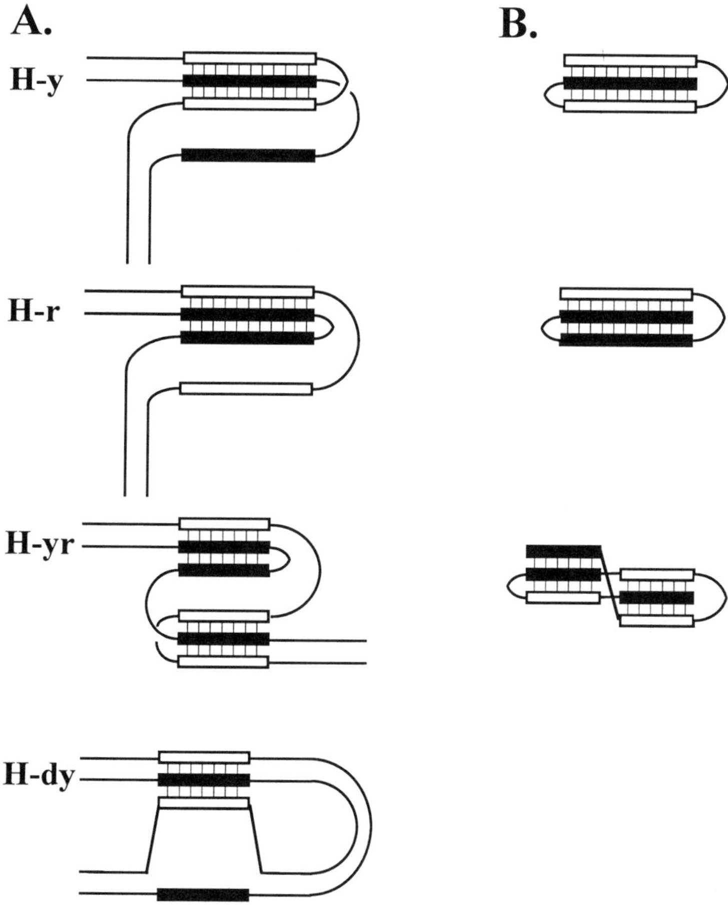

Figure 4. The H DNA family. **A**, different H forms; **B**, corresponding intrastranded triplexes. Black rectangles, purine stretches; white rectangles, pyrimidine stretches; thin lines, adjacent DNA strands.

198

stretches adopt H-y conformation under the influence of high DNA supercoiling (27).
Transition of this sequence into a triplex state might be complicated by its unusually
rigid and stable double-helical configuration characterized by propeller twist and
formation of extra hydrogen bonds (28).

Using a variety of different methods including electrophoretic analysis of
topoisomers, affinity cleavage approaches, DNA melting experiments, *etc.*, it was
demonstrated that incorporation of any of the 14 remaining triads into the H-y triplex
destabilizes it by 3 kcal mol^{-1} (29-35). This triad vocabulary restricts H-y-forming
sequences to perfect homopurine-homopyrimidine mirror repeats (H palindromes),
though individual mismatches can be tolerated.

Structural features of H DNA in supercoiled plasmids were analyzed by indirect
approaches including electrophoresis, combined with chemical (17-21, 36),
enzymatic (13, 17, 37) and photo (38-40) probing. These methods showed that in
H-y conformation, half of the homopurine strand and several bases in the center of
the homopyrimidine strand are single-stranded, consistent with the model.
Methylation protection experiments revealed that the third strand is located in the
major groove and is paired with purines of the duplex (18, 21). Finally, based on
dramatically decreased incidence of photoproduct formation in H DNA compared to
duplex DNA (38-40), it was suggested that the duplex portion of the triplex differs
from the orthodox Watson-Crick (WC) duplex.

By far more refined knowledge of triplex structure came from the NMR analysis
of intrastranded H-y-like triplexes (Figure 4B). In this case, a triplex is formed upon
folding of a single DNA chain containing a homopurine stretch followed by a
complementary homopyrimidine stretch, which is, in turn, succeeded by its
homopyrimidine mirror image (41). These studies, carried out primarily by Feigon
and co-authors (41-43) and Patel and co-authors (44-46), demonstrated the enormous
power of two-dimensional NMR in resolving complex nucleic acid structures. Most
of these results were reviewed in (47, 48), and here I will only outline their main
conclusions.

First, it was directly proven that structures of TA*T and CG*C$^+$ triads correspond
to those shown in Figure 2A. Each triad is stabilized by two Hoogsteen hydrogen
bonds between the third strand and the purine strand of the duplex, which does not
affect the WC base pairing. Two pyrimidines in a triad are antiparallel. The two
triads are very isomorphous, which allows for good stacking interactions in a triple
helix.

Second, incorporation of the third strand substantially changed the helical
parameters of the WC duplex. This is most notably illustrated by a   -2Å
x displacement. This negative value means that the WC base pairs are displaced
towards the minor groove, for the optimal accommodation of the third strand into the
major groove. The x-displacement is accompanied by unwinding of the duplex
resulting in approximately 11.6 base pairs per helical turn. Though the x-
displacement value places H-y triplexes somewhere between the A- and B-DNA
families, sugars in the duplex part of a triplex are definitely of the S-type (in the

C2'-endo range), which is characteristic of B-DNA. Overall, the duplex part of a triplex can be considered as a distorted and unwound B-helix.

The stability of H-y structures was studied by a wide range of approaches. The energetics of B-to-H transitions in supercoiled DNA was analyzed by 2-D electrophoresis of DNA topoisomers (49).

Simple thermodynamic treatment of H-y DNA formation in superhelical DNA (reviewed in 22) shows that superhelical density required for this transition is:

$$-\sigma_{tr} = n/2N + \Delta G/20RT + (G_j - m\Delta G)/20nRT$$

where n is the number of base pairs adopting H-y structure, N is the total number of base pairs in DNA, ΔG is the free energy of a Hoogsteen interaction, G_j is the energy of the B-H junctions, m is the number of bases in the central loop of H-y, R is the gas constant and T is temperature.

This allows one to estimate G_j by comparing the superhelical densities of B-to-H transition for regular homopurine-homopyrimidine repeats of varying lengths. For both $d(G-A)_n{\cdot}d(T-C)_n$ and $d(G)_n{\cdot}d(C)_n$ repeats, the energy of the H-y junctions appeared to be 18 kcal mol^{-1}. However, attempts to assign any ΔG value to individual base triads failed, indicating that nearest neighbor effects must be taken into consideration.

The energetic characteristics of intrastranded and intermolecular H-y-like triplexes were studied by calorimetry, quantitative affinity cleavage, electrophoretic mobility assays, UV absorbance spectroscopy, etc. (33, 34, 50-56). The relative advantages and disadvantages of these approaches with regard to triplex studies were discussed in an excellent review by Breslauer and co-authors (57). Below, I will summarize the conclusions from these studies and discuss major recent developments.

Analysis of enthalpy of triplex formation gave ΔH values ranging between -2 kcal mol^{-1} to -8 kcal mol^{-1} per Hoogsteen interaction. This deviation may not be that surprising, considering that these experiments were done for different sequences and in different ambient conditions. Since enthalpy reflects the stacking interactions between adjacent DNA bases, the nearest neighbor effect should be taken into consideration in assigning enthalpy values to specific DNA sequences. An attractive model was described in (54). The ΔH of an H-y triplex was calculated as the simple sum of enthalpies for dinucleotide steps CC, TC/CT and TT. The predicted ΔH values for different triplexes closely matched ΔH values determined experimentally.

The entropy penalty for a triad formation is around 13 cal mol^{-1} K^{-1}. The ratio of the ΔS values for duplexes and triplexes was close to 2:3, reflecting the strandedness of these two states. This supports the idea that the entropy change is caused primarily by changes in the conformation entropy of the sugar phosphate backbone, rather than by the DNA sequence.

As mentioned above, the free energy of an H-y triplex can not be predicted without taking into consideration the nearest neighbor effects. A satisfactory model predicting the free energy of this structure depending on its sequence has been developed recently (54), based on the analysis of 23 different intermolecular triplexes. The free energy (kcal mol^{-1}) at 37°C was calculated as:

$$\Delta G = -3.00(C) - 0.65(T) + 1.65(CC) + (C)(pH - 5.0)(1.26 - 0.08(CC)) + 6.0$$

where (C) and (T) are the number of cytosines and thymines, respectively, and (CC) is the number of CC dinucleotides. The physical sense of the above equation is that at pH 5.0, the $CG*C^+$ triad is very energetically favorable ($\Delta G = -3$ kcal mol^{-1}), while the TA*T triad is not ($\Delta G = -0.65$ kcal mol^{-1}). However, triplex stability for C-rich sequences is undermined by two factors: (i) there is a 1.65 kcal mol^{-1} penalty for each CC-contact, and (ii) the stability of $CG*C^+$ triads decreases by roughly 1 kcal mol^{-1} per pH unit above 5.0. This, paradoxically, makes $d(C-T)_n \cdot d(A-G)_n$ repeats the best H-y-forming sequences at physiological pH.

As mentioned above, the H-y form is topologically equivalent to unwound DNA. It can exist in two isoforms, H-y5 and H-y3, depending on the part of the homopyrimidine strand donated to the triplex. The isoforms for the H-DNA family will be discussed at the end of this section.

H-r form

The H-r structure (Figure 4A), first described by Kohwi and Kohwi-Shigematsu in 1988 (*19*), is also known as *H (*58, 59*) and H' DNA (*19*) in the literature. The structure originally described was formed by $d(G)_n \cdot d(C)_n$ repeats where the $d(G)_n$ strand folded back to form a triplex. Consequently, this structure was made of the CG*G triads (Figure 2B) where guanines in the third strand formed reverse Hoogsteen hydrogen bonds (*25*) with the GC base pairs. Further work showed that two other triads could easily be incorporated into the H-r triplex: TA*A (*58, 60*) and, surprisingly, TA*T (*61, 62*) (Figure 2B). In those triads, adenines and thymines of the third strand form reverse Hoogsteen hydrogen bonds with AT base pairs.

Different approaches, including affinity cleavage and DNA melting experiments, revealed that all other triads destabilize H-r triplexes by approximately 4 kcal mol^{-1} (*63, 64*). This results in rather peculiar sequence requirements for H-r DNA: guanines in the purine-rich strand should be arranged in a mirror-repeated way, while adenines in one half of the purine-rich strand could be reflected by either adenines or thymines in its other half (*61*). Therefore, unlike H-y triplexes, the H-r conformation can be adopted by sequences that are neither homopurine-homopyrimidine, nor mirror repeats. The fact that the third strand in H-r DNA does not have to be entirely homopurine makes the term H-r somewhat awkward. However it could withstand, assuming that *r* stands for *reverse* Hoogsteen.

Unlike H-y DNA, H-r structure is stable at neutral pH in the presence of divalent metal cations. Surprisingly, different divalent metal cations are required to stabilize H-r structures depending on their triad composition (*19, 58, 59, 65-67*). It is believed that these cations coordinate with the N7 position of purines in the third strand, rather than serve as trivial counterions (*67, 68*). Such coordination might lead to the differential stabilization of different reverse Hoogsteen bonds by different cations.

The structure of H-r DNA in supercoiled plasmids was analyzed using the same approaches that were utilized for studying the H-y form. These approaches genuinely confirmed that half of the homopyrimidine strand and the center of the purine-rich strand are single-stranded (*19, 58-60, 69, 70*).

The understanding of the fine structure of an H-r DNA came from NMR analysis of intrastranded H-r-type triplexes (Figure 4B) (*71-74*). To form such a structure, a single DNA chain must have a tricky design, which is due to the higher sequence versatility of H-r triplexes. Specifically, a homopyrimidine stretch must be followed by its homopurine complement, and then by a purine-rich stretch. Guanines in this last stretch are mirror reflections of their counterparts in the central homopurine segment, while either adenines or thymines reflect adenines in the central homopurine segment. Since these studies have been reviewed extensively (*48*), I will only briefly summarize their conclusions.

The third strand of an H-r triplex is anti-parallel to the homopurine strand of the duplex. Each of CG*G, TA*A and TA*T triads is stabilized by two reverse Hoogsteen hydrogen bonds (Figure 2B). The three triads of this triplex are not isomorphous, which is evident from the very different positions of the C1' atoms of the third base in the different triads.

The features of a duplex within the H-r triplex are very similar to those in the H-y triplex. It is characterized by the negative x-displacement value (-2Å) and significant unwinding (approximately 12 bp per helical turn). As in the H-y structure, sugars in the duplex part of the H-r triplex are of the S-type as in B-DNA.

However, there are important differences between the structures of the two triplexes that are due to the lack of isomorphism between the H-r-forming triads, contrasting with perfectly isomorphous H-y triads. In fact, the neighboring bases in the third strand of the H-r structure overlap so poorly that this causes profound structural perturbations in the phosphodiester torsion angles between neighboring bases of the third strand. The extent of these perturbations depends on the base composition of the triplex. TpG steps are shown to be the most constrained.

The energetics of H-r triplexes is less well studied. The energy of H-r junctions in supercoiled DNA was never determined, although it is not expected to differ substantially from that of the H-y structure.

There are only a limited number of studies on the thermodynamics of intra- and intermolecular H-r triplexes. The general consensus is that these triplexes are more stable than their H-y counterparts at near physiological conditions. A strong argument favoring this conclusion comes from the comparison of melting curves of H-y and H-r structures. H-y triplexes usually melt in a biphasic manner (except under extremely acidic conditions), where the third strand dissociates at temperatures far below those needed for duplex dissociation (*51*). H-r triplexes, in contrast, melt as a single transition at temperatures which are higher than or equal to the duplex T_m. This persists over a broad range of Mg^{2+} concentrations (*64, 75, 76*). Thus, these triplexes must be at least as stable as the corresponding duplexes. It is also apparent from melting studies that CG*G triads are the main players in stabilizing H-r triplexes, since the stability increases dramatically with CG*G triad incidence (*75*).

Several studies on the stability of intermolecular H-r triplexes were performed using calorimetric and spectroscopic approaches (*76-79*). Most of these results, however, were obtained at a rather high Mg^{2+} concentration, making their extrapolation to the physiological situation difficult. Nevertheless, the conclusions were that ΔH -4 kcal mol^{-1} and ΔS 7 cal mol^{-1} K^{-1} for a reverse Hoogsteen interaction. Consistent with the notion of elevated stability of H-r triplexes, in the same set of experiments the enthalpy of Hoogsteen interaction for an H-y triplex was significantly lower, -2.4 kcal mol^{-1}.

Many issues concerning the stability of H-r triplexes remain to be addressed. It is obvious that the comparative analysis of different sequences is crucial to address the nearest neighbor effects. Also, the relative contribution of TA*A and TA*T triads to triplex stability remains controversial. While calorimetric and spectroscopic analysis indicated that TA*T triads are less stable (77), the quantitative affinity cleavage experiments indicated the opposite (62, 63).

Similarly to H-y, H-r DNA is topologically equivalent to unwound DNA (80). It can also exist in two isoforms, which will be discussed below.

H-yr form

The third member of the H-family might be called H-yr DNA (Figure 4A), though it is also called "nodule" DNA in the literature (69, 81). It is a chimera of H-y and H-r triplexes. To form an H-yr structure, 2/3 of the pyrimidine strand must form an H-y triplex with 1/3 of the purine strand, while the remaining third of the pyrimidine strand forms an H-r triplex with 2/3 of the purine strand. The advantage of this structure over both H-y and H-r forms is that it contains a very few unpaired bases: only those at the tips of the triplexes. The disadvantage is that it has three energetically costly junctions: two triplex-to-duplex and one triplex-to-triplex. This consideration explains the experimental facts that the H-yr conformation was only detected for very long homopurine-homopyrimidine stretches (69, 81). Indeed, the lack of single-stranded regions inevitable in canonical H DNA is a big gain for long repeats, and the energy cost for junctions is easily compensated by the release of a large number of DNA supercoils. The ambient conditions required for H-yr formation are dictated by its integral parts: it is stable under mild acidic pH in the presence of divalent cations (69).

H-dy form

The last member of the H-family might be called H-dy, where d stands for *distance*. Currently it is also called a "tethered loop" in the literature (Figure 4A). Formation of an H-dy structure in supercoiled DNA has been detected electrophoretically and by electron microscopy (82, 83). The proposed structure involves two separate homopurine-homopyrimidine clusters, and the triplex is formed between a homopyrimidine strand of one cluster and both strands of the distant cluster. This structure is reminiscent of an early model for S1 hypersensitivity in the human thyroglobulin gene (9, 84). The mechanisms of H-dy formation are unknown, but the structure is topologically feasible. H-dr structure could, in principle, be formed, but it has never been detected experimentally.

Isomerization

As mentioned above, all members of the H DNA family can exist in two isoforms, named by the 5' or 3' half of a strand donated to the triplex (which can be either homopyrimidine in H-y structures or purine-rich in H-r structures). Most of the data were obtained for the H-y structure. It was realized very early on that the H-y3 isoform dominates in DNA with physiological superhelical density (σ -0.05) (*17-21, 36, 37*). An understanding of this phenomenon came from the pioneering study of Htun and Dahlberg (*85*). They found that the H-y3 isoform is energetically more favorable under high torsional stress because it releases one extra supercoil as compared to the H-y5 structure. At low superhelical stress, in contrast, the H-y5 isoform predominates. To explain this observation, it was suggested that the nucleation topology of the two H-y isoforms is unequal. In the absence of torsional stress, H-y5 is sterically more favorable, whereas at high torsional stress significant unwinding of the prenucleation center both promotes H-y3 formation and causes the release of an extra supercoil.

Subsequently, it was found that H DNA isomerization could be affected by many factors. For example, the H-y5 isoform is best stabilized by divalent cations, whereas the H-y3 isoform favors monovalent cations (*86, 87*). Further, the loop sequence plays an important role in determining isomerization direction: GC-rich loops favor H-y5 formation (*88*).

Recent progress in understanding the isoform paradox came from kinetic studies of bi-molecular triplexes (*89*). These structures were formed between a mirror repeated homopyrimidine sequence and a homopurine fragment complementary to one half of it. Depending on whether the homopurine stretch was complementary to the 5'- or the 3'-half of the homopyrimidine sequence, H y3- or H y5-like triplexes, respectively, were formed. Careful analysis of these structures revealed that their thermodynamic stability is identical within experimental error. At the same time, they are quite different kinetically: the H y3 isomer forms more than an order of magnitude faster than its H y5 counterpart. Remarkably, however, the addition of extra bases to the homopurine segment, extending past the pyrimidine loop (much as seen by the purine stretch in H DNA), slowed down formation of the H y3 structure 200-fold.

An elegant mechanism explaining all the above observations was proposed by Roberts and Crothers (*89*) and is shown in Figure 5. At low superhelical stress, when the DNA duplex is not grossly unwound, H-y5 nucleation can be achieved by a simple overwinding kink of the homopyrimidine strand. At the same time, H-y3 nucleation is difficult, since extending the purine strand obstructs the path for the triplex-forming half of the pyrimidine strand to the major groove of a duplex. At high torsional stress, however, the situation changes. Duplex unwinding clears a path to the major groove for the third strand in the H-y3 conformation, and the first triad can be formed upon a simple underwinding kink. The necessity for path clearing in H-y3 formation nicely explains its prevalence at high superhelicity, as well as its exclusion under conditions of increased duplex stability.

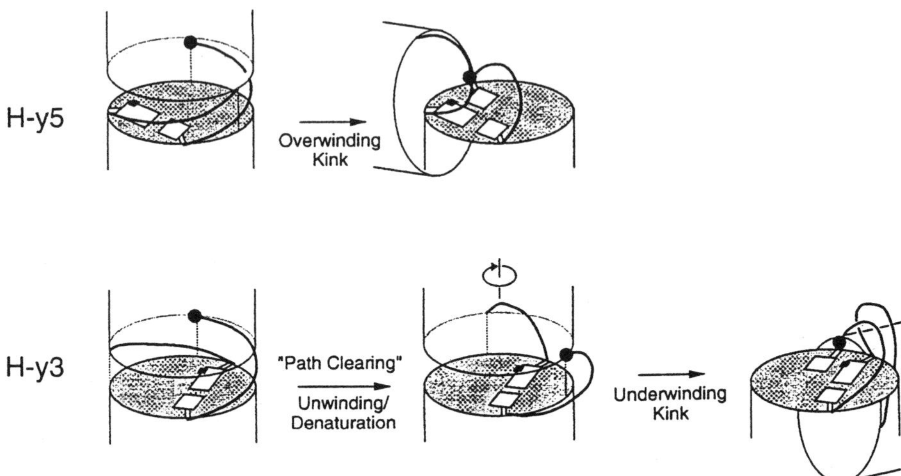

Figure 5. Proposed mechanism for H-y3 and H-y5 formation. Reprinted from (*89*) by permission of Academic Press.

The H-r structure can also exist in two isoforms, H-r5 and H-r3, depending on which half of the purine-rich strand is donated to the triplex. For H-r triplexes built of CG*G and TA*A triads, H-r3 isoform forms predominantly at physiological levels of DNA supercoiling (19, 59, 60, 70). This can probably be attributed to the same factors that determine H-y3 isoform prevalence at high supercoiling density. Note, however, that for the H-r structures containing TA*T triads, only one isoform is allowed, since only the third (but not the central) strand of a triplex can contain thymines (61).

H-yr form can, in principle, exist in two isoforms. This question was never addressed experimentally in supercoiled DNA. Analysis of this question using the intrastrand H-yr-like triplexes (Figure 4B) (90) revealed that these two isoforms are energetically very unequal (91). In this case it might primarily depend on factors other than prenucleation topology. To form such a triplex, the third strand must cross the major groove on its way from the H-y to the H-r half or vice versa. This strand switching going in the direction 3'-Rn-Yn-5' along the third strand is allowed, whereas switching in the direction 3'-Yn-Rn-5' is very unfavorable (91).

THE BIOLOGY OF H DNA

Polymerization suicide at H-forming sequences

During the last several years substantial attention has been paid to the peculiar interplay between H-forming sequences and different polymerases. In their pioneering studies, Manor and co-authors found that d(G-A)$_n$ or d(C-T)$_n$ repeats in single-stranded

DNA templates cause partial termination of DNA polymerization at the center of those repeats (*92, 93*). It was hypothesized that when the newly synthesized DNA chain reaches the center of the homopolymer sequence, the remaining half of the homopolymer stretch can fold back, forming a stable triplex (Figure 6A). As a

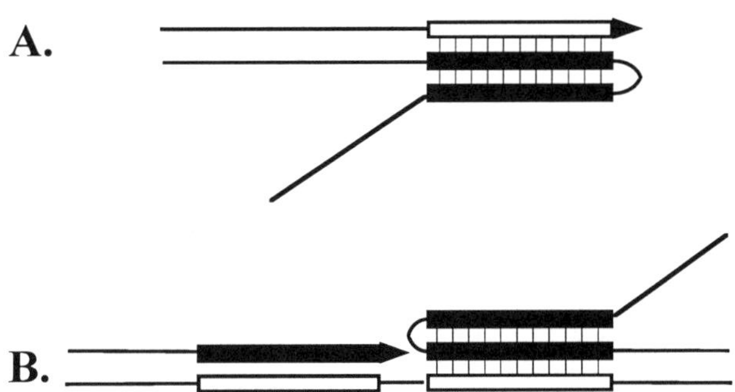

Figure 6. Conformational suicide of DNA polymerization at triplex-forming sequences. **A**, single-stranded template; **B**, double-stranded template. Black rectangles, purine stretches; white rectangles, pyrimidine stretches; arrows show 3'-ends of growing DNA chains.

result, DNA polymerase is trapped and further elongation stops. Similar blockage of DNA polymerization was shown for other H-forming sequences, most recently $d(T)_n$ clusters (*94*).

A different mechanism of polymerization blockage was discovered in our lab (*95*). In open circular DNA templates, DNA polymerization stopped almost completely at the center of H-r-forming sequences. This only happened when the homopyrimidine strand served as a template. Using the mutational approach described above, we showed that the symmetry of these sequences which is required for H-r DNA formation was also required for the polymerization block. Thus, the formation of the H-r structure during DNA polymerization caused polymerization arrest. The mechanism is presented in Figure 6B. During DNA synthesis on a double-stranded template, the DNA polymerase must displace the non-template DNA strand. When the displaced segment contains the purine-rich half of an H-r motif, it can fold back to form an intramolecular triplex downstream of the polymerase. This, in turn, stops further polymerization.

These two examples illustrate a crucial feature of polymerase-triplex interplay. While the original DNA template may not even contain any triplex structure, the action of the polymerase, itself, can create one. And as soon as this structure is formed, it will block all subsequent polymerization. This interplay can be called conformational suicide.

There are also important kinetic differences between the two mechanisms. Polymerization can only be blocked if a triplex is formed prior to the polymerization of the 5'-half of an H-forming sequence. Thus, on single-stranded DNA templates, where polymerization rate is high, one would expect only partial termination, whereas on double-stranded DNA templates, where the polymerization rate is at least 10-fold slower, one would expect a near complete blockage. This prediction has been confirmed experimentally (*93, 95*).

Obviously, in supercoiled DNA templates, where H-r DNA exists prior to polymerization, one should expect the polymerase to stop upon encountering this structure. This was shown in (*61*). When the polymerase approached different isoforms of H-r DNA from different directions, termination always occurred at the triplex edge.

Several possible mechanisms for triplex-caused polymerization blockage have been discussed. First, under polymerization conditions, triplexes may be so much more stable than duplexes that the triplex blockage of polymerization is a simple reflection of their persistence. Second, the kinetics of polymerase passage through triplexes may be much slower than through duplexes, simulating polymerization blockage. Finally, DNA polymerases, while capable of dismantling duplexes, may be unable to do so with triplexes.

Our recent study (*75*) distinguished among these possibilities. We used single-stranded DNA templates containing intrastranded H-r triplexes or control duplexes and studied the efficiency of Vent DNA polymerase passage at different temperatures and time intervals. In parallel, the stabilities of different triplex and duplex structures were determined in DNA melting experiments. Comparisons of the data revealed that, although triplexes were slightly more stable than the corresponding duplexes, this effect was unlikely to account for the dramatic differences in temperatures (20°C to 40°C) at which the polymerase traverses these structures. Projection of polymerase passage temperatures onto melting curves for different structures showed that the polymerase passes triplex barriers at temperatures where they start to dissociate, whereas duplexes are overcome far below their dissociation temperatures (Figure 7). This shows that the DNA polymerase can actively untangle duplexes in DNA templates but lacks such an ability for triplex structures.

It is not unreasonable to expect similarly complex relationships between H-forming sequences and other polymerizing enzymes, especially RNA polymerases. This question has been less extensively pursued, but there are some provocative data. RNA polymerase was shown to stall immediately after several H-forming repeats (*96-99*). This stalling was profoundly orientation dependent, occurring only when the transcript carried an oligopurine or an oligopyrimidine stretch for G- and A-rich H palindromes, respectively. Further, a portion of the RNA transcript corresponding to the H palindrome was apparently involved in some kind of interaction with its DNA template: it was resistant to RNase A but cleaved by RNase H (*97, 98*). Finally, the formation of a stalled complex was accompanied by the release of supercoils equivalent to the unwinding of an H palindrome (*96, 98*). Therefore, it is believed that transcriptional stall is caused by the formation of a stable complex between the H palindrome and its transcript.

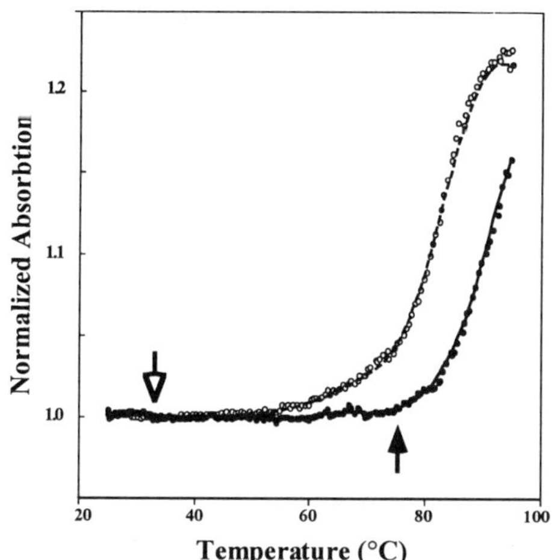

Figure 7. Comparison of melting curves for a triplex and a duplex with the temperatures of DNA polymerase passage through these structures. Filled circles, triplex melting data; open circles, duplex melting data; arrows, temperatures of polymerase passage (filled, triplex template; unfilled, duplex template).

The exact nature of this complex remains unclear, and several possibilities are currently considered. Figure 8 shows two structures that seem the most adequate. First, since RNA polymerization generates negative supercoiling upstream of the enzyme, H DNA could be formed and additionally stabilized by an RNA transcript which might bind to its single-stranded portion (*96, 98*). Formation of such complex immediately upstream of the RNA polymerase might attenuate its propagation. The second model, called collapsed R-loop, proposes that transcription through a homopurine-homopyrimidine sequence creates an unusually long and stable R-loop. The non-template DNA strand can then associate with this loop, forming additional hydrogen bonds with either the DNA or the RNA strand as possible (*97*). Current data are insufficient to distinguish between those structures or to explain why RNA polymerization stalls upon their formation. This is such an important problem, however, that major breakthroughs might be expected in the near future.

Representation of H-forming sequences in natural DNAs

It was quickly recognized that certain homopurine-homopyrimidine repeats, for example $d(G-A)_n \cdot d(T-C)_n$, are very abundant in eukaryotic DNA (*100-102*).

208

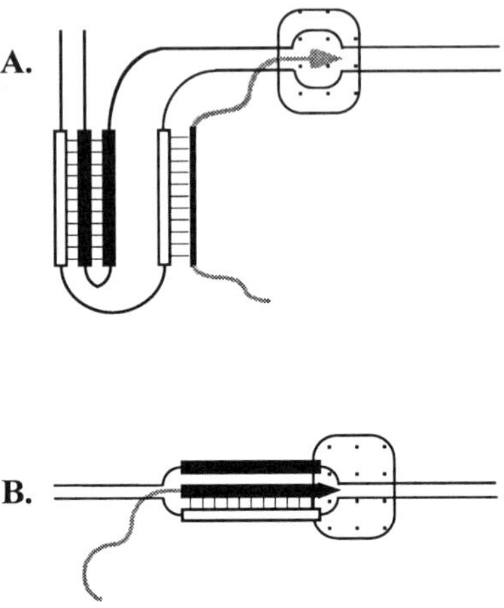

Figure 8. Hypothetical structures responsible for transcription blockage at homopurine-homopyrimidine sequences. **A**, transcript-stabilized H DNA; **B**, collapsed R-loop. Black rectangles, purine stretches; white rectangles, pyrimidine stretches; arrows, 3'-ends of growing RNA chains; patterned rectangle, RNA polymerase; dotted line, RNA transcript; thin lines, DNA strands.

However, the first systematic study of this question was carried out only recently by Schroth and Ho (*103*). They found that H palindromes are overrepresented in eukaryotes, but occur at chance frequencies in prokaryotes. They also concluded that overrepresentation of H palindromes increases from yeast to human DNA, where they occur approximately once per 40 kb.

A potential problem with the above study was that the chance occurrence for H palindromes was calculated assuming that natural DNAs are homogeneously 50% GC-rich, which is far from true. In order to address this problem, we compared the frequencies of H-y and H-r forming sequences with their chance occurrence calculated on the basis of local GC-content in many different DNAs (*104*). In agreement with the previous study, we found that both H-y- and H-r-forming sequences are over-represented in eukaryotes, but not in bacteria. Unlike in the previous study, however, the adjusted overrepresentation of H palindromes appeared to be very similar in all eukaryotes studied. One might hypothesize, therefore, that H palindromes are widespread in the eukaryotic branch of evolution.

In supercoiled DNA, the H-forming ability of a homopurine-homopyrimidine repeat exponentially depends on its length (reviewed in *22*). It was important, therefore, to evaluate the frequency of H-palindrome occurrence depending on their lengths. Our analysis showed that long H palindromes are the most overrepresented (up to 10^9 fold over chance) repeats in eukaryotes. Remarkably, normalizing the frequencies of H palindrome occurrence to their expected incidence gave a nearly perfect exponential dependence on their length. It is plausible to speculate, therefore, that the overrepresentation of H palindromes may correspond simply to their probability of H DNA formation.

Existence of H DNA in vivo

Since H palindromes are only enriched in eukaryotes, it would be most relevant to analyze H DNA formation in eukaryotic cells. However, the direct detection of H DNA in those cells is very difficult due to the high complexity of genomic DNA. Therefore, most of the data addressing this question were obtained for *E.coli*.

These studies gave several important conclusions. First, it appeared that the level of DNA supercoiling *in vivo* is the major limiting factor for H DNA formation. The transient appearance of H DNA was observed in exponentially growing *E.coli* cells (*105*), but was much more pronounced when intracellular DNA supercoiling increased due to mutations in the topo I gene (*105*), or due to treatment of cells with chloramphenicol (*106, 107*). Second, environmental conditions during *E. coli* growth also significantly contributed to the formation of triplexes. H-y form was greatly favored when cells were growing in mildly acidic media (*106*), where H-r structures were mainly observed in cells growing in media with a high concentration of Mg^{2+} (*107*). Third, besides the steady-state level of DNA supercoiling, the local level of supercoiling depends on transcription, since elongating RNA polymerases create domains of high negative and positive supercoiling upstream and downstream of them, respectively (reviewed in *108*). It was found that transcription-driven negative supercoiling promotes H-r-DNA formation in $d(G)_n \cdot d(C)_n$ stretches (*109*). These results unequivocally show that H DNA can be formed in normal intracellular conditions.

As mentioned above, the detection of H DNA in eukaryotic cells is very difficult. Substantial progress in this direction was achieved using the triplex-specific monoclonal antibodies Jel 318 and Jel 466 developed by Lee and his colleagues (*110-112*). These antibodies were found to interact with eukaryotic chromosomes so that their staining pattern correlated with G and Hoechst 33258 banding. The interpretation of these results in terms of H DNA existence *in vivo* is complicated by the potentially severe chromatin distortions caused by the antibody staining procedure. More recently the same group addressed this concern by showing that triplex-specific antibodies bind to agarose encapsulated, permeabilized nuclei, where the chromatin is presumably intact (*113*).

Altogether the above data show that potential H-forming sequences are very common elements of eukaryotic genomes. There are no significant impediments to H DNA formation *in vivo* given some supercoiling. In fact, triple-stranded DNA seems to exist in eukaryotic chromosomes *in vivo*. It is reasonable to believe, therefore, that H DNA could be involved in some processes of DNA metabolism.

Credible data attributing H DNA to different genetic processes are still scarce, but there are several convincing and interesting cases which are discussed below.

H DNA in transcription

Historically, interest in H DNA first arose from studies of eukaryotic promoters. This led to many studies of H DNA in transcription. Over the last ten years, the formation of H DNA in many eukaryotic promoters *in vitro* was unequivocally proven. Among those promoters are *Drosophila hsp*26 (*114*), chicken β^A-globin (*65*) and malik enzyme (*115*), mouse c-Ki-*ras* (*116*), metallothionein-I (*117*) and TGF-β3 (*118*), human EGFR (*119*), γ-globin (*120*), *ets*-2 (*121*), *MUC* 1 (*122*), and Na,K-ATPase α2 (*123*), and many others. In most cases, the H-forming elements were essential for the promoter function (*115, 119, 124, 125*), though in some promoters they were dispensable (*126*). It is far from clear, however, whether a homopurine-homopyrimidine sequence per se, or its ability to form H DNA is required in a given promoter.

For several promoters, including *Drosophila hsp*26 (*127*), and particularly mouse c-Ki-*ras* (*124*), careful mutational analysis revealed that the primary sequence of the homopurine-homopyrimidine stretch, rather than its mirror symmetry and H-forming ability, is essential for transcription. These results represent a powerful argument against the role of H DNA in promoter functioning. However, their interpretation may not be so simple.

The main complication is that homopurine-homopyrimidine sequences in eukaryotic promoters serve as binding sites for nuclear proteins involved in transcription regulation. Several transcription factors that bind to homopurine-homopyrimidine sequences were described, including BPG1 (*128*), NSEP-1 (*129*), MAZ (*130*), nm23-H2 (*131*), PYBP (*132*), Pur-1 (*133*), BP-8 (*134*), etc. Peculiarly, these proteins often prefer to bind to the separate purine- or pyrimidine strands of an H motif rather than to double-stranded DNA (*128-130, 132-134*). Therefore, these proteins can bind to H DNA or even promote its formation. For example, BP-8 protein was shown to bind H DNA in a supercoiled plasmid (*134*). At the same time, the above proteins are, at least to some extent, sequence specific. This unusual combination of sequence and structure specificity could make the above mutational analysis rather misleading. In addition, there are several reports describing proteins which bind to the putative triplex part of H DNA (*135-137*), although their primary structure and function are unknown.

This consideration led to the popular hypothesis that H palindromes in eukaryotic promoters could represent molecular switches between the "on" and "off" states of those promoters, where the equilibrium between B and H conformations is largely determined by conformation-specific transcription factors (reviewed in *23, 24*).

One of the most interesting examples supporting this idea is the human fetal γ-globin promoter. In this promoter, an H-forming element is located 200 bp upstream of the transcriptional start site (*120*) and serves as a binding site for factor BP-8 (*134*). This promoter is normally active during embryonic development but turns off in adulthood. In the disease called hereditary persistence of fetal hemoglobin

(HPFH), this developmental switching does not occur, leading to the production of fetal globin in adults. Several human mutations leading to HPFH are located within an H-forming element and destroy its H-forming ability (*120, 138*). Therefore, it was suggested that H DNA might be responsible for the repression of the fetal gene. Another very recent example is the promoter of the human androgen receptor gene (*139*). Here the H-forming element could interact with two proteins: Sp1, which binds a double-stranded H palindrome, and ssPyrBF, which recognizes its single pyrimidine strand. The Sp1 factor is crucial for the activation of this promoter. It is believed, therefore, that ssPyrBF can switch this promoter into an H conformation, preventing Sp1 binding and thus leading to transcriptional activation.

Though the above data and hypotheses are very provocative and may have profound biological implications, they still do not constitute proof for a role for H DNA in promoter function.

As discussed in the previous section, there are interesting indications that H-forming sequences block transcription elongation *in vitro*. There are several reports showing that these sequences block transcription *in vivo*, as well. Brahmachari and his colleagues created constructs where different H palindromes were incorporated in frame into the 5'-part of the lacZ gene. These led to an approximately 2-fold decrease in lacZ expression in *E. coli* cells, due to partial transcription blockage (*140, 141*). The extent of transcription blockage slightly depended on the H-palindrome orientation within the transcription unit. A more extreme case of orientation-dependent transcription blockage by H-forming elements in bacteria was described in our recent study (*142*). We found that $d(G)_n \cdot d(C)_n$ stretches block transcription elongation with 30%-to-80% efficiency when the $r(G)_n$ stretch, but not the $r(C)_n$ stretch, was a part of the RNA transcript. A transcriptional stall occurred immediately after the repeated segment and was very likely due to the formation of a three-stranded structure between the $r(G)_n$ sequence in the RNA transcript and its DNA template. The exact nature of this structure remains to be established.

H DNA in replication

In 1987 we suggested that H DNA could be formed in eukaryotic replication origins, providing the initial unwinding needed for replication initiation (*15*). The support of this hypothesis was lacking until a recent persuasive study of Epstein-Barr virus replication (*143*). It was found that one of the viral replication origins, OriLyt, consists of two parts: an upstream part essential for replication, and a downstream part which contains an H palindrome. *In vitro*, the latter sequence preferentially adopts an H-r DNA conformation. To understand the role of this element in OriLyt-dependent replication *in vivo*, Portes-Sentes *et al.* (*143*) used the mutational approach described above. Remarkably, it was found that mutations destroying the sequence's H-forming ability impair replication, while the compensatory mutations restoring H DNA potential restore viral replication as well.

As discussed in the previous section, H DNA arrests DNA polymerization *in vitro*. Thus it was very tempting to speculate that H-forming sequences in eukaryotic genomes can serve as pause sites for replication *in vivo*. There are still only fragmentary data supporting this idea. In a model system using episomal replication driven by the SV40 origin, several homopurine-homopyrimidine inserts inhibited replication (*144, 145*). Direct visualization of replication fork pausing within a $d(G\text{-}A)_n \cdot d(T\text{-}C)_n$ insert *in vivo* was demonstrated using two-dimensional electrophoresis of replication intermediates (*146*).

Though these data makes H DNA involvement in the replication elongation control look promising, they don't prove it. The problem is that other factors, such as protein binding, changes in the chromatin structure (*147*), or the interplay of several DNA metabolic processes might be responsible for replication inhibition. Our recent study on the effects of $d(G)_n \cdot d(C)_n$ repeats on DNA replication in bacterial cells is a good illustration of the complexity of the situation (*142*). Using two-dimensional separation of replication intermediates, it was directly shown that $d(G)_n \cdot d(C)_n$ repeats block replication elongation in a length- and orientation-dependent manner. It appeared, however, that the replication blockage was primarily dependent on transcription through the repeated DNA. When the long $r(G)_n$ stretch existed in an RNA transcript, the RNA polymerase was stalled (see above) and this, in turn, abrogated replication fork progression.

Very recently, another exciting possibility linking H DNA to DNA replication and genome instability was described by McMurray and co-authors (*148*). This group studies a new type of mutations, called dynamic mutations, caused by trinucleotide repeats which are responsible for many human neurological and neuromuscular diseases (reviewed in *149-151*). Briefly, when the number of those repeats exceeds a threshold of 30, they start to expand in length, and each expansion makes the likelihood of a subsequent expansion higher (anticipation phenomenon). Eventually, repeated DNA stretches become so long that they shut down the expression of the genes carrying them. Although the exact mechanism of expansion is unknown, it is believed to be caused by an abnormal replication of the repeated DNAs (*152*). One disease, Friedreich Ataxia, is caused by expansion of a $d(GAA)_n$ repeat situated in the intron of the frataxin gene (*153*). It was found that, *in vitro*, this repeat preferentially forms the H-y structure even at neutral pH. Further, formation of the H-y-like triplex by this repeat halts DNA polymerization *in vitro* (*148*). The authors suggested a model for repeat expansion (Figure 9) based on an assumption that the replication fork may pause within the long $d(GAA)_n \cdot d(TTC)_n$ stretches *in vivo*. This might promote dissociation of template and daughter DNA strands, and lead to the formation of an H-y triplex. If this structure is formed by folding back the newly synthesized pyrimidine-rich strand, continuation of DNA replication would lead to repeat expansion.

H DNA in recombination

There are a few indications that H DNA might affect genetic recombination *in vivo*. In *E. coli* cells, it was found that long $d(G)_n \cdot d(C)_n$ stretches located upstream of an inducible promoter formed H-r DNA upon transcriptional induction. This, in turn,

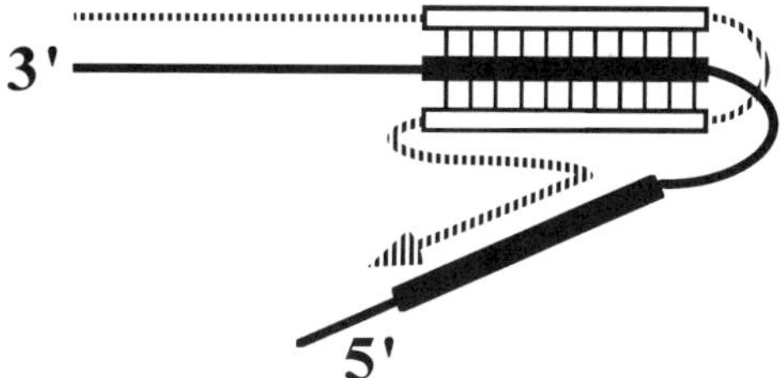

Figure 9. Model for (GAA)$_n$-repeat expansion via triplex formation (adopted from (*148*) with modifications). Black rectangles, purine stretches; white rectangles, pyrimidine stretches; solid line, DNA template; dotted line, daughter DNA chain.

stimulated recombination between short direct repeats flanking the d(G)$_n$·d(C)$_n$ sequences (*109*). In cultured mammalian cells, H-forming repeats were found to increase the level of homologous recombination by 3 to 5-fold (*154*). The molecular mechanisms responsible for those effects are unknown.

CONCLUDING REMARKS

As one can see from this chapter, the structural features of the H DNA family are understood in substantial detail. The definite proof for the biological function of H DNA is, however, still lacking. Summarizing numerous studies on the biology of H DNA, one can see a change in their direction. While originally most studies attempted to define a unique biological function specific for H DNA (with a transcription regulatory switch as the prime suspect), this does not look feasible anymore. The three-stranded nature of H DNA provides rich biological opportunities and makes the formation of this structure possible in the course of many different genetic processes. This plasticity of H DNA could be utilized by different machines involved in DNA metabolism. It seems hopeful that many more interesting studies on the biology of H DNA are to follow.

ACKNOWLEDGMENTS

I thank my wife, Elena, for many years of unconditional support that made my studies of H DNA possible; my principal collaborators in Moscow, Victor Lyamichev, Oleg Voloshin and Maxim Frank-Kamenetskii for their pivotal role in unraveling H DNA structure; the current and former members of my lab in Chicago, Andrew Dayn, George Samadashwily, Sergei Tevosian, Ganapathirama Raghu,

Randal Cox, Andrei Krasilnikov, Maria Krasilnikova and Gordana Raca for their invaluable role in studying the biology of H DNA; Randal Cox for critical reading of the manuscript, Donald Crothers for his kind permission to use one of his figures, and Gerald Buldak for editorial help. Supported by grants from the National Science Foundation (MCB-9723924) and the National Institutes of Health (GM 54247).

REFERENCES

1. Lyamichev, V. I., Mirkin, S. M. and Frank-Kamenetskii, M. D. (1985). A pH-dependent structural transition in the homopurine-homopyrimidine tract in superhelical DNA. *J. Biomol. Struct. Dyn.* **3**, 327-338.

2. Wells, R. D., Collier, D. A., Hanvey, J. C., Shimizu, M. and Wohlrab, F. (1988).The chemistry and biology of unusual DNA structures adopted by oligopurine·oligo-pyrimidine sequences. *FASEB. J.* **2**, 2939-2949.

3. Hentchel, C. C. (1982). Homocopolymer sequences in the spacer of a sea urchin histone gene repeat are sensitive to S1 nuclease. Nature **29,** 714-716.

4. Mace, H. A. F., Pelham, H. R. B. and Travers, A. (1983). Association of an S1 nuclease-sensitive structure with short direct repeats 5' of Drosophila heat shock genes. *Nature* **304**, 555-557.

5. McKeon, C., Schmidt, A. and de-Crombrugghe, B. A. (1984). A sequence conserved in both the chicken and mouse $\alpha 2(I)$ collagen promoter contains sites sensitive to S1 nuclease. *J. Biol. Chem.* **259**, 6636-6640.

6. Shen, C. K. (1983). Superhelicity induces hypersensitivity of a human polypyrimidine.polypurine DNA sequence in the human $\alpha 2$-$\alpha 1$ globin intergenic region to S1 nuclease digestion. *Nucleic Acids Res.* **11**, 7899-7910.

7. Cantor, C. R. and Efstratiadis, A. (1984). Possible structures of homopurine-homopyrimidine S1-hypersensitive sites. *Nucleic Acids Res.* **12**, 8059-8072.

8. Margot, J. B. and Hardison, R. C. (1985). DNase I and nuclease S1 sensitivity of the rabbit β-globin gene in nuclei and in supercoiled plasmids. *J. Mol. Biol.* **184**, 195-210.

9. Christophe, D., Cabrer, B., Bacolla, A., Targovnik, H., Pohl, V. and Vassart, G. (1985). An unusually long poly(purine)-poly(pyrimidine) sequence is located upstream from the human thyroglobulin gene. *Nucleic Acids Res.* **13**, 5127-5144.

10. Lee, J. S., Johnson, D. A. and Morgan, A. R. (1979). Complexes formed by (pyrimidine)$_n$·(purine)$_n$ DNAs on lowering the pH are three-stranded. *Nucleic Acids Res.* **6**, 3073-3091.

11. Johnson, D. and Morgan, A. R. (1978). Unique structures formed by pyrimidine-purine DNAs which may be four-stranded. *Proc. Natl. Acad. Sci. USA* **75**, 1637-1641.

12. Pulleyblank, D. E., Haniford, D. B. and Morgan, A. R. (1985). A structural basis for S1 nuclease sensitivity of double-stranded DNA. *Cell* **42**, 271-280.

13. Htun, H., Lund, E. and Dahlberg J. E. (1984). Human U1 RNA genes contain an unusually sensitive nuclease S1 cleavage site within the conserved 3' flanking region. *Proc. Natl. Acad. Sci. USA* **81**, 7288-7292.

14. Lyamichev, V. I., Mirkin, S. M. and Frank-Kamenetskii M. D. (1986). Structures of homopurine-homopyrimidine tract in superhelical DNA. *J. Biomol. Struct. Dynam.* **3**, 667-669.

15. Mirkin, S. M., Lyamichev, V. I., Drushlyak, K. N., Dobrynin, V. N., Filippov, S. A. and Frank-Kamenetskii, M. D. (1987). DNA H form requires a homopurine-homopyrimidine mirror repeat. *Nature* **330**, 495-497.

16. Hanvey, J. C., Shimizu, M. and Wells, R. D. (1988). Intramolecular DNA triplexes in supercoiled plasmids. *Proc. Natl. Acad. Sci. USA* 85, 6292-6296.

17. Htun, H. and Dahlberg, J. E. (1988). Single strands, triple strands and kinks in H-DNA. *Science* 241, 1791-1796.

18. Johnston, B. H. (1988). The S1-sensitive form of $d(C-T)_n \cdot d(A-G)_n$: chemical evidence for a three-stranded structure in plasmids. *Science* 241, 1800-1804.

19. Kohwi, Y. and Kohwi-Shigematsu, T. (1988). Magnesium ion-dependent triple-helix structure formed by homopurine-homopyrimidine sequences in supercoiled plasmid DNA. *Proc. Natl. Acad. Sci. USA* 85, 3781-3785.

20. Vojtiskova, M., Mirkin, S., Lyamichev, V., Voloshin, O., Frank-Kamenetskii, M. and Palecek, E. (1988). Chemical probing of the homopurine·homopyrimidine tract in supercoiled DNA at single-nucleotide resolution. *FEBS Lett.* 234, 295-299.

21. Voloshin, O. N., Mirkin, S. M., Lyamichev, V. I., Belotserkovskii, B. P. and Frank-Kamenetskii, M. D. (1988). Chemical probing of homopurine-homopyrimidine mirror repeats in supercoiled DNA. *Nature* 333, 475-476.

22. Mirkin, S. M. and Frank-Kamenetskii, M. D. (1994). H-DNA and related structures. *Annu. Rev. Biophys. Biomol. Struct.* 23, 541-576.

23. Sinden, R. R. DNA structure and function. San Diego: Academic Press, 1994.

24. Soyfer, V. N. and Potaman, V. N. Triple-helical nucleic acids. New York: Springer-Verlag, 1996.

25. Hoogsteen, K. (1963). The crystal and molecular structure of a hydrogen-bonded complex between 1 methylthymine and 9 methyladenine. *Acta Cryst.* 16, 907-916.

26. Lyamichev, V. I., Mirkin, S. M. and Frank-Kamenetskii, M. D. (1987). Structure of $(dG)n \cdot (dC)n$ under superhelical stress and acid pH. *J. Biomol. Struct. Dynam.* 5, 275-282.

27. Fox, K. R. (1990). Long $(dA)_n \cdot (dT)_n$ tracts can form intramolecular triplexes under superhelical stress. *Nucleic Acids Res.* 18, 5387-5391.

28. Nelson, H. C., Finch, J. T., Luisi, B. F. and Klug, A. (1987). The structure of an oligo(dA)·oligo(dT) tract and its biological implications. *Nature* 330, 221-226.

29. Fossella, J. A., Kim, Y. J., Shih, H., Richards, E. G. and Fresco, J. R. (1993). Relative specificities in binding of Watson-Crick base pairs by 3rd strand residues in a DNA pyrimidine triplex motif. *Nucleic Acids Res.* 21, 4511-4515.

30. Belotserkovskii, B. P., Veselkov, A. G., Filippov, S. A., Dobrynin, V. N., Mirkin, S. M. and Frank-Kamenetskii, M. D. (1990). Formation of intramolecular triplex in homopurine-homopyrimidine mirror repeats with point substitutions. *Nucleic Acids Res.* 18, 6621-6624.

31. Macaya, R. F., Gilbert, D. E., Malek, S., Sinsheimer, J. S. and Feigon, J. (1991). Structure and stability of XGC mismatches in the third strand of intramolecular triplexes. *Science* 254, 270-274.

32. Mergny, J. L., Sun, J. S., Rougee, M., Montenay-Garestier, T., Barcelo, F., Chomilier, J. and Helene, C. (1991). Sequence specificity in triple-helix formation, experimental and theoretical studies of the effect of mismatches on triplex stability. *Biochemistry* 30, 9791-9798.

33. Roberts, R. W. and Crothers, D. M. (1991). Specificity and stringency in DNA triplex formation. *Proc. Natl. Acad. Sci. USA* 88, 9397-9401.

34. Singleton, S. F. and Dervan, P. B. (1992). Thermodynamics of oligodeoxyribonucleotide-directed triple-helix formation at single DNA sites. *J. Am. Chem. Soc.* 114, 6957-6965.

35. Yoon, K., Hobbs, C. A., Koch, J., Sardaro, M., Kutny, R. and Weis, A. L. (1992). Elucidation of the sequence-specific third-strand recognition of four Watson-Crick base pairs in a pyrimidine triple-helix motif, T·AT, C·GC, T·CG, and G·TA. *Proc. Natl. Acad. Sci. USA* 89, 3840-3844.

36. Hanvey, J. C., Klysik, J. and Wells, R. D. (1988). Influence of DNA sequence on the formation of non-B right-handed helices in oligopurine·oligopyrimidine inserts in plasmids. *J. Biol. Chem.* **263**, 7386-7396.

37. Hanvey, J. C., Shimizu, M. and Wells, R. D. (1990). Site-specific inhibition of EcoRI restriction/modification enzymes by a DNA triple helix. *Nucleic Acids Res.* **18**, 157-161.

38. Tang, M.-S., Htun, H., Cheng, Y. and Dahlberg, J. E. (1991). Supression of cylobutane and <6-4> dipyrimidines formation in triple-stranded H-DNA. *Biochemistry* **30**, 7021-7026.

39. Lyamichev, V. I., Voloshin, O. N., Frank-Kamenetskii, M. D. and Soyfer, V. N. (1991). Photofootprinting of DNA triplexes. *Nucleic Acids Res.* **19**, 1633-1638.

40. Lyamichev, V. I., Frank-Kamenetskii, M. D. and Soyfer, V. N. (1990). Protection against UV-induced pyrimidine dimerization in DNA by triplex formation. *Nature* **344**, 568-570.

41. Sklenar, V. and Feigon, J. (1990). Formation of a stable triplex from a single DNA strand. *Nature* **345**, 836-838.

42. Wang, E., Malek, S. and Feigon, J. (1992). Structure of a G.T.A triplet in an intramolecular DNA triplex. *Biochemistry* **31**, 4838-4846.

43. Macaya, R., Wang, E., Schultze, P., Sklenar, V. and Feigon, J. (1992). Proton nuclear magnetic resonanse assignments and structural characterization of an intramolecular DNA triplex. *J. Mol. Biol.* **225**, 755-773.

44. Radhakrishnan, I., Gao, X., de-los-Santos, C., Live, D. and Patel, D. J. (1991). NMR structural studies of intramolecular $(Y+)_n(R+)_n(Y-)_n$ DNA triplexes in solution: imino and amino proton and nitrogen markers of GTA base triple formation. *Biochemistry* **30**, 9022-9030.

45. Radhakrishnan, I. and Patel, D. J. (1994). Solution structure of a pyrimidine· purine·pyrimidine DNA triplex containing T·AT, C^+·GC and G·TA triples. *Structure* **2**, 17-32.

46. Radhakrishnan, I. and Patel, D. J. (1994). Solution structure and hydration patterns of a pyrimidine·purine·pyrimidine DNA triplex containing a novel T·CG base-triple. *J. Mol. Biol.* **241**, 600-619.

47. Feigon, J., Koshlap, K. M. and Smith, F. W. (1995). 1H NMR spectroscopy of DNA triplexes and quadruplexes. *Meth. Enzymol.* **261**, 225-255.

48. Radhakrishnan, I. and Patel, D. J. (1994). DNA triplexes: solution structures, hydration sites, energetics, interactions, and function. *Biochemistry* **33**, 11405-11416.

49. Lyamichev, V. I., Mirkin, S. M., Kumarev, V. P., Baranova, L. V., Vologodskii, A. V. and Frank-Kamenetskii, M. D. (1989). Energetics of the B-H transition in supercoiled DNA carrying $d(CT)_x \cdot d(AG)_x$ and $d(C)_n \cdot d(G)_n$ inserts. *Nucleic Acids Res.* **17**, 9417-9423.

50. Plum, G. E., Park, Y. W., Singleton, S. F., Dervan, P. B. and Breslauer, K. J. (1990). Thermodynamic characterization of the stability and the melting behavior of a DNA triplex: a spectroscopic and calorimetric study. *Proc. Natl. Acad. Sci. USA* **87**, 9436-9440.

51. Plum, G. E. and Breslauer, K. J. (1995). Thermodynamics of an intramolecular DNA triple helix: a calorimetric and spectroscopic study of the pH and salt dependence of thermally induced structural transitions. *J. Mol. Biol.* **248**, 679-695.

52. Manzini, G., Xodo, L. E., Gasparotto, D., Quadrifoglio, F., van-der-Marel, G. A. and van-Boom, J. H. (1990). Triple helix formation by oligopurine-oligopyrimidine DNA fragments. Electrophoretic and thermodynamic behavior. *J. Mol. Biol.* **213**, 833-843.

53. Pilch, D. S., Brousseau, R. and Shafer, R. H. (1990). Thermodynamics of triple helix formation: spectrophotometric studies on the $d(A)_{10} \cdot 2d(T)_{10}$ and $d(C^+_3T_4C^+_3) \cdot d(G_3A_4G_3) \cdot d(C_3T_4C_3)$ triple helices. *Nucleic Acids Res.* **18**, 5743-5750.

54. Roberts, R. W. and Crothers, D. M. (1996). Prediction of the stability of DNA triplexes. *Proc. Natl. Acad. Sci. USA* **93**, 4320-4325.

55. Volker, J., Botes, D. P., Lindsey, G. G. and Klump, H. H. (1993). Energetics of a stable intramolecular DNA triple helix formation. *J. Mol. Biol.* **230**, 1278-1290.

56. Xodo, L. E., Manzini, G. and Quadrifoglio, F. (1990). Spectroscopic and calorimetric investigation on the DNA triplex formed by d(CTCTTCTTTCTTTTCTTTCTTCTC) and d(GAGAAGAAAGA) at acidic pH. *Nucleic Acids Res.* **18**, 3557-3564.

57. Plum, G. E., Pilch, D. S., Singleton, S. F. and Breslauer, K. J. (1995). Nucleic acid hybridization: triplex stability and energetics. *Annu. Rev. Biophys. Struct. Biol.* **24**, 319-350.

58. Bernues, J., Beltran, R., Casasnovas, J. M. and Azorin, F. (1990). DNA-sequence and metal-ion specificity of the formation of *H-DNA. *Nucleic Acids Res.* **18**, 4067-4073.

59. Beltran, R., Martinez-Balbas, A., Bernues, J., Bowater, R. and Azorin, F. (1993). Characterization of the zinc-induced structural transition to *H-DNA at a $d(GACT)_{22}$ sequence. *J. Mol. Biol.* **230**, 966-978.

60. Bernues, J., Beltran, R., Casasnovas, J. M. and Azorin, F. (1989). Structural polymorphism of homopurine-homopyrimidine sequences: the secondary DNA structure adopted by a $d(GACT)_{22}$ sequence in the presence of zinc ions. *EMBO J.* **8**, 2087-2094.

61. Dayn, A., Samadashwily, G. M. and Mirkin, S. M. (1992). Intramolecular DNA triplexes: unusual sequence requirements and influence on DNA polymerization. *Proc. Natl. Acad. Sci. USA* **89**, 11406-11410.

62. Beal, P. A. and Dervan, P. B. (1991). Second structural motif for recognition of DNA by oligonucleotide-directed triple-helix formation. *Science* **251**, 1360-1363.

63. Beal, P. A. and Dervan, P. B. (1992). The influence of single base triplet changes on the stability of a purpurpyr triple helix determined by affinity cleaving. *Nucleic Acids Res.* **20**, 2773-2776.

64. Vo, T., Wang, S. and Kool, E. T. (1995). Targeting pyrimidine single strands by triplex formation: structural optimization of binding. *Nucleic Acids Res.* **23**, 2937-2944.

65. Kohwi, Y. (1989). Cationic metal-specific structures adopted by the poly(dG) region and the direct repeats in the chicken adult βA globin gene promoter. *Nucleic Acids Res.* **17**, 4493-4502.

66. Malkov, V. A., Voloshin, O. N., Soyfer, V. N. and Frank-Kamenetskii, M. D. (1993). Cation and sequence effects on stability of intermolecular pyrimidine-purine-purine triplex. *Nucleic Acids Res.* **21**, 585-591.

67. Martinez-Balbas, A. and Azorin, F. (1993). The effect of zinc on the secondary structure of d(GATC)n DNA sequences of different length: a model for the formation *H-DNA. *Nucleic Acids Res.* **21**, 2557-2562.

68. Potaman, V. N. and Soyfer, V. N. (1994). Divalent metal cations upon coordination to the N7 of purines differentially stabilize the PyPuPu DNA triplex due to unequal hoogsteen-type hydrogen bond enhancement. *J. Biomol. Struct. Dynam.* **11**, 1035-1040.

69. Kohwi-Shigematsu, T. and Kohwi, Y. (1991). Detection of triple-helix related structures adopted by poly(dG)-poly(dC) sequences in supercoiled plasmid DNA. *Nucleic Acids Res.* **19**, 4267-4271.

70. Kohwi, Y. and Kohwi-Shigematsu, T. (1993). Structural polymorphism of homopurine-homopyrimidine sequences at neutral pH. *J. Mol. Biol.* **231**, 1090-1101.

71. Radhakrishnan, I., de-los-Santos, C. and Patel, D. J. (1991). Nuclear magnetic resonance structural studies of intramolecular purine·purine·pyrimidine DNA triplexes in solution. Base triple pairing alignments and strand direction. *J. Mol. Biol.* **221**, 1403-1418.

72. Radhakrishnan, I. and Patel, D. J. (1993). Solution structure of a purine·purine· pyrimidine DNA triplex containing G·GC and T·AT triples. *Structure* **1**, 135-152.

73. Radhakrishnan, I., de los Santos, C. and Patel, D. J. (1993). NMR structural studies of A·AT base triple alignments in intramolecular purine·purine·pyrimidine DNA triplexes in solution. *J. Mol. Biol.* **234**, 188-197.

74. Radhakrishnan, I. and Patel, D. J. (1994). Hydration sites in purine.purine. pyrimidine and pyrimidine·purine·pyrimidine DNA triplexes in aqueous solution. *Structure* **2**, 395-405.

75. Krasilnikov, A. S., Panyutin, I. G., Samadashwily, G. M., Cox, R., Lazurkin, Y. S. and Mirkin, S. M. (1997). Mechanisms of triplex-caused polymerization arrest. *Nucleic Acids Res.* **25**, 1339-1346.

76. Scaria, P. V., Will, S., Levenson, C. and Shafer, R. H. (1995). Physicochemical studies of the $d(G_3T_4G_3){*}d(G_3A_4G_4){\cdot}d(C_3T_4C_3)$ triple helix. *J. Biol. Chem.* **270**, 7295-7303.

77. He, Y., Scaria, P. V. and Shafer, R. H. (1997). Studies on formation and stability of the $d[G(AG)_5]{*}d[G(AG)_5]{\cdot}d[C(TC)_5]$ and $d[G(TG)_5]{*}d[G(AG)_5]{\cdot}d[C(TC)_5]$ triplex helices. *Biopolymers* **41**, 431-441.

78. Scaria, P. V. and Shafer, R. H. (1996). Calorimetric analysis of triple helices targeted to the $d(G_3A_4G_3){\cdot}d(C_3T_4C_3)$ duplex. *Biochemistry* **35**, 10985-10994.

79. Pilch, D. S., Levenson, C. and Shafer, R. H. (1991). Structure, stability, and thermodynamics of a short intermolecular purine-purine-pyrimidine triple helix. *Biochemistry* **30**, 6081-6088.

80. Panyutin, I. G., Kovalsky, O. I. and Budowsky, E. I. (1989). Magnesium-dependent supercoiling-induced transition in $(dG)_n{\cdot}(dC)_n$ stretches and formation of a new G-structure by $(dG)_n$ strand. *Nucleic Acids Res.* **17**, 8257-8271.

81. Panyutin, I. G. and Wells, R. D. (1992). Nodule DNA in the $(GA)_{37}{\cdot}(CT)_{37}$ insert in superhelical plasmids. *J. Biol. Chem.* **267**, 5495-5501.

82. Hampel, K. J., Ashley, C. and Lee, J. S. (1994). Kilobase-range communication between polypurine·polypyrimidine tracts in linear plasmids mediated by triplex formation - a braided knot between two linear duplexes. *Biochemistry* **33**, 5674-5681.

83. Lee, J. S., Ashley, C., Hampel, K. J., Bradley, R. and Scraba, D. G. (1995). A stable interaction between separated pyrimidine.purine tracts in circular DNA. *J. Mol. Biol.* **252**, 283-288.

84. Christophe, D. (1988). H-form DNA and the hairpin-triplex model [letter]. *Nature* **333**, 214.

85. Htun, H. and Dahlberg, J. E. (1989). Topology and formation of triple-stranded H-DNA. *Science* **243**, 1571-1576.

86. Kang, S., Wohlrab, F. and Wells, R. D. (1992). Metal ions cause the isomerization of certain intramolecular triplexes. *J. Biol. Chem.* **267**, 1259-1264.

87. Kang, S. and Wells, R. D. (1992). Central non-pur.pyr sequences in oligo (dG·dC) tracts and metal ions influence the formation of intramolecular DNA triplex isomers. *J. Biol. Chem.* **267**, 20887-20891.

88. Shimizu, M., Kubo, K., Matsumoto, U. and Shindo, H. (1994). The loop sequence plays crucial roles for isomerization of intramolecular DNA triplexes in supercoiled plasmids. *J. Mol. Biol.* **235**, 185-197.

89. Roberts, R. W. and Crothers, D. M. (1996). Kinetic discrimination in the folding of intramolecular triple helices. *J. Mol. Biol.* **260**, 135-146.

90. Jayasena, S. D. and Johnston, B. H. (1992). Intramolecular triple-helix formation at $(Pu_nPy_n)\cdot(Pu_nPy_n)$ tracts: recognition of alternate strands via Pu·PuPy and Py·PuPy base triplets. *Biochemistry* **31**, 320-327.

91. Jayasena, S. D. and Johnston, B. H. (1993). Sequence limitations of triple helix formation by alternate-strand recognition. *Biochemistry* **32**, 2800-2807.

92. Lapidot, A., Baran, N. and Manor, H. (1989). $(dT-dC)_n$ and $(dG-dA)_n$ tracts arrest single stranded DNA replication in vitro. *Nucleic Acids Res.* **17**, 883-900.

93. Baran, N., Lapidot, A. and Manor, H. (1991). Formation of DNA triplexes accounts for arrests of DNA synthesis at $d(TC)_n$ and $d(GA)_n$ tracts. *Proc. Natl. Acad. Sci. USA* **88**, 507-511.

94. Mikhailov, V. S. and Bogenhagen, D. F. (1996). Termination within oligo(dT) tracts in template DNA by DNA polymerase gamma occurs with formation of a DNA triplex structure and is relieved by mitochondrial single-stranded DNA-binding protein. *J. Biol. Chem.* **271**, 30774-30780.

95. Samadashwily, G. M., Dayn, A. and Mirkin, S. M. (1993). Suicidal nucleotide sequences for DNA polymerization. *EMBO J.* **12**, 4975-4983.

96. Reaban, M. E. and Griffin, J. A. (1990). Induction of RNA-stabilized DNA conformers by transcription of an immuniglobulin switch region. *Nature* **348**, 342-344.

97. Reaban, M. E., Lebowitz, J. and Griffin, J. A. (1994). Transciption induces the formation of a stable RNA·DNA hybrid in the immunoglobulin α switch region. *J. Biol. Chem.* **269**, 21850-21857.

98. Grabczyk, E. and Fishman, M. C. (1995). A long purine-pyrimidine homopolymer acts as a transcriptional diode. *J. Biol. Chem.* **270**, 1791-1797.

99. Kiyama, R. and Oishi, M. (1996). In vitro transcription of a poly(dA)·poly(dT)-containing sequence is inhibited by interaction between the template and its transcripts. *Nucleic Acids Res.* **24**, 4577-4583.

100. Beasty, A. M. and Behe, M. J. (1988). An oligopurine sequence bias occurs in eukaryotic viruses. *Nucleic Acids Res.* **16**, 1517-1528.

101. Manor, H., Sridhara-Rao, B. and Martin, R. G. (1988). Abundance and degree of dispersion of genomic $d(GA)_n\cdot d(TC)_n$ sequences. *J. Mol. Evol.* **27**, 96-101.

102. Tripathi, J. and Brahmachari, S. K. (1991). Distribution of simple repetitive $(TG/CA)_n$ and $(CT/AG)_n$ sequences in human and rodent genomes. *J. Biomol. Struct. Dynam.* **9**, 387-397.

103. Schroth, G. P. and Ho, P. S. (1995). Occurrence of potential cruciform and H-DNA forming sequences in genomic DNA. *Nucleic Acids Res.* **23**, 1977-1983.

104. Cox, R. and Mirkin, S. M. (1997). Characteristic enrichment of DNA repeats in different genomes. *Proc. Natl. Acad. Sci. USA* **94**, 5237-5242.

105. Ussery, D. W. and Sinden, R. R. (1993). Environmental influences on the in vivo level of intramolecular triplex DNA in Escherichia coli. *Biochemistry* **32**, 6206-6213.

106. Karlovsky, P., Pecinka, P., Vojtiskova, M., Makaturova, E. and Palecek, E. (1990). Protonated triplex DNA in E. coli cells as detected by chemical probing. *FEBS Lett.* **274**, 39-42.

107. Kohwi, Y., Malkhosyan, S. R. and Kohwi-Shigematsu, T. (1992). Intramolecular dG·dCdC triplex detected in Escherichia coli cells. *J. Mol. Biol.* **223**, 817-822.

108. Wang, J. C. (1996). DNA topoisomerases. Annu. Rev. Biochem. **65**, 635-692.

109. Kohwi, Y. and Panchenko, Y. (1993). Transcription-dependent recombination induced by triple-helix formation. *Genes Dev.* **7**, 1766-1778.

110. Burkholder, G. D., Latimer, L. J. and Lee, J. S. (1988). Immunofluorescent staining of mammalian nuclei and chromosomes with a monoclonal antibody to triplex DNA. *Chromosoma* **97**, 185-192.

111. Lee, J. S., Latimer, L. J., Haug, B. L., Pulleyblank, D. E., Skinner, D. M. and Burkholder, G. D. (1989). Triplex DNA in plasmids and chromosomes. *Gene* **82**, 191-199.

112. Agazie, Y. M., Lee, J. S. and Burkholder, G. D. (1994). Characterization of a new monoclonal antibody to tripler DNA and immunofluorescent staining of mammalian chromosomes. *J. Biol. Chem.* **269**, 7019-7023.

113. Agazie, Y. M., Burkholder, G. D. and Lee, J. S. (1996). Triplex DNA in the nucleus: direct binding of triplex-specific antibodies and their effect on transcription, replication and cell growth. *Biochem. J.* **316**, 461-466.

114. Gilmour, D. S., Thomas, G. H. and Elgin, S. C. R. (1989). Drosophila nuclear proteins bind to regions of alternaring C and T residues in gene promoters. *Science* **245**, 1487-1490.

115. Xu, G. and Goodridge, A. G. (1996). Characterization of a polypyrimidine/ polypurine tract in the promoter of the gene for chicken malic enzyme. *J. Biol. Chem.* **271**, 16008-16019.

116. Pestov, D. G., Dayn, A., Siyanova, E., George, D. L. and Mirkin, S. M. (1991). H-DNA and Z-DNA in the mouse c-Ki-*ras* promoter. *Nucleic Acids Res.* **19**, 6527-6532.

117. Bacolla, A. and Wu, F. Y.-H. (1991). Mung bean nucelase cleavage pattern at a polypurine-polypyrimidine sequence upstream from the mouse metallothionein gene. *Nucleic Acids Res.* **19**, 1639-1647.

118. Lafyatis, R., Denhez, F., Williams, T., Sporn, M. and Roberts, A. (1991). Sequence specific protein binding to and activation of the TGF-β3 promoter through a repeated TCCC motif. *Nucleic Acids Res.* **19**, 6419-6425.

119. Johnson, A. C., Jinno, Y. and Merlino, G. T. (1988). Modulation of epidermal growth factor receptor proto-oncogene transcription by a promoter site sensitive to S1 nuclease. *Mol. Cell. Biol.* **8**, 4174-4184.

120. Ulrich, M. J., Gray, W. J. and Ley, T. J. (1992). An intramolecular DNA triplex is disrupted by point mutations associated with hereditary persistence of fetal hemoglobin. *J. Biol. Chem.* **267**, 18649-18658.

121. Mavrothalassitis, G. L., Watson, D. K. and Papas, T. S. (1990). Molecular and functional characterization of the promoter of ETS2, the human *c-ets-2* gene. *Proc. Natl. Acad. Sci. USA* **87**, 1047-1051

122. Nelson, K. L., Becker, N. A., Pahwa, G. S., Hollingsworth, M. A. and Maher, L. J. (1996). Potential for H-DNA in the human MUC1 mucin gene promoter. *J. Biol. Chem.* **271**, 18061-18067.

123. Potaman, V. N., Ussery, D. W. and Sinden, R. H. (1996). Formation of combined H-DNA/open TATA box structure in the promoter sequence of the human Na,K-ATPase α2 gene. *J. Biol. Chem.* **271**, 13441-13447.

124. Raghu, G., Tevosian, S., Anant, S., Subramanian, K., George, D. L. and Mirkin, S. M. (1994). Transcriptional activity of the homopurine-homopyrimidine repeat of the c-Ki-*ras* promoter is independent of its H-forming potential. *Nucleic Acids Res.* **22**, 3271-3279.

125. Tewari, D. S., Cook, D. M. and Traub, R. (1990). Characterization of the promoter region and 3' end of the human insulin receptor gene. *J. Biol. Chem.* **264**, 16238-16245.

126. Becker, N. A. and Maher, L. J. (1998). Characterization of a polypurine/polypyrimidine sequence upstream of the mouse metallothionein-I gene. *Nucleic Acids Res.* **26**, 1951-1958.

127. Glaser, R. L., Thomas, G. H., Siegfried, E., Elgin, S. C. and Lis, J. T. (1990). Optimal heat-induced expression of the Drosophila hsp26 gene requires a promoter sequence containing $(CT)_n \cdot (GA)_n$ repeats. *J. Mol. Biol.* **211**, 751-761.

128. Clark, S. P., Lewis, C. D. and Felsenfeld, G. (1990). Properties of BGP1 a poly(dG)-binding protein from chicken erythrocytes. *Nucleic Acids Res.* **18**, 5119-5126.

129. Kolluri, R., Torrey, T. A. and Kinniburgh, A. J. (1992). A CT promoter element binding protein: definition of a double-strand and a novel single-strand DNA binding motif. *Nucleic Acids Res.* **20**, 111-116.

130. Bossone, S. A., Asselin, C., Patel, A. J. and Marcu, K. B. (1992). MAZ, a zinc finger protein, binds to c-MYC and C2 gene sequences regulating transcriptional initiation and termination. *Proc. Natl. Acad. Sci. USA* **89**, 7452-7456.

131. Postel, E. H., Berberich, S. J., Flint, S. J. and Ferrone, C. A. (1993). Human *c myc* transcription factor PuF identified as nm23-H2 nucleoside diphosphate kinase, a candidate supressor of tumor metastasis. *Science* **261**, 478-483.

132. Brunel, F., Alzari, P. M., Ferrara, P. and Zakin, M. M. (1991). Cloning and sequencing of PYBP, a pyrimidine-rich specific single strand DNA-binding protein. *Nucleic Acids Res.* **19**, 5237-5245.

133. Kennedy, G. C. and Rattner, J. B. (1992). Pur-1 P a zinc-finger protein that binds to purine-rich sequences, transactivates an insulin promoter in heterologous cells. *Proc. Natl. Acad. Sci. USA* **89**, 11498-11502.

134. Horwitz, E. M., Maloney, K. A. and Ley, T. J. (1994). A human protein containing a "cold shock" domain binds specifically to H-DNA upstream from the human gamma-globin genes. *J. Biol. Chem.* **269**, 14130-14139.

135. Kiyama, R. and Camerini-Otero, D. (1991). A triplex DNA-binding protein from human cells: purification and characterization. *Proc. Natl. Acad. Sci. USA* **88**, 10450-10454.

136. Guieysse, A.-L., Praseuth, D. and Helene, C. (1997). Identification of a triplex DNA-binding protein from human cells. *J. Mol. Biol.* **267**, 289-298.

137. Musso, M., Nelson, L. D. and Van Dyke, M. W. (1998). Characterization of purine-motif triplex DNA-binding proteins in HeLa extracts. *Biochemistry* **37**, 3086-3095.

138. Bacolla, A., Ulrich, M. J., Larson, J. E., Ley, T. J. and Wells, R. D. (1995). An intramolecular triplex in the human gamma-globin 5'-flanking region is altered by point mutations associated with hereditary persistence of fetal hemoglobin. *J. Biol. Chem.* **270**, 24556-24563.

139. Chen, S., Supakar, P. C., Vellanoweth, R. L., Song, C. S., Chatterjee, B. and Roy, A. K. (1997). Functional role of a conformationally flexible homopurine/homopyrimidine domain of the androgen receptor gene promoter interacting with Sp1 and a pyrimidine single strand DNA-binding protein. *Mol. Endocrinol.* **11**, 3-15.

140. Sarkar, P. S. and Brahmachari, S. K. (1992). Intramolecular triplex potential sequence within a gene down regulates its expression in vivo. *Nucleic Acids Res.* **20**, 5713-5718.

141. Brahmachari, S. K., Sarkar, P. S., Raghavam, S., Narayan, M. and Maiti, A. K. (1997). Polypurine/polypyrimidine sequences as cis-acting transcriptional regulators. *Gene* **190**, 17-26.

142. Krasilnikova, M. M., Samadashwily, G. M., Krasilnikov, A. S. and Mirkin, S. M. (1998). Transcription through a simple DNA repeat blocks replication elongation. *EMBO J.* **17**, 5095-5102.

143. Portes-Sentis, S., Sergeant, A. and Gruffat, H. (1997). A particular DNA structure is required for the function of a cis-acting component of the Epstein-Barr virus OriLyt origin of replication. *Nucleic Acids Res.* **25**, 1347-1354.

144. Brinton, B. T., Caddle, M. S. and Heintz, N. H. (1991). Position and orientation-dependent effects of a eukaryotic Z-triplex DNA motif on episomal DNA replication in COS-7 cells. *J. Biol. Chem.* **266**, 5153-5161.

145. Rao, S., Manor, H. and Martin, R. G. (1988). Pausing in simian virus 40 DNA replication by a sequence containing (dG-dA)$_{27}$·(dT-dC)$_{27}$. *Nucleic Acids Res.* **16**, 8077-8094.
146. Rao, B. S. (1994). Pausing of simian virus 40 DNA replication fork movement in vivo by (dG-dA)$_n$·(dT-dC)$_n$ tracts. *Gene* **140**, 233-237.
147. Espinas, M. L., Jimenes-Garcia, E., Martinez-Balbas, A. and Azorin, F. (1996). Formation of triple-stranded DNA at d(GATC)$_n$ sequences prevents nucleosome assembly and is hindered by nucleosomes. *J. Biol. Chem.* **271**, 31807-31812.
148. Gacy, A. M., Goellner, G. M., Spiro, C., Chen, X., Gupta, G., Bradbury, E. M., Dyer, R. B., Mikesell, M. J.,Yao, J. Z., Johnson, A. J., Richter, A., Melancon, S. B. and McMurray, C. T. (1998). GAA instability in Friedreich's ataxia shares a common, DNA-directed and intraallelic mechanism with other trinucleotide diseases. *Molecular Cell* **1**, 583-593.
149. Ashley, C., Jr. and Warren, S. T. (1995). Trinucleotide repeat expansion and human disease. *Annu. Rev. Genet.* **29**, 703-728.
150. Wells, R. D. (1996). Molecular basis of genetic instability of triplet repeats. *J. Biol. Chem.* **271**, 2875-2878.
151. McMurray, C. T. (1995). Mechanisms of DNA expansion. *Chromosoma* 104, 2-13.
152. Samadashwily, G. M., Raca, G. and Mirkin, S. M. (1997). Trinucleotide repeats affect DNA replication in vivo. *Nature Genet.* **17**, 298-304.
153. Campuzano, V., Montermini, L., Molto, M. D., Pianese, L., Cossee, M., Cavalcanti, F., Monros, E., Rodius, F., Duclos, F., Monticelli, A., et al. (1996). Friedreich's ataxia: autosomal recessive disease caused by an intronic GAA triplet repeat expansion. *Science* **271**, 1423-1427.
154. Rooney, S. M. and Moore, P. D. (1995). Antiparallel, intramolecular triplex DNA stimulates homologous recombination in human cells. *Proc. Natl. Acad. Sci. USA* **92**, 2141-2144.

16 TRIPLEX-BINDING PROTEINS

Anne-Laure Guieysse and Danièle Praseuth

INTRODUCTION

Oligopurine-oligopyrimidine sequences, when they exhibit a mirror symmetry can fold under conditions of negative superhelicity into triple-helical structures and adopt the so-called H-DNA form characterized by the presence of both a triple helix and a single-strand DNA structure (see Chapter 15). The over-representation of oligo-purine-oligopyrimidine stretches in eukaryotic genomes constitutes an argument in favor of the role of these structures *in vivo (1)*. These intramolecular DNA triplexes may have a function in regulation of gene expression, and have been postulated to play a role in replication, transcription, and recombination, and also in chromatin structure and condensation. Proteins could be involved in the biological functions of these sequences by promoting and stabilizing altered DNA structures. A number of non-B-DNA-structure binding proteins have been identified, arguing for the implication of such altered structures in transcriptional activity *(2)*. In this review we will discuss the potential role of H-DNA binding proteins.

SINGLE-STRAND-BINDING PROTEINS

Several proteins specific for natural sequences that can form H-DNA structures have been described and are presented in Table 1. All these proteins are specific for the single-strand polypyrimidine or polypurine loop and/or the duplex and at the present time there is no example of protein specific for the triple helix part of the structure. Most of the H-DNA binding proteins described to date are specific for natural sequences present in gene promoters and might act as positive or negative transcriptional regulators by promoting or stabilizing the H-DNA structure. In all cases there is no direct demonstration of a clear relationship between protein binding to the structure *in vivo* and transcriptional regulation.

Several authors have investigated proteins specific for the 5' region of the *c-myc* gene containing a polypyrimidine tract corresponding to a DNase I hypersensitive

Origin	Name	Sequence binding	Comments	Ref.
HL-60 cell extracts	RNP factor	NSE of *c-myc*	A ribonucleoprotein (RNP) and a protein factor are suggested to act as positive transcriptional regulators.	(3)
HeLa cells cDNA library	NSEP-1	ss polypyrimidine and ds of the NSE of the *c-myc* promoter	sequence specific, also binds to the EGF-R NSE	(4, 5)
human cDNA library	THZif-1	ss pyrimidine and ds of the NSE of *c-myc*	transcription factor, zinc finger protein	(6)
Drosophila nuclear extracts	(GAGA factor)	alternating C,T regions in the promoters of the heat shock genes *hsp70* and *hsp26* and the histone genes *his4* and *his3*	the same or closely related proteins interact with the (CT)-rich regions of the four promoters. Same as or closely related to the GAGA factor.	(7)
adult mouse and human hematopoietic cells		pyrimidine-rich domain upstream from the human δ-globin gene	activity restricted to specific tissue (hematopoietic)	(8)
human adult bone marrow cDNA library	dbpB/BP8	ss polypyrimidine human γ-globin gene promoter	cold-shock protein, binds the corresponding ds with a 100-fold lower affinity	(9)
livers of chicks		ss polypyrimidine chicken malic enzyme gene promoter	a (CT)$_7$ sequence binds the protein and is essential for promoter activity	(10)

Table 1. H-DNA binding proteins (continued on next page)

Origin	Name	Sequence binding	Comments	Ref.
chick embryo fibroblast cDNA library	CHkYB-I	ss polypyrimidine $\alpha2(I)$ collagen gene promoter	cold-shock protein	(11)
HeLa nuclear extract	ssPyrBF	ss polypyrimidine androgen receptor gene promoter	binding of ssPyrBF is expected to inhibit Sp1 binding to the duplex	(12)
human, monkey, mouse cell extracts		ss $d(TC)_n$	family of proteins with similar size range (55.5 to 57 kDa)	(13)
mouse tumor cells	p70	ss $d(CCT)_n$	partial sequence specificity, binds to $d(CCCT)_8$ but not $d(CTT)_{12}$ or $d(CT)_{16}$	(14)
human fibroblasts	PGB	ss $d(GA)_n$ and $d(GT)_n$	protein has lower affinities to $d(G)_n$ tracts	(15)
human colon and pancreatic adenocarcinoma cell lines		ss polypurine human CFTR gene promoter and MUC1 promoter	binds to scrambled polypurine oligonucleotides	(16)

Table 1. H-DNA binding proteins (continued from previous page)

NSE: Nuclease Sensitive Element; ss: single-strand; ds: double-strand; CFTR: cystic fibrosis transmembrane conductance regulator; Ref.: reference.

site (NSE). This region potentially forms a H-DNA structure. An oligonucleotide homologous to this region formed several complexes when it was incubated with HL-60 cell extracts and submitted to electrophoretic mobility shift assay (EMSA) *(3)*. Experiments with mutants allowed the authors to define a consensus binding sequence for the protein factors named the "AC box". They proposed that the H-DNA structure would stimulate gene transcription, since the levels of these factors were correlated with the level of *c-myc* gene expression. The screening of a HeLa cell cDNA expression library for proteins binding to an oligonucleotide containing the NSE allowed the same group to characterize a factor termed NSEP-1 *(4, 5)*. This protein was shown to bind to both duplex and CT-rich single-strand and was specific for CT- asymmetric DNA sequences. Moreover NSEP-1 additionally recognizes the NSE of the EGF-R gene suggesting a general role in regulation of gene expression by its ability to recognize unusual DNA structures. Recently, another group showed that a zinc finger protein called THZif-1 binds to both the double-strand and the pyrimidine single-strand of the *c-myc* NSE, the binding to the triple-stranded structure being at least questionable *(6)*.

Another nuclease-sensitive element present on the fetal δ-globin gene promoter was shown by EMSA to interact with a protein factor *(8)*. Interestingly, this factor was only present in hematopoietic tissues at late states in development, suggesting a possible role in hemoglobin switching from fetal (γ) to adult (δ and β) globin gene expression. Horwitz *et al.* used a single-strand oligopyrimidine corresponding to a potentially H-DNA-forming region of the human γ-globin gene promoter to screen an adult bone marrow cDNA library *(9)*. They identified a protein, BP-8, which specifically binds to the pyrimidine-rich single-strand of the γ-globin promoter region. They studied *in vitro* the binding of the protein to the target sequence cloned in a plasmid under conditions known to stabilize H-DNA structures, such as the presence of magnesium ions, negative superhelical density and acidic pH. All these effectors enhanced BP-8 binding to the secondary structure. In addition, mutations that disrupt H-DNA structure significantly reduced BP-8 binding. It was not excluded that the protein could also recognize the intramolecular triplex part of the structure, but unfortunately this was not determined.

An interesting study has been devoted to the regulatory role of a polypyrimidine/polypurine tract in the promoter of the gene coding for chicken malic enzyme *(10)*. This sequence exhibits DNase I hypersensitivity specifically under conditions where the gene is actively transcribed. In the presence of nuclear extracts three complexes were identified by EMSA. Two of them were specific for pyrimidine-rich single-strand and one for the duplex. Competition experiments allowed the authors to determine the specificity of the proteins involved in the complexes, and they proposed an attractive model for promoter function. Two pyrimidine-rich sequences are involved in the gene regulation, a $(CT)_7$ repeat and an upstream region containing two C-rich elements (C-rich region). They suggested that in the active promoter the $(CT)_7$ repeat is in a melted form stabilized by a $(CT)_n$-specific protein. In contrast, in an inactive form, the formation of a triplex structure by the C-rich region and the binding of a poly(dC)-specific protein to the single-strand part of the H-DNA might prevent the formation of the open conformation of the $(CT)_7$ tract. Finally, the binding of a double-stranded specific protein as third partner to the C-rich region might prevent triplex formation and relieve triplex mediated inhibition of transcription initiation. Chen *et al.* *(12)* again detected a complex in the presence of HeLa extracts with a pyrimidine-rich probe corresponding to one strand of a long polypyrimidine/polypurine stretch present in the promoter of the androgen receptor gene. This single-stranded binding protein, termed SSpYrBF, was shown to be sequence-specific and expressed in diverse mammalian cell types. This S1-nuclease sensitive region additionally contained a site for Sp1 transcription factor. They suggested a negative regulatory role in gene transcription for H-DNA formation and binding of the protein to the resulting pyrimidine-rich single-strand.

In summary, in all these examples a correlation between the existence of a nuclease-sensitive region present in a gene promoter and the level of gene expression stimulated the search for a sequence-specific binding protein. Single-stranded binding proteins specific for the pyrimidine-rich strand were generally found, consistent with their possible role in stabilization of H-DNA structures harboring a PY•PU x PU triplex. In some cases a double-stranded specific protein was also identified; and in general the mechanism proposed is that H-DNA formation results in transcription inhibition. Besides studies showing a correlation between the expression of single-strand-binding protein at different developmental stages and the level of gene transcription, there is, for now, no determinant element that definitely demonstrates the role of H-DNA binding factors in transcriptional activity.

TRIPLE-HELIX BINDING PROTEINS

An interesting approach was developed by Lee *et al.* to study the possible existence and distribution of triplexes *in vivo* *(17, 18)*. They showed by immunofluorescence that two triplex-specific monoclonal antibodies bind to specific sites on cell nuclei and chromosomes. More recently, they demonstrated that these antibodies inhibit replication and transcription in cell nuclei, and cause myeloma cell growth *(19)*.

Three studies deal with triple helix recognition by specific proteins, where triplex-motif-binding proteins were identified from HeLa cells; they are presented in Table 2. The first one was partially purified on the basis of its affinity for an intermolecular $(dT)_{34}•(dA)_{34}$ x $(dT)_{34}$ triplex *(20)*. The protein has an apparent molecular mass of 55 kDa and exhibited also an affinity 2 to 5 times lower for the duplex sequence. In contrast, the binding to the triplex was not affected in the presence of an excess of $(dT)_{34}$. The authors suggested that the protein facilitates triplex formation by promoting the dismutation of two duplexes into the formation of the triple helix and single-strand structures.

More recently, we identified an intramolecular-triplex-binding protein from HeLa cells *(21)*. For this study, we used 55-nucleotide-long oligomers forming intramolecular PY•PU x PY and PY•PU x PU triplexes (Table 2). We showed by EMSA and thermic denaturation experiments that this triple helix was very stable.

Origin	Oligonucleotides, sequence	Comments	Ref.
monoclonal antibody	mice immunized with poly[d(TC)]•poly[d(GA)]	no binding to ssDNA or to duplexes; bind to cell nuclei	*(17, 19)*
HeLa cells	34-mer forming intermolecular triplex: $(dT)_{34}$ • $(dA)_{34}$ x $(dT)_{34}$	affinity 2 to 5-fold lower for the duplex partial purification: apparent MW: 55 kDa	*(20)*
HeLa cells	55-mer forming intramolecular triplexes 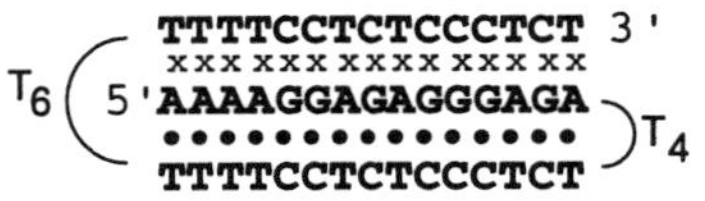 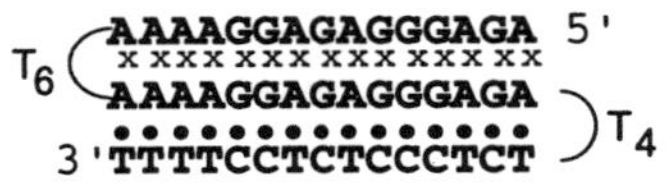 TTTTCCTCTCCCTCT 3' AAAAGGAGAGGGAGA 5' TTTTCCTCTCCCTCT T_6, T_4 AAAAGGAGAGGGAGA 5' AAAAGGAGAGGGAGA 3'TTTTCCTCTCCCTCT T_6, T_4	very low affinity for ss polypyrimidine, no affinity for ds apparent MW of the oligonucleotide-protein complex: 55 kDa	*(21)*
HeLa cells	19-mer forming covalent triplex (5' psoralen conjugate) 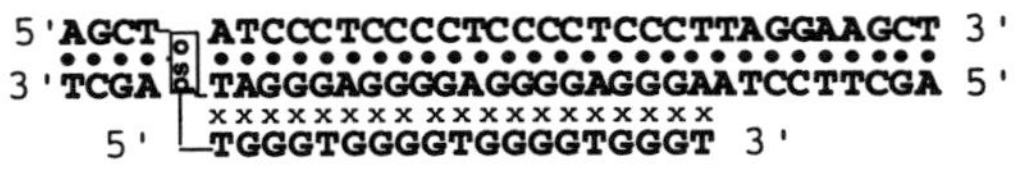 5'AGCT ATCCCTCCCCTCCCCTCCCTTAGGAAGCT 3' 3'TCGA TAGGGAGGGGAGGGGAGGGAATCCTTCGA 5' 5' TGGGTGGGGTGGGGTGGGT 3'	affinity for ss(GA) and ss(GT) 3 polypeptides identified apparent MW: 100 kDa, 60 kDa, 15 kDa.	*(22)*

Table 2. Triplex-binding proteins

These oligonucleotides allowed us to identify, by EMSA, a protein that specifically recognizes the triplex structures. The 55-mer-binding protein activity was not inhibited by an excess of double-strand nor single-strand competitor, but was affected by an excess of intermolecular triplex. No binding was detected on the single or double strand, demonstrating the triplex specificity of the protein. UV cross-linking between the protein and the oligonucleotide allowed us to determine an apparent molecular mass of 55 kDa. Interestingly, experiments using a mutated oligonucleotide forming a partial triple helix seemed to indicate that the binding of the protein might stabilize this partial triplex. The same complex was found in various human nuclear and cytoplasmic extracts. A protein with a similar activity was also found in yeast extracts with an apparent molecular mass of 80 kDa as determined by UV cross-linking (Guieysse, unpublished data); the purification of the protein is currently in progress. To study the possibility that the protein we identified might be identical to the one partially purified by Kiyama and Camereni-Ottero *(20)*, we performed cross-competition experiments between intramolecular 55-mer and intermolecular $(dT)_{34} \cdot (dA)_{34} \times (dT)_{34}$ triplexes. Major differences in specificities between the two proteins indicated that the two proteins were different. Further purification and sequencing will be required to fully characterize them.

Lastly, Musso *et al.* reported the identification of several HeLa proteins that recognize a covalent PY•PU x PU triplex *(22)*. They identified, by EMSA, four major protein-DNA complexes, none of which was inhibited by the corresponding duplex. In contrast, oligonucleotides corresponding to (GT) or (GA) third strand competed with protein binding to the covalent triplex. The authors favored the idea that the putative proteins could also recognize structures formed by these G-rich oligonucleotides. Whether this interpretation is correct has yet to be determined. To study the PY•PU x PU-motif triplex binding specificity, a PY•PU x PY-motif triplex of distinct sequence was used as competitor but did not inhibit complex formation, leading to the premature author's conclusion that the protein was specific for triplexes formed with the purine motif. Furthermore, results obtained with different types of purine triplexes indicated that they were not equally recognized by the putative binding proteins. Southwestern blotting allowed the authors to identify three major polypeptides that preferentially bound to the covalent PY•PU x PU triplex, and whose apparent molecular weights were 100, 60 and 15 kDa.

PERSPECTIVES

The growing number of proteins specific for triplex and H-DNA structures identified to date tends to argue for the existence of such structures *in vivo*. However, the actual role of these factors in transcription regulation has not been yet clearly established. A possible mechanism is that the formation of triplex structures might inhibit or enhance transcription factor binding. The oligopurine-oligopyrimidine sequences potentially forming H-DNA could then act as regulatory units controlling gene activity by modulating the binding of transcription factors and by interacting with specific DNA-binding proteins. Interestingly, two recent studies report the implication of an oligopurine-oligopyrimidine domain in transcription regulation of two TATA-box-less promoters, the androgen receptor and the chicken malic enzyme *(10, 12)*. In both cases the domains have been shown to be essential for promoter activity, and single and double-stranded DNA proteins that interact with these sequences are expected to regulate transcription. The authors suggested that the domains can form H-DNA structures and that the binding of polypyrimidine single-strand-specific proteins negatively regulates gene transcription. The characterization of H-DNA binding proteins in the future, and particularly proteins specific for the triplex, will hopefully give essential information concerning the role of these factors in transcription regulation. They could enhance triplex formation or stabilize triple helices; and such proteins could prove to be important partners for the development of the anti-gene strategy if they are able to stabilize intermolecular triplexes. Furthermore, the characterization of such factors will help to shed light on the role of H DNA in essential processes.

ABBREVIATIONS

PY•PU x Z, base triplet where PY•PU represents the Watson-Crick base-pair (PY=pyrimidine, PU=purine) and Z the base in the third strand forming Hoogsteen or reverse Hoogsteen hydrogen bonds with the purine base of the Watson-Crick duplex.

REFERENCES

1. Behe, M. J. (1995). An overabundance of long oligopurine tracts occurs in the genome of simple and complex eukaryotes. *Nucleic Acid Res.* **23**, 689-695.
2. Van Holde, K. and Zlatanova, J. (1994). Unusual DNA structures, chromatin and transcription. *Bioessays* **16**, 59-68.
3. Davis, T. L., Firulli, A. B. and Kinniburgh, A. J. (1989). Ribonucleoprotein and protein factors bind to an H-DNA-forming c-myc DNA element: possible regulators for the c-myc gene. *Proc. Natl. Acad. Sci. USA* **86**, 9682-9686.
4. Kolluri, R. and Kinniburgh, A. J. (1991). Full length cDNA sequence encoding a nuclease-sensitive element DNA binding protein. *Nucleic Acids Res.* **19**, 4771.
5. Kolluri, R., Torrey, T. A. and Kinniburgh, A. J. (1992). A CT promoter element binding protein: definition of a double-strand and a novel single-strand binding motif. *Nucleic Acids Res.* **20**, 111-116.
6. Sakatsume, O., Tsutsui, H., Wang, Y. F., Gao, H., Tang, X. R., Yamauchi, T., Murata, T., Itakura, K. and Yokoyama, K. K. (1996). Binding of THZif-1, a MAZ-like zinc finger protein to the nuclease-hypersensitive element in the promoter region of the c-myc protooncogene. *J. Biol. Chem.* **271**, 31322-31333.
7. Gilmour, D. S., Thomas, G. H. and Elgin, S. C. (1989). Drosophila nuclear proteins bind to regions of alternating C and T residues in gene promoters. *Science* **245**, 1487-1490.

8. O'Neil, D., Bornschlegel, K., Flamm, M., Castle, M. and Bank, A. (1991). A DNA-binding factor in adult hematopoietic cells interacts with a pyrimidine-rich domain upstream from the human d-globin gene. *Proc. Natl. Acad. Sci. USA* **88**, 8953-8957.

9. Horwitz, E., Maloney, K. and Ley, T. (1994). A human protein containing a "cold shock" domain binds specifically to H-DNA upstream from the human gamma-globin genes. *J. Biol. Chem.* **269**, 14130-14129.

10. Xu, G. and Goodridge, A. G. (1996). Characterization of a polypyrimidine/polypurine tract in the promoter of the gene for chicken malic enzyme. *J. Biol. Chem.* **271**, 16008-16019.

11. Bayarsaihan, D., Enkhmandakh, B. and Lukens, L. N. (1996). Y-box proteins interact with the S1 nuclease-sensitive site in the chicken a2(I) collagen gene promoter. *Biochem. J.* **319**, 203-207.

12. Chen, S., Supakar, P. C., Vellanoweth, R. L., Song, C. S., Chatterjee, B. and Roy, A. K. (1997). Functional role of a conformationally flexible homopurine/homopyrimidine domain of the androgen receptor gene promoter interacting with SP1 and a pyrimidine single strand DNA-binding protein. *Molecular Endocrinology* **11**, 3-15.

13. Yee, H. A., Wong, A. K. C., Van de Sande, J. H. and Rattner, J. B. (1991). Identification of a novel single-stranded d(TC)n binding protein in several mammalian species. *Nucleic Acids Res.* **19**, 949-953.

14. Muraiso, T., Nomoto, S., Yamazaki, H., Mishima, Y. and Kominami, R. (1992). A single-stranded DNA binding protein from mouse tumor cells specifically recognizes the C-rich strand of the (AGG:CCT)n repeats that can alter DNA conformation. *Nucleic Acids Res.* **20**, 6631-6635.

15. Aharoni, A., Baran, N. and Manor, H. (1993). Characterization of a multisubunit human protein which selectively binds single stranded d(GA)n and d(GT)n sequence repeats in DNA. *Nucleic Acids Res.* **21**, 5221-5228.

16. Hollingsworth, M. A., Closken, C., Harris, A., McDonald, C. D., Pahwa, G. S. and Maher, L. J. (1994). A nuclear factor that binds purine-rich, single-stranded oligonucleotides derived from S1-sensitive elements upstream of the CFTR gene and the Muc1 gene. *Nucleic Acids Res.* **22**, 1138-1146.

17. Lee, J. S., Burkholder. G. D., Latimer, L. J. P., Haug, B. L. and Braun, R. P. (1987). A monoclonal antibody to triplex DNA binds to eukaryotic chromosomes. *Nucleic Acids Res.* **15**, 1047-1061.

18. Agazie, Y. M., Lee, J. S. and Burkholder, G. D. (1994). Characterization of a new monoclonal antibody to triplex DNA and immunofluorescent staining of mammalian chromosomes. *J. Biol. Chem.* **269**, 7019-7023.

19. Agazie, Y. M., Burkholder, G. D. and Lee, J. S. (1996). Triplex DNA in the nucleus: direct binding of triplex-specific antibodies and their effect on transcription, replication and cell growth. *Biochem. J.* **316**, 461-466.

20. Kiyama, R. and Camerini-Otero, R. D. (1991). A triplex DNA-binding protein from human cells: Purification and characterization. *Proc. Natl. Acad. Sci. USA* **88**, 10450-10454.

21. Guieysse, A. L., Praseuth, D. and Hélène, C. (1997). Identification of a triplex-DNA binding protein from human cells. *J. Mol. Biol.* **267**, 289-298.

22. Musso, M., Nelson, L. D. and Van Dyke, M. W. (1998). Characterization of purine-motif triplex DNA-binding proteins in HeLa extracts. *Biochemistry* **37**, 3086-3095.

17 RNA AND TRIPLE HELICES

Kyonggeun Yoon, Jörg Jendis and Karin Moelling

INTRODUCTION

Biological processes such as transcription and recombination may involve triple-stranded nucleic acid structures *in vivo* (*1*). Triple helices have also been used as biological tools like artificial nucleases or as highly sequence-specific repressors which modulate protein recognition of DNA (*1*). Polypurine and polypyrimidine sequences form either pyrimidine motif (Py*Pu•Py) or purine motif (Pu*Pu•Py) triplexes, where • and * indicate the Watson-Crick and Hoogsteen base pairing, respectively. In the pyrimidine motif, a pyrimidine oligonucleotide binds to the major groove of a double helix in a parallel orientation with respect to the purine strand, through formation of Hoogsteen hydrogen bonds between T*A•T and protonated C^+*G•C. In the purine motif, oligonucleotides bind antiparallel to the purine strand of the duplex by reverse Hoogsteen hydrogen bonds, A*A•T and G*G•C. The Pu-motif is generally regarded as more applicable under physiological conditions since it does not involve protonation of cytidine.

Application of triple-helix-forming oligonucleotides (TFOs) has mostly been in the regulation of transcription by binding of the TFO to duplex DNA in a sequence-specific manner. Thus, TFOs can compete with the binding of transcription factors to DNA and affect transcription initiation or elongation. However, single-stranded DNA or RNA can be targeted by an oligonucleotide which can form both Watson-Crick base pairing and Hoogsteen base pairing with the target sequence. A foldback triple helix forming oligonucleotide (FTFO) and a circular oligonucleotide (CTFO) have been designed to bind to a single-stranded target sequence (*2-5*). An increase in the specificity and affinity in the binding was observed (*2-5*). When a single-stranded RNA is targeted, an FTFO or a CTFO can be utilized as an antisense oligonucleotide. In other applications, RNA can be used to target other duplexes such as double-helical RNA, RNA hairpins, or RNA-DNA hybrids which are involved in biological processes. Thus, there has been considerable interest in the stability and specificity of recognition in triplexes consisting of both RNA and DNA strands.

These triplex structures can be made from eight combinations of RNA and DNA strands. In the Py-motif (Py*Pu•Py DNA-RNA complexes), where the third strand is bound in a parallel manner, studies have shown that a triplex composed of an RNA strand bound to a duplex DNA by Hoogsteen base pairing (R*D•D) was more stable than the complex consisting only of DNA strands (D*D•D). The stability provided by an RNA strand was consistent among different triplex structures such as a single-strand and a hairpin (6), a single-strand and a double-strand (7, 8), and a circular and a single-strand oligonucleotide (9). An extremely unstable complex was observed when the third strand consisted of DNA bases, which were involved in Hoogsteen pairing with the RNA strand residing in the duplex (D*R•R and D*R•D). The exact orders of stability for the other five hybrid DNA-RNA triplexes were somewhat different in these studies, reflecting variations in sequences, duplex conformation, and experimental conditions such as pH and salt concentration (Table 1). It is important to note that the listed stability reflects different thermodynamic transitions: a duplex to triplex transition was found in the case of SS+DS and SS+HP, whereas a single-strand to triplex transition occurred in the case of SS+CTFO. Only a single transition was observed due to the cooperative nature of the binding by the Hoogsteen and Watson-Crick domains in the circle complexes. Nevertheless, similar stability rules can be applied: if the duplex contains the D purine strand, the R or D third strand will bind; if the duplex contains the R purine strand, only the R third strand will bind.

	Triplex	*Ref.*
R*DD>R*DR>D*DD>R*RD=D*DR=R*RR>>D*RR,D*RD	SS + DS, pH 7	(7)
D*DD,R*DD,R*DR,D*DR>R*RD,R*RR>>D*RR,D*RD	SS + DS, pH 7	(8)
R*DR>R*DD>R*RR>D*DR>D*DD=R*RD>>D*RR,D*RD	SS + HP, pH 5.5	(6)
R*DD>D*DD>R*RR=R*RD=R*DR>D*DR>>D*RR,D*RD	SS +CTFO, pH 5.5	(4, 9)

* Hoogsteen base pairs

R: RNA strand; D: DNA strand; SS: single strand; DS: double strand; HP: hairpin; CTFO: circular triplex-forming oligonucleotide; Ref.: references.

Table 1. Order of stabilities

Stability in the RNA-DNA complex (R*D•D) was attributed to extra hydrogen bonds between the 2' OH of the RNA strand and the phosphate oxygen of the DNA purine strand (10-13). In contrast, an RNA strand in the duplex cannot make this hydrogen bond to the third strand DNA (D*R•R and D*R•D), due to the different sugar conformations of RNA and DNA. Structural characterization of the ribose conformation of DNA and RNA in triplexes was shown to be C2'-endo and C3'-endo, respectively, as determined by NMR and infrared spectroscopy (11, 14, 15). Replacing a DNA purine strand with RNA changes the sugar conformation into a C3'-endo, making a structure which may be unfavorable for triplex formation. However, influences of the nature of the sugars on the triplex remain to be explored.

For example, an RNA strand containing 2'-O-methyl groups was more stable than unmodified RNA even though it cannot make a hydrogen bond to the phosphate oxygen of the DNA strand (*9, 11, 16*). In this case, the hydrophobic effects or van der Waals interactions may also play an important role.

In the Pu-motif (Pu*Pu•Py DNA-RNA complexes) where the third purine strand is bound in an antiparallel orientation to a purine strand of the hairpin, a stable triplex forms only when all three of the strands are composed of DNA (*17*). Exclusion of RNA in a Pu-motif contrasts with the stable triplex formed between RNA and DNA strands in a Py-motif (*6-9*). A molecular modeling study suggested that the 2' hydroxyl groups in the reverse Hoogsteen strand of the complex were unfavorably oriented with respect to the adjacent (5') sugar residues (*17*). In contrast, a stable complex of RNA and DNA was shown to exist when a double-length purine-rich oligonucleotide (either as a hairpin or as a circular oligonucleotide) binds to a SS pyrimidine strand utilizing the Pu-motif (*18*). In this case, one of the purine strands participates in Watson-Crick base pairing with the SS DNA or RNA, while the other purine strand participates in Hoogsteen base pairing with the purine strand which is linked covalently. The binding of SS RNA occurs with the same affinity as the binding of SS DNA. Differences between the two systems indicate that the nature of the linkage may be important for the stability of the triplex.

TARGETING OF A PURINE SINGLE STRAND BY FTFOs OR CTFOs

A single-stranded DNA or RNA was targeted using either FTFOs or CTFOs (*2-5*). These oligonucleotides contain two major domains, namely the Watson-Crick and Hoogsteen domains. The former domain is a complementary sequence to the SS homopurine target that binds in an antiparallel mode through Watson-Crick hydrogen bond formation. The latter is a mirror image sequence to the Watson-Crick domain and binds in a parallel orientation to the target sequence through Hoogsteen hydrogen bonding utilizing T*A•T (U*A•U) and C^+*G•C triplets, where C^+ is protonated. Thermal melting of these triplexes indicates a single transition at a much higher temperature than transitions observed for three separate strands. Binding of the Watson-Crick domain to the target increases effective local concentration and reorganization of the Hoogsteen domain so as to promote a tight binding of the third strand (*2, 3, 19*). It appears that the binding of the two domains is a cooperative phenomenon that contributes to the entropic benefit of the foldback triplex, observed both in FTFOs and CTFOs (*2-5*). However, most physicochemical studies were carried out under acidic conditions, since formation of a C^+*G•C triplet requires protonation of cytidine. In order to target a polypurine RNA strand such as the PPT region of HIV, a cytidine residue was substituted by a guanine (G*G•C triplet), since the Py-motif (C^+*G•C triplet) requires an acidic condition which is not physiological.

TARGETING OF A PYRIMIDINE SINGLE STRAND
BY FTFOs OR CTFOs

A double-length purine-rich oligonucleotide, a hairpin or a circular oligonucleotide, was shown to bind to a target pyrimidine sequence, folding back to form an antiparallel Pu*Pu•Py triple helix (*18*). The resulting complex between a SS target and a hairpin or a circular oligonucleotide was shown to be much more stable than the duplex. Both complexes contain identical Watson-Crick hydrogen bonding to the target sequence. However, a purine strand of a hairpin or a circular oligonucleotide folds back and participates in the antiparallel Pu*Pu•Py triple helix, contributing higher stability. Mismatches in the purine strand making a reverse Hoogsteen hydrogen bond were shown to be detrimental to the overall stability of the complex, indicating the importance of foldback structure (*18*).

CHARACTERIZATION OF DNA*RNA•DNA TRIPLE HELICES
AT THE POLYPURINE TRACT (PPT)
IN THE 3-STRAND SYSTEM

The polypurine tract of HIV and other retroviruses is essential during retroviral replication and is a potential target for triple-helix approaches. The PPT resides in the viral RNA which is transcribed into DNA by the viral reverse transcriptase. The RNA is digested by the viral RNase H except at the PPT where an RNA-DNA hybrid resists hydrolysis and the RNA serves as primer for the second strand DNA synthesis (*20, 21*). Triple helix formation is conceivable with the various intermediates, SS RNA, SS DNA, RNA-DNA hybrids, or DS DNA. Triple helices have previously been shown to protect the HIV-PPT against nuclease digestion in a D*D•D-system (*22*). A triple helix consisting of an RNA-DNA hybrid has been shown to prevent cleavage of the RNA by the retroviral RNase H. For the analysis, a 134 nucleotide long viral RNA was labeled at its 5' end and was annealed to a 40-mer oligodeoxynucleotide (DNA H40) spanning the PPT sequence (Table 2). This heteroduplex structure serves as substrate for RNase H *in vitro*. PY 1 which is parallel to the RNA purine strand exhibited a better protection of the RNA from RNase H cleavage than did PY 2, which has an antiparallel orientation (Table 2) (*23*). In another set of experiments, GT purine-pyrimidine mixed DNA oligonucleotides were used (GT1 and GT4, Table 2). They can form Hoogsteen and reverse Hoogsteen bonds resulting in the formation of G*G•C and T*A•T base triplets (*24, 25*). With the 25-mer GT1 and GT2, a strong protection (50 %) was observed against RNase H cleavage, independent of their orientations. This is surprising since more stable triplex has been shown to form when a TFO binds to duplex in an antiparallel orientation. The degree of protection was dependent on the length of the TFO, because the two 16-mer oligonucleotides GT3 and GT4 exhibited a reduced protection (Table 2). The RNase H protection results suggest that D*R•D triple helix formation occurred at the PPT. Addition of the RNA oligonucleotide (rGU) to the preformed heteroduplex results in the formation of an R*R•D complex.

It exhibited a higher thermodynamic stability than the D*R•D complex and a higher protection of RNA from RNase H cleavage, approaching 80%.

```
DNA H 40  3'....CGG TGA AAAA T T  T TCT T T T TCCCCCCT GACC TT CCC___5'
RNA    5'.......UAGCCACU UUUU A A  A AGA A AAGGGGGGAC  UGG AA GGGCU AAUUCAC A___3'
```

Sequence of deoxynucleotides	RNA uncut (%)
D*R•D:	
PY 1 5' - TTTTCTTTTCCCCCCCTGACCTTCCC - 3'	38.5
PY 2 5' - CCCTTCCAGTCCCCCCCTTTTCTTTT - 3'	19.5
GT 1 5' - TTTTGTTTTGGGGGGGTGTGGTTGGG - 3'	54.0
GT 2 5' - GGGTTGGTGTGGGGGGGTTTTGTTTT - 3'	52.0
GT 3 5' - TGGGGGGGTTTTGTTTT - 3'	25.0
GT 4 5' - TTTTGTTTTGGGGGGGT - 3'	41.0
GT 5' - TTTTGTTTTGGGGGGGTTTGGTTGGG - 3'	22.0
R*R•D:	
rGU 5' - UUUUGUUUUGGGGGGGUUUGGUUGGG - 3'	80.0

Table 2. 3-strand system

CHARACTERIZATION OF DNA*RNA•DNA TRIPLE HELICES AT THE PPT IN THE 2-STRAND SYSTEM

The data presented in Table 3 demonstrate the possibility of triple helix formation with an RNA target sequence. The "two-strand foldback" or "sandwich" system consists of a SS RNA and a FTFO containing the Watson-Crick and the Hoogsteen base-pairing domains covalently linked by T4 residues. All of the TFOs contain an identical domain for Watson-Crick base pairs with the RNA but differ in the Hoogsteen domain (Table 3). Mutations were introduced in some FTFOs in order to determine the sequence-specificity in triplex formation. The *in vivo* target for the FTFOs is a SS RNA. Therefore, we determined whether FTFOs can interfere with the enzyme activities of the reverse transcriptase or of RNase H. The RNase H cleavage assays were performed by preincubation of a radiolabelled RNA containing the PPT target sequence with increasing amounts of FTFO. After addition of RNase H, reaction products were analyzed by gel electrophoresis and autoradiography. Table 3 shows the result of RNase H cleavage reactions. The sequence specificity of several FTFOs in the RNase H inhibition was evaluated.

TFO A exhibited the greatest inhibition of RNase H because it can form triplex with SS RNA by utilizing both Watson-Crick and Hoogsteen base pairing. TFO B, which cannot form a triplex but rather a duplex with a SS RNA target, exhibited some RNase H cleavage resistance. Another TFO contained a single mismatch in its Hoogsteen strand (TFO D, underlined C) but still protected RNA against RNase H cleavage. In contrast, TFO C was not able to protect the RNA due to three base mismatches (underlined). The oligonucleotide (KO A) which cannot form Watson-Crick base pairs with the target RNA exhibited no protection of RNA. Therefore, these results indicate that FTFOs can make a D*R•D triplex with the single-stranded RNA in a sequence-specific manner. Although D*R•D triplexes have been shown to be unstable, sensitive enzyme assays such as RNase H seem to be able to detect such a complex. Furthermore, the effect of TFOs was tested in cell culture, where HIV was used to infect C81-66/45 T cells, and the effect of replicating virus was evaluated by a cytopathic effect (CPE) and by p24 synthesis (26). TFO A is the only TFO which inhibits both RNase H and p24 production, leading to an inhibition of CPE. TFO A also inhibits viral replication of drug-resistant reverse transcriptases (27).

RNA 5' - AAAAGAAAAGGGGGGACUGGAAGGG - 3'

	Hoogsteen	linker	Watson-Crick	RNase H protection uncut (%)	CPE	p24
D*R•D:						
TFO A	5'-TTTTCTTTTGGGGGGTTTGGTTGGGTTTTCCCTTCCAGTCCCCCCTTTTCTTTT-3'			> 95	–	–
TFO B	TGTGTGTGTGTGTGTGTGTGTGTGTGTTTTTCCCTTCCAGTCCCCCCTTTTCTTTT			> 50	+	+
TFO C	TTTT<u>A</u>TTTT<u>A</u>GGGG<u>A</u>TTTGGTTGGGTTTTCCCTTCCAGTCCCCCCTTTTCTTTT			32	+	+
TFO D	TTTTCTTTTGGGGGGT<u>C</u>TGGTTGGGTTTTCCCTTCCAGTCCCCCCTTTTCTTTT			76	+	+
TFO E	TTTTCTTTTCCCCCCTTTTTTCCCCCCTTTTCTTTT			24	+	+
TFO F	TTTTCTTTTGGGGGGTTTTTTCCCCCCTTTTCTTTT			31	+	+
KO A	TTTTGTTTTGGGGGGTTTTTTGGGGGGTTTTGTTTT			0	+	+
Antisense: GT			CCCTTCCAGTCCCCCCTTTTCTTTT	22	+	+
TFO A Δ5	_________GGGGGGTTTGGTTGGGTTTTCCCTTCCAGTCCCCCCTTTTCTTTT				+	+
TFO A Δ3	TTTTCTTTTGGGGGGTTTGGTTGGGTTTTCCCTTCCAGTCCCCCC_________				+	+
TFO A ΔI	TTTTCTTTTGGGGGGTTGG_____TTTT_____CCAGTCCCCCCTTTTCTTTT				+	+
TFO A rev	TTTTCTTTTCCCCCCTGACCTTCCCTTTTGGGTTGGTTGGGGGGTTTTCTTTT				+	+
TFO A anti	GGGAAGGAGAGGGGGGAAAAGAAAATTTTTTTTCTTTTCCCCCCTCTCCTTCCC				+	+
TFO NT	TTCTTTTTTGCGCGCTGCTGTTCGCTTTTGCGTTCAGTTGCGCGCTTTTTTTCT				+	+
TFO SC	TTTGGGGGGTTCTTCCTCCTTTCCTTTTTCGCCCGTCCGTTGCGTTGATTTTTT				+	+
R*R•D:						
TFO A RNA	uuuucuuuuggggggguuugguugggTTTTGGGTTCCAGTCCCCCCTTTTCTTTT				+	+

CPE: cytopathic effect

Table 3. 2-strand system

Several other TFO mutants were tested (Table 3). TFO A Δ5 and TFO A Δ3 have a deletion of nine nucleotides at the 5'- and the 3'-end, respectively. TFO ΔI contains an internal deletion of five nucleotides on both sides of the T4 linker. In "TFO A rev" oligonucleotide, the order of Hoogsteen and Watson-Crick forming domains is reversed. The oligonucleotide designated as "TFO NT" is 'not targeted' to the PPT sequence. TFO A RNA contains RNA bases in the Hoogsteen hydrogen bond-forming domain. Thus, TFO A RNA binding to the target RNA would result in the formation of an R*R•D complex. Finally, "TFO anti" is designed to bind in an antiparallel orientation to the purine target strand. When these oligonucleotides were tested for their antiviral efficacy it was found that none of these molecules were able to interfere with the replication of the laboratory propagated HIV-1 IIIB strain. This result suggests that the observed biological effect of TFO A is due entirely to its specific sequence and its structure. Any changes within these requirements render the TFO ineffective.

ALTERNATIVE EXPLANATIONS FOR THE INHIBITORY EFFECT OF TFO A

The question then arises about how the inhibitory effect is achieved by the oligonucleotide, which was originally constructed so as to form a triplex with the viral RNA at the PPT site. Since D*R•D complexes have been shown to be unstable (Table 1), we hypothesize four possible explanations for the inhibitory capacity of TFO A.

I. Inhibition of HIV-1 replication can take place at the level of newly formed proviral DNA. TFO A can make a triplex with this SS DNA at the PPT site (D*D•D).

II. Nucleic acid binding proteins can stabilize a structure which is energetically unfavorable and usually not formed without the help of such proteins. For example, the three-stranded joint molecule was formed by RecA and was shown to be a key intermediate in homologous recombination (28, 29). In another example, it has been shown that the nucleocapsid of HIV enhances the catalytic properties of ribozymes (30, 31). In these cases, the ribonucleoproteins improved annealing of RNA strands by acting as a chaperone or matchmaker (32). We have also observed a stimulatory effect of the nucleocapsid on reverse transcription. Therefore, it is conceivable that the proposed D*R•D triplex may be stabilized by RNA binding proteins such as the nucleocapsid of HIV.

III. A different cause of the observed inhibitory effect would be an induced mutation by triplex formation. In this case, the triplex would exist for a limited period of time but sufficient to induce mutations either in or in close proximity to the targeted viral sequence. Triple helix induced mutagenesis has been demonstrated in mammalian cells, albeit at a low frequency (33).

IV. The antiviral effect of TFO A could also be attributed to mutations in the PPT sequence caused by interaction between cellular proteins involved in recombination or DNA repair and the three-stranded structure. Recently, a unique hybrid oligonucleotide composed of both RNA and DNA has been shown to correct a point mutation in a site-specific and inheritable manner in extrachromosomal and chromosomal targets (*34-38*). The RNA-DNA hybrids were shown to be highly active in homologous pairing reactions *in vitro* (*39-41*). It has been hypothesized that the perturbation and partial destacking of the RNA-DNA base pairing in the oligonucleotide may facilitate recombination, followed by a subsequent recognition of a mismatch and repair by the DNA repair system. One can extend such hypotheses to the PPT sequence, where a SS RNA is involved in the three-stranded structure with a folded duplex, TFO A. Any mutations in the PPT sequence would finally result in a premature block of HIV-1 replication. Although the mechanism of the inhibition of retroviral replication by TFO A is still not elucidated, this work will hopefully contribute to a deeper understanding of the biological effects of TFOs.

REFERENCES

1. Vasquez, K. M. and Wilson J. H. (1998). Triplex-directed modification of genes and gene activity. *TIBS* **23**, 4-9.
2. Giovannangeli, C., Montenay-Garestier, T., Rougée, M., Chassignol, M., Thuong, N. T. and Hélène, C. (1991). Single-stranded DNA as a target for triple-helix formation. *J. Am. Chem. Soc.* **113**, 7775-7777.
3. Kool, E. T. (1991). Molecular recognition of circular oligonucleotide increasing the selectivity of DNA binding. *J. Am. Chem. Soc.* **113**, 6265-6266.
4. Wang, S. and Kool, E. T. (1994 a). Circular RNA oligonucleotides. Synthesis, nucleic acid binding properties and a comparison with circular DNAs. *Nucleic Acids Res.* **22**, 2326-2333.
5. Wang, S. and Kool, E. T. (1994 b). Recognition of single-stranded nucleic acids by triplex formation: the binding of pyrimidine-rich sequences. *J. Am. Chem. Soc.* **116**, 8857-8858.
6. Roberts, R. W. and Crothers, D. M. (1992). Stability and properties of double and triple helices: dramatic effects of RNA or DNA backbone composition. *Science* **258**, 1463-1466.
7. Escudé, C., François, J. C., Sun, J. S., Ott, G., Sprinzl, M., Garestier, T. and Hélène, C. (1993). Stability of triple helices containing RNA and DNA strands: experimental and molecular modeling studies. *Nucleic Acids Res.* **21**, 5547-5553.
8. Han, H. and Dervan, P. B. (1993). Sequence-specific recognition of double helical RNA and RNA-DNA by triple helix formation. *Proc. Natl. Acad. Sci. USA* **90**, 3806-3810.
9. Wang, S. and Kool, E. T. (1995). Relative stabilities of triple helices composed of combinations of DNA, RNA and 2'-O-methyl-RNA backbones: chimeric circular oligonucleotides as probes. *Nucleic Acids Res.* **23**, 1157-1164.
10. Holland, J. and Hoffman, D. W. (1996). Structural features and stability. *Nucleic Acids Res.* **14**, 2841-2848.
11. Dagneaux, C., Liquier, J. and Taillandier, E. (1995). Sugar conformations in DNA and RNA-DNA triple helices determined by FTIR spectroscopy: role of backbone composition. *Biochem.* **34**, 16618-23.

12. Van Dongen, M. J., Heus, H. A., Wymenga, S. S., van der Marel, G. A., van Boom, J. H. and Hilbers, C. W. (1996). Unambiguous structure characterization of a DNA-RNA triple helix by 15N- and 13C-filtered NOESY spectroscopy. *Biochem.* **35**, 1733-1739.

13. Liquier, J., Taillandier, E., Klinck, R., Guittet, E., Gouyette, C. and Huynh-Dinh, T. (1995). Spectroscopic studies of chimeric DNA-RNA and RNA 29-base intramolecular triple helices. *Nucleic Acids Res.* **23**, 1722-1728.

14. Macaya, R., Wang, S., Schultze, P., Sklenar, V. and Feigon, J. (1992). Proton Nuclear magnetic resonance assignments and structural characterization of an intramolecular DNA triplex. *J. Mol. Biol.* **225**, 755-773.

15. Klinck, R., Liquier, J., Taillandier, E., Gouyette, C., Huynh-Dinh, T. and Guittet, E. (1995). Structural characterization of an intramolecular RNA triple helix by NMR spectroscopy. *Eur. J. Biochem.* **233**, 544-553.

16. Shimizu, M., Konishi, A., Shimada, Y., Inoue, M. and Ohtsuka, E. (1992). Oligo(2'-0-methyl)ribonucleotides. Effective probes for duplex DNA. *FEBS Lett.* **302**, 155-158.

17. Semerad, C. L. and Maher, L. J. (1994). Exclusion of RNA strands from a purine motif triple helix. *Nucleic Acids Res.* **22**, 5321-5325.

18. Vo, T., Wang, S. and Kool, E. T. (1995). Targeting pyrimidine single strands by triplex formation: structural optimization of binding. *Nucleic Acids Res.* **23**, 2937-2944.

19. Prakash, G. and Kool, E. T. (1991). Molecular recognition by circular oligonucleotides. Strong binding of single-stranded DNA and RNA. *J. Chem. Soc., Chem. Commun.* **17**, 1161-1163.

20. Hansen, J., Schulze, T., Mellert, W. and Moelling, K. (1988). Identification and characterization of HIV-specific RNase H by monoclonal antibody. *EMBO J.* **7**, 239-243.

21. Wöhrl, B. and Moelling, K. (1990). Interaction of HIV-1 RNase H with polypurine tract containing RNA-DNA hybrids. *Biochem.* **29**, 10141-10147.

22. Sun, J. S., Giovannangeli, C., François, J. C., Kurfurst, R., Montenay-Garestier, T., Asseline, U., Saison-Behmoaras, T., Thuong, N. T. and Hélène, C. (1991). Triple-helix formation by oligodeoxynucleotides and oligodeoxynucleotide-intercalator conjugates. *Proc. Natl. Acad. Sci. USA* **88**, 6023-6027.

23. Volkmann, S., Dannull, J. and Moelling, K. (1993). The polypurine tract, PPT, of HIV as target for antisense and triple-helix-forming oligonucleotides. *Biochimie* **75**, 71-78.

24. Hoogsteen, K. (1959). The structure of crystals containing a hydrogen-bonded complex of 1-methylthymine and 9-methyladenine. *Acta Crystallogr.* **12**, 822-823.

25. Giovannangeli, C., Rougee, M., Garestier, T., Thuong, N. T., Hélène C. (1992). Triple-helix formation by oligonucleotides containing the three bases thymine, cytosine, and guanine. *Proc. Natl. Acad. Sci. USA* **89**, 8631-8635.

26. Volkmann, S., Jendis, J., Frauendorf, A. and Moelling, K. (1995). Inhibition of HIV-1 reverse transcription by triple-helix-forming oligonucleotides with viral RNA. *Nucleic Acids Res.* **23**, 1204-1212.

27. Jendis, J., Strack, B. and Moelling, K. (1998). Inhibition of replication of drug-resistant HIV type 1 isolates by polypurine tract specific oligodeoxynucleotide TFO A. *AIDS Res. Hum. Retroviruses* **14**, 999-1005.

28. Bianchi, M., DasGupta, C. and Radding, C. M. (1983). Synapsis and the formation of paranemic joints by *E. coli* RecA protein. *Cell* **34**, 931-939.

29. Camerini-Otero, R. D. and Hsieh, P. (1993). Parallel DNA triplexes, homologous recombination, and other homology-dependent DNA interactions. *Cell* **73**, 217-223.

30. Müller, G., Strack, B., Dannull, J., Sproat, B. S., Surovoy, A., Jung, G. and Moelling, K. (1994). Amino acid requirements of the nucleocapsid protein of HIV-1 for increasing catalytic activity of a Ki-ras ribozyme *in vitro*. *J. Mol. Biol.* **242**, 422-429.

31. Tsuchihashi, Z., Khosla, M. and Herschlag, D. (1993). Protein enhancement of hammerhead ribozyme catalysis. *Science* **262**, 99-102.

32. Portman, D.S. and Dreyfuss, G. (1994). RNA annealing activities in HeLa nuclei. *EMBO J.* **13**, 213-221.

33. Wang, G., Seidman, M. M. and Glazer, P. M. (1996). Mutagenesis in mammalian cells induced by triple helix formation and transcription-coupled repair. *Science* **271**, 802-805.

34. Yoon, K., Cole-Strauss, A. and Kmiec, E. B. (1996). Targeted gene correction of episomal DNA in mammalian cells mediated by a chimeric RNA/DNA oligonucleotide. *Proc. Natl. Acad. Sci. USA* **93**, 2071-2076.

35. Cole-Strauss, A., Yoon, K., Xiang, Y., Byrne, B. C., Rice, M. C., Gryn, J., Holloman, W. K. and Kmiec, E. B. (1996). Correction of the mutation responsible for sickle cell anemia by an RNA-DNA oligonucleotide. *Science* **273**, 1386-1389.

36. Xiang, Y., Cole-Strauss, A., Yoon, K., Gryn, J. and Kmiec, E. B. (1997). Targeted gene conversion in a mammalian CD34[+]-enriched cell population using a chimeric RNA/DNA oligonucleotide. *J. Mol. Med.* **75**, 829-835.

37. Kren, B. T., Cole-Strauss, A., Kmiec, E. B. and Steer, C. J. (1997). Targeted nucleotide exchange in the alkaline phosphatase gene of Hul-I-7 cells mediated by a chimeric RNA/DNA oligonucleotide. *Hepatology* **25**, 1462-1468.

38. Kren, B. T., Bandyopadhyay, P. and Steer, C. J. (1998). *In vivo* site-directed mutagenesis of the factor IX gene by chimeric RNA/DNA oligonucleotides. *Nat. Med.* **4**, 285-290.

39. Kotani, H. and Kmiec, E B. (1994a). A role for RNA synthesis in homologous pairing events. *Mol. Cell Biol.* **14**, 6097-6106.

40. Kotani, H. and Kmiec, E B. (1994b). Transcription activates RecA-promoted homologous pairing of nucleosomal DNA. *Mol. Cell Biol.* **14**, 1949-1955.

41. Kotani, H., Germann, M. W., Andrus, A., Vinayak, R., Mullah, B. and Kmiec, E. B. (1996). RNA facilitates RecA-mediated DNA pairing and strand transfer between molecules bearing limited regions of homology. *Mol. Gen. Genet.* **250**, 626-634.

NEW DEVELOPMENTS AND APPLICATIONS

18 TRIPLEXES INVOLVING PNA

Thomas Bentin and Peter E. Nielsen

INTRODUCTION

Peptide nucleic acids (PNAs) are synthetic oligonucleotide mimics containing a pseudopeptide backbone to which the nucleobases are attached. Although PNAs can be degraded during prolonged incubation under alkaline conditions (*1*), these molecules are generally not degraded in serum, in microbial extracts, or by proteases (*2*), and PNAs should therefore be sufficiently stable for therapeutic applications. Homopyrimidine PNAs form very stable and sequence selective complexes with complementary nucleic acids by formation of an unusual triplex structure. Due to the extraordinary properties of such PNAs, they have found several relevant biological (*3-13*) and biotechnological (*14-19*) applications.

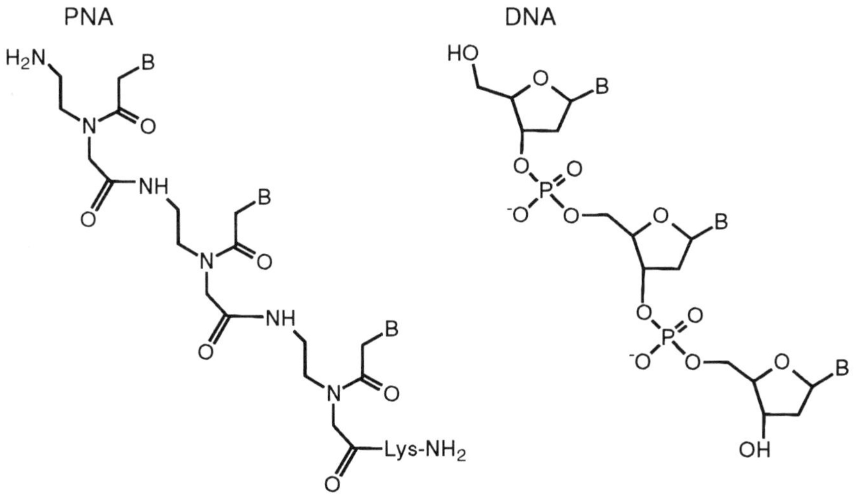

Figure 1. Chemical structure of PNA compared to DNA (B = Nucleobase).

PNA CHEMICAL STRUCTURE

PNA contains an uncharged and achiral backbone composed of aminoethyl glycine units to which the nucleobases are attached by methylene carbonyl linkers (Figure 1). Despite having only faint chemical resemblance to standard nucleic acids, the PNA backbone retains the same number of covalent inter-nucleobase bonds (20). This feature, in combination with restrained flexibility, is a key to the remarkable hybridization properties of PNA (21-23).

PNA-NUCLEIC ACID TRIPLEXES

PNA hybridization to nucleic acids obeys the traditional base pairing rules: adenine is recognized by thymine and guanine by cytosine. At mixed purine-pyrimidine sequence targets, PNA recognition is accomplished by duplex formation involving Watson-Crick hydrogen bonds (24). In the case of homopurine targets, however, remarkably thermostable complexes with a 2:1 PNA to nucleic acid stoichiometry are formed (21, 25, 26), and such triplexes adopt a helical structure (27). This PNA to DNA ratio is compatible with a complex in which one PNA strand employs regular Watson-Crick hydrogen bonds while another binds its DNA complement *via* Hoogsteen hydrogen bonds (Figure 2) (21). Below, these are termed H-PNA and WC-PNA, respectively. Furthermore (·) implies Hoogsteen hydrogen bonds, whereas (-) refers to Watson-Crick hydrogen bonds.

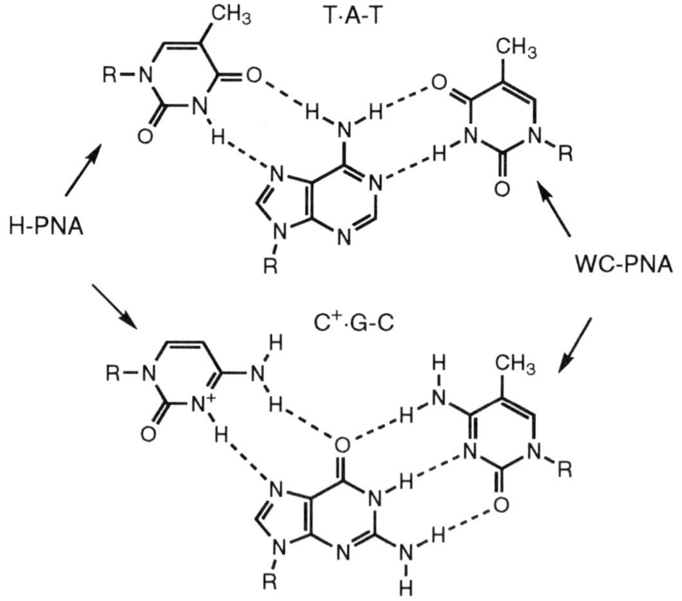

Figure 2. Watson-Crick and Hoogsteen hydrogen bonds in the T·A-T and C^+·G-C nucleobase triplets. R = backbone attachment bond.

PNA·DNA-PNA triplex formation is sensitive to pH, and this further substantiated participation of Hoogsteen hydrogen bonds because they require protonization of the cytosine N3 atom. Accordingly, substituting pseudoisocytosine for cytosine markedly reduces the pH sensitivity (28).

As a consequence of the 2:1 PNA to DNA ratio in the PNA·DNA-PNA triplex, linked bis-PNAs were generated using flexible linkers, thereby simplifying binding stoichiometry. The most favorable binding orientation for such bis-PNAs is clear: antiparallel (the PNA NH$_2$-terminus aligns with the 3'-DNA) for the WC-PNA strand, and parallel (the PNA NH$_2$-terminus aligns with the 5'-DNA) for the H-PNA strand (28). In contrast, the binding affinity for monomeric PNAs (mono-PNAs) appears insensitive to strand orientation (4, 28). This is because the preferred arrangement, in which the two PNAs are antiparallel to each other, cannot be accommodated without obtaining unfavorable binding orientation for one of the PNAs.

The crystal structure of a PNA·DNA-PNA triplex directly confirms and extends the above statements (Figure 3). In this so called "P-form" (29), the WC-PNA strand is antiparallel to the DNA strand, whereas the H-PNA is parallel. The P-form

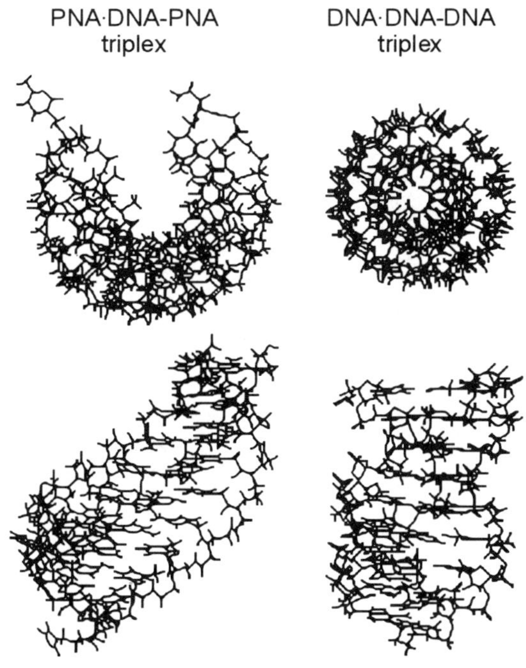

Figure 3. Structure of a PNA·DNA-PNA triplex compared to a conventional "A type" DNA·DNA-DNA triplex (29).

helix is wide (diameter, d = 26 Å) with a large central cavity (average base displacement, D = 6.8 Å) and 16 bp per helical turn – features which distinguish it from other helical nucleic acid structures, including standard B-DNA (d = 20 Å, D = -0.2 Å, and 10.5 bp/helical turn). The high resolution structure provides information regarding the unexpected high stability of PNA·DNA-PNA triplexes. For example, the H-PNA backbone, maybe due to its lack of charge, lies in close proximity to the DNA backbone, where hydrogen bonds form directly between the PNA amide protons and one of the DNA phosphate oxygen atoms.

The stability of PNA·DNA-PNA triplexes is severely compromised by the introduction of extended PNA backbones (*21, 22*) or by replacing the methylene carbonyl nucleobase linker with the more flexible ethylene unit (*23*). These observations emphasize the importance of having both the correct inter-nucleobase distance and restrained flexibility, *i.e.*, a limited number of possible conformations. This outlines a framework for introduction of backbone substitutions (*30, 31*), and thus opens ways for fine tuning the interactions of PNA with DNA

PNA BINDING TO DOUBLE-STRANDED DNA

Types of complexes

Unlike conventional triple helix structures in which triplex forming oligonucleotides (TFOs) bind double-stranded DNA (dsDNA) as a third strand in the major groove, PNA binding results in displacement of the non-complementary DNA strand, while the target DNA strand is occupied by two PNAs (Figure 4A). Upon binding,

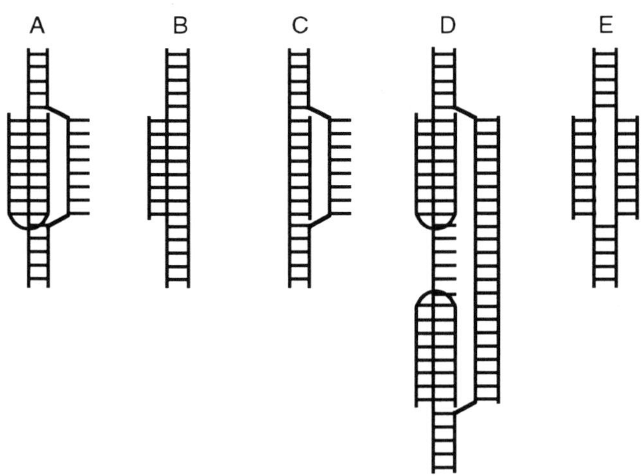

Figure 4. Complexes of PNA with dsDNA. **A**, PNA·DNA-PNA/DNA triplex invasion complex. **B**, PNA·DNA-DNA triplex. **C**, PNA-DNA/DNA duplex invasion complex. **D**, PD-loop. **E**, PNA-DNA/PNA-DNA Double duplex invasion complex.

PNA·DNA-PNA/DNA triplex invasion complexes are formed, where / refers to the displaced DNA strand. Similar to the binding to single-stranded nucleic acids, this involves both WC-PNA (20) and H-PNA (20, 32). Such complexes exhibit remarkable sequence selectivity (33, 34) and stability properties (35).

The common PNA·DNA-PNA/DNA triplex invasion complex occurs with thymine-rich PNAs such as PNA(T_8C_2). However, alternative motifs have been reported. With PNAs rich in cytosines, e.g., mono-PNA(C_{10}) and mono-PNA(CT)$_5$, formation of "Hoogsteen type" PNA·DNA-DNA triplexes involving H-PNA only, was reported (Figure 4B). These complexes are comparable to conventional TFO-dsDNA triplexes (36).

A third type of complex involves homopurine PNAs recognizing a homopyrimidine dsDNA target (Figure 4C). A single mono-PNA(A_6G_4) strand bound its DNA complement by duplex invasion, thereby forming a PNA-DNA/DNA complex involving Watson-Crick hydrogen bonds and displacement of the non-complementary DNA strand (37). Although the thermal stability of the oligonucleotide duplex with this PNA compared well with that of an ordinary decamer thymine-rich PNA·DNA-PNA triplex (Figure 3), duplex invasion complexes could not be detected by electrophoretic mobility shift analysis. This presumably indicates that the dissociation rate is significantly higher as compared to triplex invasion complexes. Presently, PNA·DNA-DNA triplexes (Figure 4B) and PNA-DNA/DNA duplex invasion complexes (Figure 4C) should be regarded as special cases.

Recently a novel complex termed "PD-loop" was reported (Figure 4D) (38). The PD-loop is based on the regular PNA·DNA-PNA/DNA triplex invasion complex with the added feature that the displaced DNA strand is hybridized with a complementary oligonucleotide by Watson-Crick hydrogen bonds. This allows for recognition of mixed purine-pyrimidine sequences located amidst adjacent complexes, because the DNA duplex is dissociated between the PNA targets. Although this PNA mediated "duplex opening effect" is expected to propagate for only a few base pairs per triplex invasion complex (13), formation of PD-loops was possible using as much as 11 bp mixed purine-pyrimidine sequences between the two PNA target sites, although at reduced efficiency (38).

The requirement for homopurine stretches limits the versatility of recognition involving Hoogsteen hydrogen bonds. To overcome this limitation a new strategy termed double duplex invasion has been developed (Figure 4E). Two pseudo-complementary PNAs containing diaminopurine-thiouracil base pairs substituting adenine-thymine base pairs bind complementary dsDNA target sequences selectively. In this PNA-DNA/PNA-DNA double duplex invasion complex, both DNA complements are bound by PNA via Watson-Crick hydrogen bonds. The PNA strands do not hybridize to each other due to steric occlusion in the diaminopurine-thiouracil base pair (39; Peter E. Nielsen, unpublished). In contrast to PNA-DNA/DNA duplex invasion complexes, double duplexes are sufficiently stable to be resolved using electrophoretic mobility shift analysis. Importantly, it is estimated that any sequence containing at least 40% adenine-thymine base pairs can be targeted by this strategy. Because this statistically is the case in >80% of all decameric sequences, double duplex invasion greatly expands the number of sequences which can be targeted by PNA.

Sequence selectivity

Triplex invasion is a highly sequence-selective process, and sequence discrimination is kinetically controlled (*33, 34*). Once formed, however, both correctly matched and (singly) mismatched complexes are thermodynamically very stable under the experimental conditions usually employed. For example, Griffith *et al.* (1995) reported <10% dissociation in 100 hours for a decamer bis-PNA (*35*). The combination of high stability and sequence selectivity is explained by the observation that perfectly matched complexes are formed at a rate 1-3 orders of magnitude faster as compared to singly mismatched complexes (*33, 34*).

Triplex invasion by homopyrimidine PNAs is severely compromised by increased salt concentrations (*13, 34*), presumably due to DNA duplex stabilization. For instance, the rate is approximately 100 fold decreased when comparing binding at 10 mM and 40 mM of NaCl (*13*), and binding is further inhibited at higher salt concentrations (*4, 40, 41*). This indicates that application in a physiological context requires improved PNAs. Development of positively charged (lysine containing) bis-PNAs has partly solved this problem, because such PNAs can exhibit saturation binding at elevated ionic strength, although with slow association kinetics (*35*). In fact it is found that triplex invasion is enhanced by increasing the number of positive charges (*34*), at least up to 5 lysine residues per decamer bis-PNA (Peter E. Nielsen, unpublished). This improved affinity does not necessarily occur at the expense of sequence selectivity, because positively charged bis-PNAs have been found to bind two orders of magnitude faster to a perfect match site as compared to a singly mismatched site. However, the position of the charge in the PNA molecule may affect both sequence discrimination and binding rate (*34*), thereby adding complexity.

Mechanism of triplex invasion

The mechanism of PNA·DNA-PNA/DNA triplex invasion is not fully resolved. It is not known if the PNA strands bind simultaneously or if one strand binds before the other. Intermediates that exclusively involve either H-PNA or WC-PNA have not been detected. However, as described above, such complexes may form as end products (see Figure 4B-C), and thus all possibilities remain open.

What is the rate-limiting step for triplex invasion? The present data are not conclusive, but transient dissociation of a few nucleobase pairs has been proposed (*41, 42*). Indeed, events leading to unwinding or other destabilization of dsDNA, including negative supercoiling (*41*) and passage by RNA polymerases over the PNA target (*12*) dramatically enhance PNA binding. Also, recent cooperativity studies are in accord with this hypothesis. Importantly, there was correlation between accelerated binding of PNA adjacent to an existing triplex invasion complex and increased sensitivity to permanganate probing of thymidyl residues in the inter-target region (*13*). This is in fact the first structural data linking local dsDNA destabilization with increased PNA binding.

APPLICATION OF TRIPLEX-FORMING PNA

Regulation of gene expression by exogenous molecules has been a long-standing goal in medicinal chemistry, because this could enable treatment at the genetic level.

Basic science may benefit from such developments, *e.g.*, in the elucidation of gene function. Finally, there is an obvious potential for PNA-based diagnostic tools. Thus molecules which recognize RNA and DNA digitally are of prime interest, and indeed PNAs have been employed in both anti-gene (dsDNA targeted) and antisense (RNA targeted) experimental strategies.

Anti-gene

PNA·DNA-PNA/DNA triplex invasion complexes have been shown to arrest transcription elongation by the phage T3 and T7 RNA polymerases (*5*) and by RNA polymerase II (*3, 4*). As predicted from the sequence selectivity of binding, transcription inhibition occurred exclusively with templates containing a target. Notably, high efficiency was obtained when using the PNA bound strand as a template, whereas reduced (RNA polymerase II) or virtually no (T3 and T7 RNA polymerases) transcription inhibition was seen with the displaced DNA strand. As an alternative to PNA-mediated elongation arrest, transcription down regulation also can be achieved using homopyrimidine PNAs targeted to regulatory promoter sequences. This provides a strategy for interference with transcription factor binding, as in the case of NF-κB (*7, 10*). Whereas the anti-gene approach seems quite efficient and mechanistically well-documented as regards transcription inhibition *in vitro* (*3-5*), present data are scarce and not fully consistent with respect to whole-cell systems (*7, 10*).

It has also been shown that triplex invasion complexes arrest T7 DNA polymerase elongation (*10*), and inhibit restriction enzyme cleavage (*43*). Thus a picture emerges in which tracking processes seem obstructed by triplex invasion involving the template strand, and where sequence-specific protein-DNA interactions are prevented. However, this is not the full picture because, *e.g.*, *E. coli* RNA polymerase may employ a PNA·DNA-PNA/DNA complex for transcription initiation *in vitro* (*44*), and mouse fibroblasts treated with triplex-forming PNAs exhibited an elevated mutation frequency primarily in the PNA target region (*45*). Thus a complex series of events may be expected to take place upon exposure of triplex invasion complexes to a cellular environment.

Antisense

Due to the single-stranded nature of mRNA, both duplex and triplex forming PNAs may be used for antisense-mediated translation inhibition. In the original report (*3*), it was shown that PNAs with propensity for triplex formation exhibited antisense activity in a cell-free system, as evidenced by a reduction in the formation of full-length translation product with concomitant appearance of a truncated version which corresponded to termination at the PNA target site. Furthermore, upon microinjection into cells of PNAs targeted to SV40 large T Ag, expression was specifically reduced. It now seems clear that the translation start region is a markedly susceptible target for both duplex- and triplex-forming PNAs. Targeting of 3'-translated sequences *in vitro* using triplex-forming PNAs has been efficient (*6, 8, 9*). Whereas duplex-forming PNAs targeted to the 3'-region could perform well upon microinjection (*6*), such PNAs did not cause translation inhibition *in vitro* (*9*). More data are required before general rules may be deduced to identify the most sensitive sites, and therefore "scanning" of the relevant mRNA is appropriate. Additionally, the additive effects observed when simultaneously using two or three PNAs targeted to different sites suggest that maximal efficiency could require a combination of PNAs (*46*). Finally, homopyrimidine PNAs were shown to inhibit reverse transcriptase elongation (*3, 11*), again conforming to the principle that PNA triplexes with nucleic acids function as roadblocks for enzymes.

PNA in biotechnology

Techniques for DNA cleavage of specific sequences are of importance in DNA technology. A wide variety of restriction endonucleases are now available for this purpose, and these enzymes mostly recognize 4-6 bp sequences. Alternative techniques exist, but these require predetermined sites of more than 15 bp (*47*). Thus a gap of sequences 7-15 bp in length remains in which DNA cannot be cleaved. This is important because rare cleavage of genomic sequences requires recognition sites in this range. To address the issue, Achilles heel type DNA cleavage employing bis-PNAs has been developed (*15, 16*). The strategy is termed PNA assisted rare cleavage (PARC); and bis-PNAs containing sequences with binding overlap to methylase/restriction endonuclease sites were used. The resulting recognition sites (PNA target + methylase/restriction endonuclease site) were in the range 10-13 bp. In λ-DNA and *Saccharomyces cerevisiae* genomic DNA, methylase activity was inhibited upon triplex invasion. After dissociation of the PNA·DNA-PNA/DNA complexes, restriction digestion of targeted sites occurred without cleavage of non-target sites. Although this approach gives a great deal of freedom in choosing the length and thus the number of cleavage sites, the technique is limited by the requirement for both a methylation/restriction site and an overlapping homopyrimidine sequence. Other PNA-assisted DNA cleavage strategies have been explored. These include recognition of triplex invasion complexes by S1 nuclease (*48*) and DNA cleavage by Fe-EDTA oxidation using EDTA conjugated PNAs (*14*).

Finally, triplex-forming PNAs have been used in diagnostically relevant technology. Seeger *et al.* (1997) reported a method for purification of bulk human genomic DNA from whole blood samples (*19*). Biotinylated bis-PNA(T7) was hybridized to genomic DNA after lysis of the blood samples, taking advantage of the fact that A7 sequences are quite frequent. Subsequent purification was accomplished using streptavidin-coated paramagnetic beads, and this affinity capture yielded DNA of sufficient quality for PCR amplification, suggesting that PNA-based diagnostic sample preparation kits can be developed.

ACKNOWLEDGMENTS

We thank Dr. Jakob Larsen and Dr. Liam Good for comments on the manuscript.

REFERENCES

1. Eriksson, M., Christensen, L. Schmidt, J., Orgel, L. and Nielsen, P. E. (1998). Sequence dependent N-terminal degradation of Peptide Nucleic Acid (PNA) in aqueous solution. *New Journal of Chemistry*, in press.

2. Demidov, V., Potaman, V. N., Frank-Kamenetskii, M. D., Buchardt, O., Egholm, M., and Nielsen, P. E. (1994). Stability of peptide nucleic acids in human serum and cellular extracts. *Biochem. Pharmacol.* **48**, 1309-1313.
3. Hanvey, J. C., Peffer, N. C., Bisi, J. E., Thomson, S. A., Cadilla, R., Josey, J. A., Ricca, D. J., Hassman, C. F., Bonham, M. A., Au, K. G., Carter, S. G., Bruckenstein D. A., Boyd, A L., Noble S. A. and Babiss, L. E. (1992). Antisense and antigene properties of peptide nucleic acids. *Science* **258**, 1481-1485.
4. Peffer, N. J., Hanvey, J. C., Bisi, J. E., Thomson, S. A., Hassman, F. C., Noble, S. A. and Babiss, L. E. (1993). Strand-invasion of duplex DNA by peptide nucleic acid oligomers. *Proc. Natl. Acad. Sci. USA* **90**, 10648-10652.
5. Nielsen, P. E., Egholm, M. and Buchardt, O. (1994). Sequence specific transcription arrest by PNA bound to the template strand. *Gene* **149**, 139-145.
6. Bonham, M. A., Brown, S., Boyd, A. L., Brown, P. H., Bruckenstein, D. A., Hanvey, J. C., Thomson, S. A., Pipe, A., Hassman, F., Bisi, J. E., Froehler, B. C., Matteucci, M. D., Wagner, R. W., Noble, S. A. and Babiss, L. E. (1995). An assessment of the antisense properties of RNase H-competent and steric-blocking oligomers. *Nucleic Acids Res.* **23**, 1197-1203.
7. Vickers, T. A., Griffith, M. C., Ramasamy, K., Risen, L. M. and Freier, S. M. (1995). Inhibition of NF-kappa B specific transcriptional activation by PNA strand invasion. *Nucleic. Acids Res.* **23**, 3003-8.
8. Gambacorti-Passerini, C., Mologni, L., Bertazzoli, C., Marchesi, E., Grignani, F. and Nielsen, P. E. (1996). In vitro transcription and translation inhibition by anti-PML/RAR and PML Peptide Nucleic Acid (PNA). *Blood* **88**, 1411-1417.
9. Knudsen, H. and Nielsen, P. E. (1996). Antisense properties of duplex and triplex forming PNA. *Nucleic Acids Res.* **24**, 494-500.
10. Praseuth, D., Grigoriev, M., Guieysse, A. L., Pritchard L. L., Harel-Bellan, A., Nielsen, P. E. and Hélène, C. (1997). Peptide nucleic acids directed to the promoter of the alpha-chain of the interleukin-2 receptor. *Biochim. Biophys. Acta* **1309**, 226-238.
11. Koppelhus, U., Zachar, V., Nielsen, P. E., Liu, X., Eugen-Olsen, J. and Ebbesen, P.(1997). Efficient in vitro inhibition of HIV-1 gag reverse transcription by peptide nucleic acid (PNA) at minimal ratios of PNA/RNA. *Nucleic Acids Res.* **25**, 2167-2173
12. Larsen, H. J. and Nielsen, P. E. (1996). Transcription-mediated binding of peptide nucleic acid (PNA) to double stranded DNA: Sequence specific suicide transcription. *Nucleic Acids Res.* **24**, 458-463.
13. Kurakin, A., Larsen, H. J. and Nielsen, P. E. (1998). Cooperative strand displacement by Peptide Nucleic Acids (PNA). *Chem. and Biol.* **5**, 81-89.
14. Lohse, J., Hui, C., Sönnichsen, S. H. and Nielsen, P. E. (1996). Sequence selective DNA cleavage by PNA-NTA conjugates. *NATO Adv. Ser. DNA and RNA Cleavers and Chemotherapy of Cancer and Viral Diseases* 133-141.
15. Veselkov, A. G., Demidov, V., Nielsen, P. E. and Frank-Kamenetskii, M. D. (1996). PNA as a rare genome-cutter. *Nature* **379**, 214.
16. Veselkov, A. G., Demidov, V. V., Nielsen, P. E. and Frank-Kamenetskii, M. (1996). A new class of genome rare cutters. *Nucleic. Acids Res.* **24**, 2483-2487.
17. Armitage, B., Koch, T., Frydenlund, H., Örum, H., Batz, H-G., and Schuster, G. (1997). Peptide nucleic acid-anthraquinone conjugates: strand invasion and photoinduced cleavage of duplex DNA. *Nucleic Acids Res.* **25**, 4674-4678.
18. Bigey, P., Sönnichsen, S. H., Nielsen, P. E., Meunier, B. (1997). DNA binding and cleavage by a cationic manganese porphyrin-peptide nucleic acid conjugate. *Bioconjugate Chem.* **8**, 267-270.
19. Seeger, C., Batz, H-G., and Örum, H. (1997). PNA-mediated purification of PCR amplifiable human genomic DNA from whole blood. *BioTechniques.* **23**, 512-517.

20. Nielsen, P. E., Egholm, M., Berg, R. H. and Buchardt, O. (1991). Sequence selective recognition of DNA by strand displacement with a thymine-substituted polyamide. *Science* **254**, 1497- 1500.
21. Hyrup, B., Egholm, M., Rolland, M., Nielsen, P. E., Berg, R. H. and Buchardt, O. (1993). Modification of the binding affinity of peptide nucleic acids (PNA). PNA with extended backbones consisting of 2-aminoethyl-alanine or 3-aminopropylglycine units. *J. Chem. Soc. Chem. Commun.* **6**, 518-519.
22. Hyrup, B., Egholm, M., Nielsen, P. E., Wittung, P., Nordén, B. and Buchardt, O. (1994). Structure-activity studies of the binding of modified peptide nucleic acids (PNA) to DNA *J. Amer. Chem. Soc.* **116**, 7964-7970.
23. Hyrup, B., Egholm, M., Buchardt, O and Nielsen, P. E. (1996). A flexible and positively charged PNA analogue with an ethylene-linker to the nucleobase: Synthesis and hybridization properties. *Bioorg. Med. Chem. Lett.* **6**, 1083-1088.
24. Egholm, M., Buchardt, O., Christensen, L., Behrens, C., Freier, S. M., Driver, D. A., Berg, R. H., Kim, S. K., Nordén, B. and Nielsen, P. E. (1993). PNA hybridizes to complementary oligonucleotides obeying the Watson-Crick hydrogen bonding rules. *Nature* **365**, 556-568.
25. Egholm, M., Buchardt, O., Nielsen, P. E. and Berg, R. H. (1992). Peptide nucleic acids (PNA). Oligonucleotide analogues with an achiral peptide backbone. *J. Amer. Chem. Soc.* **114**, 1895-1897.
26. Egholm, M., Buchardt, O., Nielsen, P. E. and Berg, R. H. (1992). Recognition of guanine and adenine in DNA by cytosine and thymine containing peptide nucleic acids (PNA). *J. Amer. Chem. Soc.* **114**, 9677-9678.
27. Kim, S. K., Nielsen, P. E., Egholm, M., Buchardt, O., Berg, R. H. and Nordén, B. (1993). Right-handed triplex formed between peptide nucleic acid PNA-T8 and poly(dA) shown by linear and circular dichroism spectroscopy. *J. Amer. Chem. Soc.* **115**, 6477-6481.
28. Egholm, M., Christensen, L., Dueholm, K., Buchardt, O., Coull, J. and Nielsen, P. E. (1995). Efficient pH independent sequence specific DNA binding by pseudoisocytosine-containing bis-PNA. *Nucleic Acids Res.* **23**, 217-222.
29. Betts, L., Josey, J. A., Veal, J. M., Jordan, S. R. (1995). A nucleic acid triple helix formed by a peptide nucleic acid-DNA complex. *Science* **270**, 1838-1841
30. Lagriffoule, P., Buchardt, O., Wittung, P., Nordén, B., Jensen, K. K. and Nielsen, P. E. (1997). Peptide Nucleic Acids (PNAs) with a conformationally constrained, chiral cyclohexyl derived backbone. *Chem. Eur. J.* **3**, 912-919.
31. Haaima, G., Lohse, A., Buchardt, O., Nielsen, P. E. (1996). Peptide nucleic acids (PNAs) containing thymine monomers derived from chiral amino acids: hybridization and solubility properties of D-Lysine PNA. *Angew. Chem. Int. Engl.* **35**, 1939-1941.
32. Nielsen, P. E., Egholm, M. and Buchardt, O. (1994). Evidence for (PNA)$_2$/DNA triplex structure upon binding of PNA to dsDNA by strand displacement. *J. Mol. Recognition* **7**, 165- 70.
33. Demidov, V. V., Yavnilovich, M. V., Belotserkovskii, B. P., Frank-Kamenetskii, M. D. and Nielsen, P. E. (1995). Kinetics and mechanism of PNA binding to duplex DNA. *Proc. Natl. Acad. Sci. USA* **92**, 2637-2641.
34. Kuhn, H., Demidov, V., Frank-Kamenetskii, M. D. and Nielsen, P. E. (1997). Kinetic sequence discrimination of bis-PNAs upon targeting of double-stranded DNA. *Nucleic Acids Res.* **26**, 582-587.
35. Griffith, M. C., Risen, L. M., Greig, M. J., Lesnik, E. A., Sprangle, K. G., Griffey, R. H., Kiely, J. S. and Freier S. M. (1995). Single and bis-peptide nucleic acids as triplexing agents: Binding and Stoichiometry. *J. Amer. Chem. Soc.* **117**, 831-832.
36. Wittung, P., Nielsen, P. E. and Nordén, B. (1997). Extended DNA-recognition repertoire of PNA. *Biochemistry* **36**, 7973-7979.

37. Nielsen, P. E. and Christensen, L. (1996). Strand displacement binding of a duplex forming homopurine PNA to a homopyrimidine duplex DNA target. *J. Amer. Chem. Soc.* **118**, 2287- 2288.
38. Bukanov, N., Demidov, V., Nielsen, P. E., Frank-Kamenetskii, M. D. (1998). PD-loop: a complex of duplex DNA with an oligonucleotide. *Proc. Natl. Acad. Sci. USA* **95**, 5516-5520.
39. Lohse, J. (1997). The principle of non-complementarity. Ph.D.-thesis, Faculty of Science, Copenhagen University, Denmark.
40. Cherny, D. Y., Belotserkovskii, B. P., Frank-Kamenetskii, M. D., Egholm, M., Buchardt, O., Berg, R. H. and Nielsen, P. E. (1993). DNA unwinding upon strand displacement binding of PNA to double stranded DNA. *Proc. Natl. Acad. Sci. USA* **90**, 1667-1670.
41. Bentin, T. and Nielsen, P. E.: Enhanced peptide nucleic acid (PNA) binding to supercoiled DNA: Possible implications for DNA "breathing" dynamics. *Biochemistry* **35**, 8863-8869.
42. Wittung, P., Nielsen, P. E. and Nordén, B. (1996). Direct Observation of PNA Strand Invasion by Circular Dichroism *J. Amer. Chem. Soc.* **118**, 7049-7054.
43. Nielsen, P. E., Egholm. M., Berg, R. H. and Buchardt, O. (1993). Sequence specific inhibition of restriction enzyme cleavage by PNA. *Nucleic Acids Res.* **21**, 197-200.
44. Möllegaard, N. E., Buchardt, O., Egholm, M. and Nielsen, P. E. (1994). PNA-DNA Strand displacement loops as artificial transcription promoters. *Proc. Natl. Acad. Sci. USA* **91**, 3892- 3895.
45. Farugi, A. F., Egholm, M., Glazer, P. M. (1998). Peptide nucleic acid-targeted mutagenesis of a chromosomal gene in mouse cells. *Proc. Natl. Acad. Sci. USA* **95**, 1398-1403.
46. Mologni, L., Nielsen, P. E., Gambacorti-Passerini, C. (1998). Additive antisense effects of different PNAs on the in vitro translation of the PML/RARα gene. *Nucleic Acids. Res.* **26**, 1934-1938.
47. Strobel, S. A., and Dervan, P. B. (1991). Single-site enzymatic cleavage of yeast genomic DNA mediated by triple helix formation. Nature. 350, 172-174.
48. Demidov, V., Frank-Kamenetskii, M. D., Egholm., M., Buchardt, O., and Nielsen, P. E. (1993). Sequence selective double strand DNA cleavage by peptide nucleic acid (PNA) targeting using nuclease S1. *Nucleic Acids Res.* **21**, 2103-2107.

19 TRIPLE HELIX STABILIZING AGENTS

Christophe Escudé and Thérèse Garestier

SUMMARY

The applications of triple helix forming oligonucleotides can be limited by the relatively low stability of triple helical complexes. Ligands which can bind more tightly to triple helical DNA than to double-helical DNA can stabilize triple helices. This chapter reviews the structural aspects of triple helix ligand complexes and briefly discusses the possible use or role of triple helix stabilizing agents in triplex-based strategies or mechanisms.

INTRODUCTION

The formation of nucleic acid triple helices has allowed the design of new DNA binding agents which could be used to develop tools for molecular biology as well as new therapeutic strategies. However, one of the limitations to these strategies is the relatively poor stability of triple helical DNA compared to double helical DNA under physiological conditions. This stability can be increased by using chemically modified oligonucleotides. Another approach which can be used to stabilize triple helices consists in using agents that bind specifically to triple helices. Indeed, triplex formation relies on binding of an oligonucleotide to a DNA duplex, according to the following equilibrium: $D + M \leftrightarrow T$ where D is the duplex, M the third strand and T is the triple helix. Thus, any ligand which would have a greater affinity for the triplex than for the duplex and third strand should displace the equilibrium to the right.

Triple helix structure

When an oligonucleotide aligns in the major groove of duplex DNA, there are three resulting grooves in the triplex (Figure 1). One of them, the Watson-Crick groove,

258

resembles the minor groove of double-stranded DNA, whereas the two other ones have completely new geometries, which are specific for triplex DNA. These grooves differ from those found in duplex DNA by their size, but also by their electrostatic as well as hydration properties. The large surface of the base triplets could be used to interact with planar chromophores which could intercalate between DNA triplets. Minor groove ligands might also be able to bind to the Watson-Crick or the Watson-Hoogsteen groove.

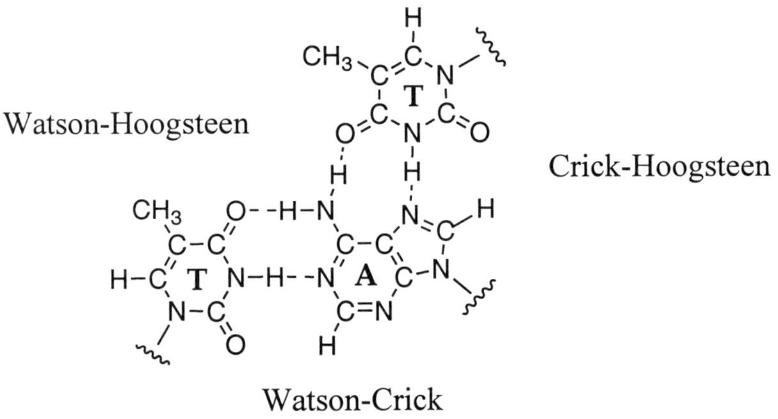

Watson-Hoogsteen

Crick-Hoogsteen

Watson-Crick

Figure 1. Structure of a T.A x T base triplet, with the names given to each of the different grooves of the triple helix.

Ligand screening strategies and binding mode studies

Thermal denaturation followed by absorption spectroscopy is a very commonly used technique to measure the stability of triple helical complexes. Quantitative footprinting experiments using DNase I can also be used to measure apparent dissociation constants (K_d) of triplex forming oligonucleotides (TFOs) from their target. Triplex stabilizing agents are expected to increase the melting temperature and to decrease apparent K_d for triplexes. Agents with these properties have been looked for, first by screening molecules that were known to interact with nucleic acids. The next two paragraphs will describe the principal results obtained for intercalators and groove binders, whose binding mode to triplex DNA can be studied by the same biophysical techniques as the ones used for characterizing duplex DNA ligands.

INTERCALATORS

The most classical duplex intercalator, ethidium bromide, was shown to have a higher affinity for poly(dA).2poly(dT) than for poly(dA).poly(dT) *(1)*, but as soon as protonated C.G x C+ base triplets are present, this compound destabilizes the triple helix *(2)*. This compound was shown to intercalate between T.A x T base triplets by fluorescence energy transfer experiments. Benzopyridoindole derivatives (BPI) were the first ligands shown to bind to triple helices containing C.G x C+ base triplets, thus leading to an important stabilization of these structures *(3)*. Next it was shown that the antitumor agent coralyne also stabilizes triple helices *(4, 5)*. Other four-membered ring chromophores can also stabilize triple helices, e.g., sanguinarine *(5)*, benzophenazine *(6)*, benzophenanthridine (S. Kukreti, unpublished results) or imidazothioxanthone *(7)*. Another series of compounds, consisting of unfused aromatic cations, stabilizes poly(dA).2poly(dT) triple helices *(8)*. More recently, various derivatives of the tricyclic amidoanthraquinones *(9)* and compounds containing five aromatic rings *(10, 11)* have also been shown to stabilize triple helices by intercalation, thus expanding the range of known structures aimed at intercalating between base triplets. Some of these ligands are described in greater detail in the following paragraphs.

Benzopyridoindoles, benzopyridoquinoxalines and derivatives

A benzo[e]pyridoindole derivative (BePI, Figure 2) has been shown to intercalate between T.A x T base triplets. Fluorescence quenching of BePI by acrylamide and Fe(CN) is inhibited in the presence of double- or triple-stranded DNA. Fluorescence energy transfer occurs between bases and intercalated BePI. The number of bases which transfer energy to the intercalated compound has been estimated to be 4.5 in the triplex and 2.8 in the duplex, which suggests that more bases are close enough to the intercalator in the triplex for the energy transfer to occur and that this transfer is more efficient. Also, viscometric experiments have shown that this ligand unwinds DNA with an angle of 21° *(12, 13)*, and linear dichroism experiments allowed to measure an angle between the planar chromophore and the triple helix helical axis close to 90°*(14)*.

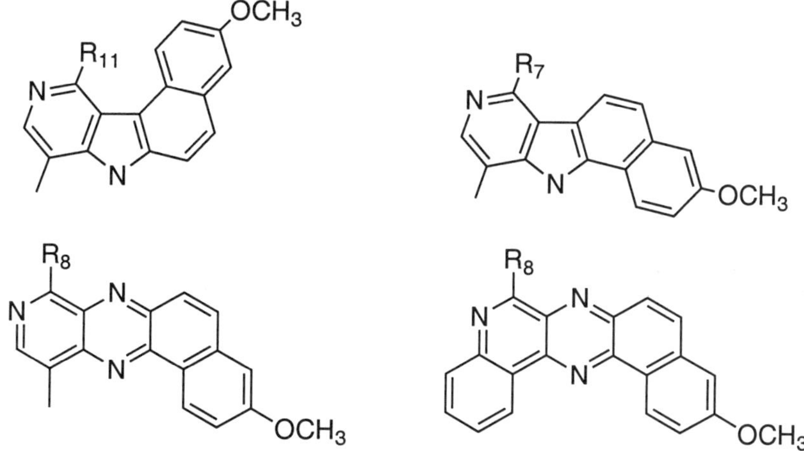

Figure 2. Structure of BePI (top left), BgPI (top right), BfPQ (bottom left) and BQQ (bottom right). For all compounds, R=NH(CH$_2$)$_3$NH$_3^+$.

260

BePI carries two positive charges, one on the alkylamine side chain at position 11 and another one on the N_{10} aromatic nitrogen, the pK_a of which is shifted from 8 to 9 when the molecule is bound to a triple helix. The crescent shape of this molecule was believed to allow good stacking interactions with base triplets, but the related BgPI (Figure 2), which has a more linear geometry, can also stabilize triple helices. As no detailed structure was available, thermal denaturation experiments and molecular modeling have been used to study the effects of the structure of BePI and BgPI derivatives on triple helix stabilization. A model was proposed in which the side chain R_{11} of BePI lies in the Watson-Hoogsteen groove whereas the side chain R_7 of BgPI lies in the Watson-Crick groove (Figure 3) *(15)*. This model also suggested that these compounds interact mostly with the Hoogsteen-paired bases within the triplet, which has led to the hypothesis, later verified, that BPI could bind to Hoogsteen-paired duplexes *(16)*.

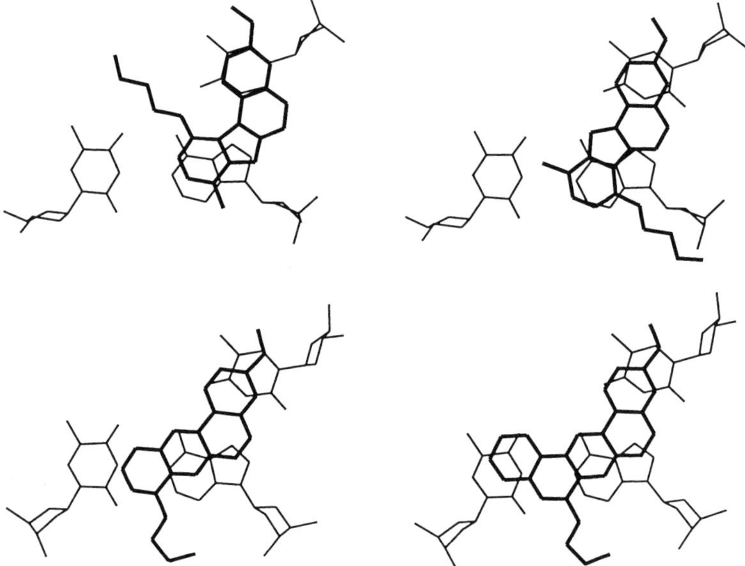

Figure 3. Model for BePI (top left), BgPI (top right), BfPQ (bottom left) and BQQ (bottom right) intercalated between T.A x T base-triplets. The ligands were drawn in bold lines. Hydrogen atoms were omitted for clarity

The validity of this model has also been checked by trying to use it to design better triplex stabilizing agents. Benzo[f]pyridoquinoxalines (BfPQ) were synthesized as analogues of benzopyridoindole derivatives, in which the indole moiety was replaced by a quinoxaline moiety *(17)*. These molecules have been shown to stabilize triple helices, probably using the same strategy as their benzopyridoindole counterpart. Based on the model which suggested that stacking interactions involved mostly the Hoogsteen-paired bases, it was predicted that a benzo[f]quino[3,4-b] quinoxaline derivative (BQQ), which possesses an additional aromatic ring, could engage additional stacking interactions with the pyrimidine strand of the Watson-

Crick double helix (Figure 3). This compound was synthesized. Thermal denaturation experiments and inhibition of restriction enzyme cleavage showed that this new compound could indeed stabilize triple helices with great efficiency and/or induce triple helix formation under physiological conditions *(18)*. Other pentacyclic compounds, derived from benzopyridoindoles, have also been synthesized and shown to induce a strong stabilization of triple helices *(19)*.

Naphtylquinoline derivatives

A series of quinoline derivatives with different alkylamine side chains at the 4-position and with different aryl substituents at the 2-position has been synthesized *(8)*. Spectral changes and energy transfer results indicated that the naphtyl compound (Figure 4) and related derivatives bind to triplex DNA by intercalation.

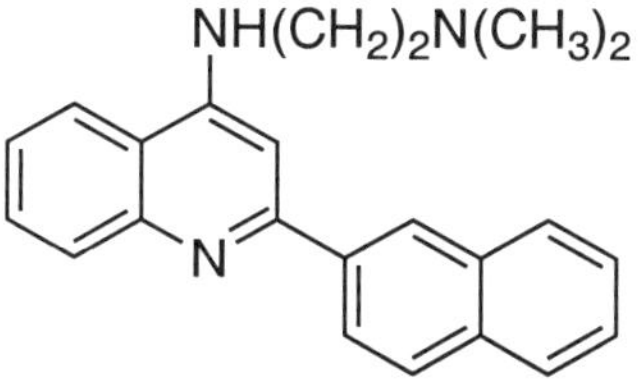

$$NH(CH_2)_2N(CH_3)_2$$

Figure 4. Structure of (N-[2-(dimethylamino)ethyl]-2-(2-naphtyl)quinolin-4-amine)

Molecular modeling results suggested good stacking of the naphtylquinoline ring system with the bases at the triplex intercalation site, whereas this ring system would be too large to stack optimally with base pairs at a duplex intercalation site. The alkylamine side chain interacts with the Watson-Crick groove as for BgPI. A structure-activity relationship study has been carried out for this series of compounds. It suggested that protonation at N_1 of the quinoline increases stability of the complex with the DNA triplex. Indeed, the measured order of pK_a for different compounds parallels the order of triplex stabilizing properties *(20)*. This series of compounds has been studied further by DNase I footprinting experiments. In the presence of the triplex-binding ligand, the concentrations of T5C5 and C5T5 required to generate DNase I footprints at the target sites A6G6.C6T6 and G6A6.T6C6, respectively, are reduced by at least 100 fold *(21)*. It should be noted that the naphtylquinoline compound has also been shown to stabilize triple helices when the third strand is modified by phosphorothioate linkages. This modification usually abolishes or strongly destabilizes triplex formation. However, phosphorothioate oligonucleotides are more resistant to intracellular nucleases than are phosphodiester oligonucleotides. It is thus interesting to stabilize triple helices with this type of third strand.

Diamidoanthraquinone derivatives

Difunctionalized bisamidoanthraquinones bearing pendant basic amine groups have been studied for their ability to interact with triple helices. It has been shown that

2,6-disubstituted amidoanthraquinones bind preferentially to triplex DNA rather than duplex DNA, whereas the 1,4-series displays opposite properties *(9)*. Molecular modeling studies suggest that the 2,6-derivatives can intercalate into triple helices with one side chain in the Watson-Crick groove and the other one in the Watson-Hoogsteen groove. For the 1,4-series however, such a geometry is not possible due to steric constraints (Figure 5). Binding from the major groove of double helical DNA would compete with binding of a triplex forming oligonucleotide, whereas binding from the minor groove would produce minimal base-chromophore stacking.

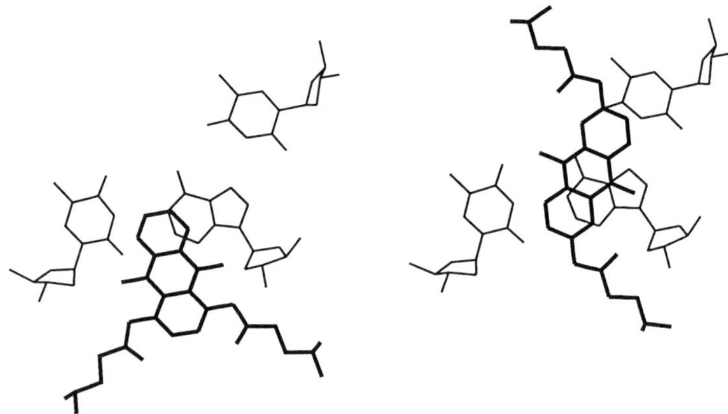

Figure 5. model for 1,4- (left) and 2,6- (right) disubstituted amidoanthraquinones

Binding of these compounds to duplex and triplex DNA has been studied by isothermal titration calorimetry, UV melting and competition dialysis techniques *(22)*. It has been shown that binding to the preferred DNA (i.e., triplex or duplex DNA) is enthalpically driven for each ligand, whereas binding to the disfavored DNA is either entropically driven or enthalpy/entropy compensated. The binding site size (3.6 base triplets) suggests DNA intercalation. A series of anthraquinone sulfonamide derivatives, which differ from the previous compounds by the chemical structure of the side chains, has also been investigated *(23)*. Phosphorescence quenching and viscosimetric titrations indicated that the quinones bind to triplex by intercalation. 2,7-derivatives provide a better stabilization than 2,6-derivatives and, interestingly, these compounds intercalate into triplex DNA whereas they bind to duplex DNA in the minor groove.

A ruthenium complex

Linear and circular dichroism and thermal denaturation experiments have been used to investigate the binding of various ruthenium complexes to double- and triple-stranded DNA *(10)*. These complexes are octahedral ruthenium complexes where ruthenium is bound by two molecules of phenanthroline and a third molecule which can be phenanthroline, dipyrido[3,2-a:2',3'-c]phenazine (DPPZ) or benzodipyrido[3,2-a:2',3'-c]phenazine (BDPPZ) (Figure 6). Linear dichroism indicated that the extended DPPZ and BDPPZ ligands lie approximately parallel to the base-pair and base-triplet planes, consistent with intercalation; this interpretation is also supported by strong hypochromism in the interligand absorption bands with either duplex or triplex. The spectral properties of the metal complexes were similar for binding to either duplex or triplex DNA, indicating that the third strand, which occupies the major groove, has little effect on the binding geometries, and hence supporting the hypothesis that the metal complexes bind from the minor groove.

Third strand stabilization depended on the nature of the third substituted phenanthroline chelate ligand but was not directly related to its size, with stabilizing power increasing in the order phen<BDPPZ<DPPZ. This observation can be explained by the fact that the extended BDPPZ ligand would protrude into the major groove and would have greater interference than DPPZ with the third strand.

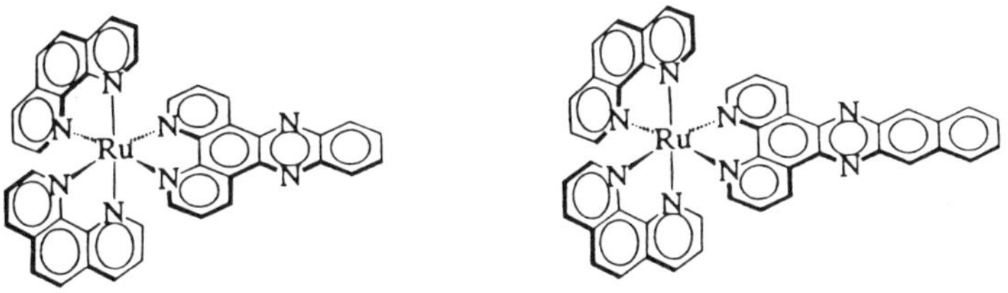

Figure 6. Molecular structure of the [Ru(II)(1,10-phenanthroline)$_2$L]$^{2+}$ complexes, where L=DPPZ (left) or L=BDPPZ (right). Only the Λ enantiomers are shown.

NON-INTERCALATING AGENTS

Minor groove binders

As the Watson-Crick groove of triple helices resembles the minor groove of double-helical DNA, minor groove binding drugs have been investigated for their ability to bind or stabilize triple helices. Netropsin was the first minor groove ligand for which binding to triplex DNA was investigated. A combination of spectroscopic and calorimetric techniques has shown that netropsin can bind in the minor groove of triple helical DNA without displacing the third strand from the major groove *(24, 25)*. Further, netropsin binding thermally destabilizes the triplex to duplex transition dramatically, contrasting with its stabilization of the duplex to single strands transition. There is also a decrease in the cooperativity of the triplex to duplex transition on binding of netropsin. These results establish major groove/minor groove crosstalk where the drug, which is bound in the minor groove of the triplex, affects the properties of this triplex as related to the equilibrium in which the third strand is expelled from the major groove. Another study has suggested that a second netropsin molecule could bind to the Watson-Hoogsteen groove of triple helical DNA *(26)*. This would account for a greater ΔH for binding to triplex compared with duplex DNA, as measured by microcalorimetric experiments.

A lot of other minor groove ligands have been studied. They include Hoechst 33258 *(27)*, berenil *(28)*, DAPI *(29)* and distamycin A *(30)*. Circular dichroism spectroscopy has shown that these molecules can bind to the minor groove of triple helices. In the case of Hoechst 33258, circular dichroism spectroscopy suggests that there might be structural differences when the ligand is complexed to double- and triple-helical DNA.

Stability of DNA triple helices is in general lower in the presence of all these minor groove binders. There are few examples in which a minor groove binding ligand can stabilize triple helices. These triple helices contain at least one RNA strand. DAPI can induce the formation of the poly(rA).poly(rA).poly(dT) triplex, and both DAPI and berenil can induce the formation of a poly(dT).poly(rA).poly(dT), an RNA-containing triple helix which does not form in the absence of any ligand *(29)*.

All these compounds bind to A.T rich sequences; thus their effect has been studied on T.A x T containing triple helices. It has also been shown that mithramycin, a ligand that binds in the minor groove of G.C rich sequences, can also interfere with the binding of T5C5 to the target A6G6.T6C6 *(31)*.

Polycations

It has been known for a long time that triple helices in the pyrimidine motif are stabilized by dications like magnesium *(32)* or polycations like spermine *(33)*. A systematic study of the effects of polyamines on triple helices has been published *(34)*. Spermine has been shown to be more efficient in stabilizing triple helices than spermidine derivatives, which are themselves better than putrescine compounds.

Chemical modifications of biogenic polyamines could improve their triplex stabilizing properties. For example, transformation of the primary amino groups in these molecules into corresponding guanidinium functions amplified the electrostatic component and resulted in better stabilization, especially at pH 7 *(35)*. Substitution of the N4-hydrogen of spermidine by a cyclohexyl ring or the fusion of the N4-hydrogen in a cyclic ring system, as in piperidine, enhances the ability of spermidine analogues to stabilize triplex DNA *(36)*. Pentamines are more efficacious than tetramines in stabilizing triplex DNA. Ethyl substitution of the pendant amino groups decreases the triplex stabilizing properties, but abolishes DNA aggregation, which in fact allows the use of higher concentrations of polyamine and thus to reach a higher overall stability for the triple helix *(37)*.

Binding of polyamines is thus governed not only by ionic effects, but also by structural specificity effects, including steric constraints and charge separation. The length of the methylene bridging region separating amino and imino groups markedly influenced their ability to stabilize triplex structures. The mechanism of polyamine-DNA interactions is not yet clearly understood. Electrostatic, major-groove-binding and minor-groove-binding modes of DNA recognition have been suggested by several investigators. Both the Watson-Hoogsteen and the Watson-Crick grooves of triple helices could be target sites for polyamines.

Peptides

The effects of basic oligopeptides on triplex DNA were studied. The peptides Lys-Lys-Lys-Lys-Lys and Lys-Arg-Lys-Arg-Lys were shown to stabilize triplex DNA with the same efficiency as spermine *(38)*. Linear polycations, such as poly(L-lysine) and poly(L-arginine), also thermally stabilize DNA triplexes, but the complexes between DNA and the polycations are liable to precipitate, which limits the use of these compounds in stabilizing triplexes. A comb-type polycation consisting of poly(L-lysine) backbone and grafted chains of hydrophilic polymers, which was designed to avoid precipitation problems, was shown to markedly stabilize triplexes *(39)*.

DESIGN AND APPLICATIONS OF TRIPLEX STABILIZING AGENTS

Design of triplex stabilizing agents

From all the available data which have been summarized above, it appears that intercalators are much more efficient triplex stabilizers than are groove binding compounds. The efficiency of triplex stabilization is the result of a higher binding affinity for the triplex than for the duplex. This structural specificity can be achieved by the shape of a polyaromatic system, but also by electrostatic interactions and steric constraints.

Stacking interactions. Triplex intercalators with 3 to 5 aromatic rings have been reported. One of the rationales used in the design of triplex specific ligands was that a too large surface area would not favor intercalation in duplex DNA. Experimental results clearly indicate that increasing the surface area of an intercalating ligand is not sufficient to improve its triplex stabilizing properties. New aromatic cycles must accommodate the geometry of the triple helix. Polyaromatic compounds with a large surface area can self-associate, a process which can compete with binding to triple helices.

Electrostatic interactions. All known triplex intercalators and groove binding ligands are positively charged. In the case of intercalators, the positive charge is carried by the polyaromatic ring system or/and by a chemical moiety located in one of the grooves, which can be a side chain or a metallic dication. The position of these side chains can drastically influence the triplex stabilizing properties. Although no detailed structure is available for such complexes, models suggest that these side chains could lie in either the Watson-Crick or the Watson-Hoogsteen groove of the triple helix (the Crick-Hoogsteen groove being to small to accommodate them). One reported ligand with two side chains would have one in each of these grooves *(9)*.

Hydrophobic interactions. Grooves in nucleic acids structure are known to be hydrated. Binding of ligands to triple helices will interfere with this hydration pattern. Hydrophobic pockets are also present within these grooves and may be responsible for the stabilizing effects of uncharged hydrophobic side chains. All these hydrophobic effects are difficult to investigate and take into account with molecular modeling techniques.

Sequence specificity. Most compounds intercalating into triple helices seem to favor T.A x T base triplets, and stabilization decreases when the number of C.G x C+ base triplets increases. This is probably explained by electrostatic repulsions between the chromophore and protonated cytosine. However, intercalation close to a C.G x C+ base triplet has not been excluded, and coralyne has been reported to stabilize the poly(dGA).2poly(dTC) triple helix *(4)*. There have been few reports regarding stabilization of triple helices with a purine-containing third strand. Binding of a 14-mer (G,T)-containing oligonucleotide which does not bind to its target in the absence of ligand could be induced by addition of BePI *(40)*. One interesting possible mechanism for this effect could be that intercalation reduces the distortion which occurs at the junction between the non-isomorphous T.A x T and C.G x G triplets. There has been only one report regarding stabilization of triple helices with a GA third strand *(41)*. BePI was also able to increase stability of the triple helix formed with the oligonucleotide T4CT4G6, in which 6 C.G x G Hoogsteen type triplets are formed *(42)*.

Stabilization of intermolecular triple helices

Formation of triple helices under physiological conditions. Most of the biophysical studies of triplex stabilizing agents involved triple helices with a pyrimidine third strand. Efficient ligands for this type of triple helices have been found, and it has been shown that some of them can induce the formation of triple helices under physiological conditions, especially the BQQ compound *(18)*. However, formation of these triple helices is still limited and pH dependent. Strong enhancement of binding of purine-containing oligonucleotides seems to be a promising approach. However, there are very few ligands reported to stabilize this type of triplexes.

Stabilization of mismatched base triplets. Effect on stringency of triple helix formation has been studied for BPI *(43)* and naphtylquinoline *(44)* derivatives. In both cases, it was shown that the most stable base triplet was the same in the absence or in the presence of the triplex specific ligand, i.e., the specificity of triple helix formation was unchanged. However, mismatched base triplets were considerably stabilized, and some of them more than other ones. This property could be used in order to recognize mismatched targets, but the use of triplex stabilizing ligands will also enhance binding of the triplex forming oligonucleotide at nonspecific target sequences, which might cause unwanted side effects.

Improvements of transcription inhibition by TFOs. Several systems have been designed to demonstrate a biological activity of TFOs. In a bacterial *in vitro* system, the sequence specific inhibition of transcription initiation by triplex formation can be strongly enhanced by stabilizing the triplex through formation of the BePI intercalation complex *(3)*. Another *in vitro* study has shown that TFOs were able to arrest transcription elongation by eukaryotic and prokaryotic polymerases. In both cases, addition of BePI dramatically increased inhibition efficiency *(42)*. Selective utilization of a polyamine was also shown to enhance the gene inhibitory effect of a TFO on the *c-myc* oncogene in MCF-7 breast cancer cells *(45)*.

Triplex stabilizing agents and H-DNA

Intramolecular triple helices can form naturally within supercoiled double-stranded DNA. Formation of this structure has been demonstrated in bacteria, whereas a lot of sequences that are likely to be able to adopt such structures exist in eukaryotic genomes, especially in the promoter regions. Only one study has addressed the question of the effect of triplex binding ligands on intramolecular triple-helices *(46)*.

A mirror repeat sequence was inserted between the *E. coli* β-lactamase gene, which confers resistance to the antibiotic ampicillin, and its promoter, in a plasmid which also expresses the resistance gene for the antibiotic tetracycline. Growth of the cells containing the mirror repeat sequence was slower in the presence of ampicillin as compared with plasmids not containing this sequence, whereas both types of cells could grow at the same rate in the presence of tetracycline. Thus, transcription of the β-lactamase gene was inhibited by the mirror repeat sequence, whereas the plasmid could replicate normally. Adding BePI was shown to increase this inhibition specifically, and chemical probes (chloroacetaldehyde) showed that BePI stabilized the H-DNA structure induced by supercoiling in the plasmid. This experiment suggests that mirror sequences that are able to form H-DNA structures could act as gene regulators, and that regulation of these genes might be influenced by triplex stabilizing agents.

Some of the triplex stabilizing agents described above have cytotoxic properties. This raises the interesting possibility of a new mechanism of action for these drugs. A triplex stabilizing agent could induce the formation of an H-DNA structure within the promoter region of a specific gene and thus downregulate expression of this gene.

CONCLUSION

From results of classical physical techniques and structure-activity relationship studies, it has been possible to start to understand how some ligands may bind specifically to triple helices. Although new very specific triple helix ligands have been designed with the help of molecular modeling calculations, getting detailed structural information about those complexes remains a challenge. Triple-helix-specific ligands can be used to enhance triple helix formation under physiological conditions. They could also potentially be used to extend the range of sequences that can be recognized by TFOs. However, binding of TFOs to nonspecific targets will also be increased. Increasing the stability of triplex structures results in an improvement of biological properties of TFOs. This has been clearly demonstrated in an *in vitro* transcription inhibition experiment. From all these studies, therapeutically relevant compounds could be selected for further *in vivo* studies, depending on the type of triple helix to be stabilized.

REFERENCES

1. Scaria, P. V. and Shafer, R. H. (1991). Binding of ethidium bromide to a DNA triple helix. *J. Biol. Chem.* **266**, 5417-5423.
2. Mergny, J. L., Collier, D., Rougée, M., Montenay-Garestier, T. and Hélène, C. (1991). Intercalation of ethidium bromide in a triple stranded oligonucleotide. *Nucl. Acids Res.* **19**, 1521-1526.
3. Mergny, J. L., Duval-Valentin, G., Nguyen, C. H., Perrouault, L., Faucon, B., Rougée, M., Montenay-Garestier, T., Bisagni, E. and Hélène, C. (1992). Triple Helix-Specific Ligands. *Science* **256**, 1681-1684.
4. Lee, J. S., Latimer, L. J. P. and Hampel, K. J. (1993). Coralyne binds tightly to both T.A.T and C.G.C+-containing DNA triplexes. *Biochemistry* **32**, 5591-5597.
5. Latimer, L. J. P., Payton, N., Forsyth, G. and Lee, J. S. (1995). The binding of analogues of coralyne and related heterocyclics to DNA triplexes. *Biochem. Cell. Biol.* **73**, 11-18.
6. Tarui, M., Doi, M., Ishida, T., Inoue, M., Nakaike, S. and Kitamura, K. (1994) DNA-binding characterization of a novel anti-tumour benzo[a]phenazine derivative NC-182: spectroscopic and viscosimetric studies. *Biochem. J.* **304**, 271-279.
7. Fox, K. R., Thurston, D. E., Jenkins, T. C., Varvaresou, A., Tsotinis, A. and Siatrapapastaikoudi, T. (1996). A novel series of DNA triple helix-binding ligands. *Biochem. Biophys. Res. Comm.* **224**, 717-720.
8. Wilson, W. D., Tanious, F. A., Mizan, S., Yao, S., Kiselyov, A. S., Zon, G. and Strekowski, L. (1993). DNA triple-helix specific intercalators as antigene enhancers: unfused aromatic cations. *Biochemistry* **32**, 10614-10621.
9. Fox, K. R., Polucci, P., Jenkins, T. C. and Neidle, S. (1995). A molecular anchor for stabilizing triple-helical DNA. *Proc. Natl. Acad. Sci. USA* **92**, 7887-7891.
10. Baudoin, O., Marchand, C., Teulade-Fichou, M.-P., Vigneron, J.-P., Sun, J. S., Garestier, T., Hélène, C. and Lehn, J. M. (1998). Stabilization of DNA triple helices by crescent-shaped dibenzophenanthrolines. *Chemistry: a European Journal* , in press.
11. Choi, S.-D., Kim, M.-S., Kim, S. K., Lincoln, P., Tuite, E. and Norden, B. (1997). Binding mode of [Ruthenium(II)(1,10-Phenanthroline)$_2$L]$^{2+}$ with poly(dT x dA-dT) triplex. ligand size effect on third-strand stabilization. *Biochemistry* **36**, 214-223.
12. Pilch, D. S., Martin, M. T., Nguyen, C. H., Sun, J. S., Bisagni, E., Garestier, T. and Hélène, C. (1993). Self-association and DNA-binding properties of two triple

helix-specific ligands: comparison of a benzo[e]- and a benzo[g]pyridoindole. *J. Am. Chem. Soc.* **115**, 9942-9951.

13. Pilch, D. S., Waring, M. J., Sun, J. S., Rougée, M., Nguyen, C. H., Bisagni, E., Garestier, T. and Hélène, C. (1993). Characterization of a triple helix-specific ligand. *J. Mol. Biol.* **232**, 926-946.

14. Kim, S. K., Sun, J. S., Garestier, T., Hélène, C., Bisagni, E., Rodger, A. and Norden, B. (1997). Binding geometries of triple helix selective benzopyrido[4,3-b]indole ligands complexed with double- and triple-helical polynucleotides. *Biopolymers* **42**, 101-111.

15. Escudé, C., Nguyen, C. H., Mergny, J. L., Sun, J. S., Bisagni, E., Garestier, T. and Hélène, C. (1995). Selective stabilization of DNA triple helices by benzopyridoindole derivatives. *J. Am. Chem. Soc.* **117**, 10212-10219.

16. Escudé, C., Mohammadi, S., Sun, J.-S., Nguyen, C.-H., Bisagni, E., Liquier, J., Taillandier, E., Garestier, T. and Hélène, C. (1996). Ligand-induced formation of Hoogsteen-paired parallel DNA. *Chem. Biol.* **3**, 57-65.

17. Nguyen, C. H., Fan, E., Riou, J.-F., Bissery, M.-C., Vrignaud, P., Lavelle, F. and Bisagni, E. (1995). Synthesis and biological evaluation of amino-substituted benzo[f]pyrido[4,3-b] and pyrido[3,4-b]quinoxalines: a new class of antineoplastic agents. *Anti-Cancer Drug Design* **10**, 277-297.

18. Escudé, C., Nguyen, C., Kukreti, S., Janin, Y., Sun, J., Bisagni, E., Garestier, T. and Hélène, C. (1998). Rational design of a triple helix-specific intercalating ligand. *Proc. Natl. Acad. Sci. U S A* **95**, 3591-3596.

19. Nguyen, C. H., Marchand, C., Delage, S., Sun, J. S., Garestier, T., Hélène, C. and Bisagni, E. (1998). Synthesis of 13H-Benzo[6,7] and 13H-Benzo[4,5]-Indolo[3,2-c]quinolines: a New Series of Potent Specific Ligands for Triplex DNA. *J. Am. Chem. Soc.* **120**, 2501-2507.

20. Strekowski, L., Gulevich, Y., Baranowski, T. C., Parker, A. N., Kiselyov, A. S., Lin, S. Y., Tanious, F. A. and Wilson, W. D. (1996). Synthesis and structure-DNA binding relationship analysis of DNA Triple-helix specific intercalators. *J. Med. Chem.* **39**, 3980-3983.

21. Cassidy, S. A., Strekowski, L., Wilson, W. D. and Fox, K. R. (1994). Effect of a triplex-binding ligand on parallel and antiparallel DNA triple helices using short unmodified and acridine-linked oligonucleotides. *Biochemistry* **33**, 15338-15347.

22. Haq, I., Ladbury, J. E., Chowdhry, B. Z. and Jenkins, T. C. (1996). Molecular anchoring of duplex and triplex DNA by disubstituted anthracene-9,10-diones: calorimetric, UV melting, and competition dialysis studies. *J. Am. Chem. Soc.* **118**, 10693-10701.

23. Kan, Y., Armitage, B. and Schuster, G. B. (1997). Selective stabilization of triplex DNA by anthraquinone sulfonamide derivatives. *Biochemistry* **36**, 1461-1466.

24. Durand, M., Thuong, N. T. and Maurizot, J. C. (1992). Binding of netropsin to a DNA triple helix. *J. Biol. Chem* **267**, 24394-24399.

25. Park, Y. W. and Breslauer, K. J. (1992). Drug binding to higher ordered DNA structures - netropsin complexation with a nucleic acid triple helix. *Proc. Natl. Acad. Sci. USA* **89**, 6653-6657.

26. Rentzeperis, D. and Marky, L. A. (1995). Ligand binding to the Hoogsteen-WC groove of TAT base triplets: Thermodynamic contribution of the thymine methyl groups. *J. Am. Chem. Soc.* **117**, 5423-5424.

27. Durand, M., Thuong, N. T. and Maurizot, J. C. (1994). Interaction of Hoechst 33258 with a DNA triple helix. *Biochimie* **76**, 181-186.

28. Durand, M., Thuong, N. T. and Maurizot, J. C. (1994). Berenil complexation with a nucleic acid triple helix. *J. Biomol. Struct. Dyn.* **11**, 1191-1202.

29. Pilch, D. S. and Breslauer, K. J. (1994). Ligand-induced formation of nucleic acid triple helices. *Proc. Natl. Acad. Sci. USA* **91**, 9332-9336.

30. Durand, M. and Maurizot, J. C. (1996). Distamycin A complexation with a nucleic acid triple helix. *Biochemistry* **35**, 9133-9139.

31. Stonehouse, T. J. and Fox, K. R. (1994). DNase I footprinting of triple helix formation at polypurine tracts by acridine-linked oligopyrimidines: Stringency, structural changes and interaction with minor groove binding ligands. *Biochimica et Biophysica Acta - Gene Structure and Expression* **1218**, 322-330.

32. Pilch, D. S., Levenson, C. and Shafer, R. H. (1990). Structural analysis of the (dA)10.2(dT10) triple helix. *Proc. Natl. Acad Sci. USA* **87**, 1942-1946.

33. Hampel, K. J., Crosson, P. and Lee, J. S. (1991). Polyamines favor DNA triplex formation at neutral pH. *Biochemistry* **30**, 4455-4459.

34. Thomas, T. and Thomas, T. J. (1993). Selectivity of polyamines in triplex DNA stabilization. *Biochemistry* **32**, 14068-14074.

35. Pallan, P. S. and Ganesh, K. N. (1996). DNA Triple Helix Stabilization by Bisguanidinyl Analogues of Biogenic Polyamines. *Biochem. Biophys. Res. Com.* **222**, 416-420.

36. Thomas, T. J., Kulkarni, G. D., Greenfield, N. J., Shirahata, A. and Thomas, T. (1996). Structural specificity effects of trivalent polyamine analogues on the stabilization and conformational plasticity of triplex DNA. *Biochem. J.* **319**, 591-599.

37. Musso, M., Thomas, T., Shirahata, A., Sigal, L. H., VanDyke, M. W. and Thomas, T. J. (1997). Effects of chain length modification and bis(ethyl) substitution of spermine analogs on purine-purine-pyrimidine triplex DNA stabilization, aggregation, and conformational transitions. *Biochemistry* **36**, 1441-1449.

38. Potaman, V. N. and Sinden, R. R. (1995). Stabilization of triple-helical nucleic acids by basic oligopeptides. *Biochemistry* **34**, 14885-14892.

39. Maruyama, A., Katoh, M., Ishihara, T. and Akaike, T. (1997). Comb-Type Polycations Effectively Stabilize DNA Triplex. *Bioconjugate Chemistry* **8**, 3-6.

40. Escudé, C., Sun, J. S., Nguyen, C. H., Bisagni, E., Garestier, T. and Hélène, C. (1996). Ligand-induced formation of triple helices with antiparallel third strands containing G and T. *Biochemistry* **35**, 5735-5740.

41. Keppler, M. D. and Fox, K. R. Proceedings of the 4th international meeting on recognition studies in nucleic acids (NACON IV), Sheffield, April 1998.

42. Giovannangeli, C., Perrouault, L., Escudé, C., Thuong, N. and Hélène, C. (1996). Specific inhibition of *in vitro* transcription elongation by triplex-forming oligonucleotide-intercalator conjugates targeted to HIV proviral DNA. *Biochemistry* **35**, 10539-10548.

43. Kukreti, S., Sun, J., Loakes, D., Brown, D., Nguyen, C., Bisagni, E., Garestier, T. and Hélène, C. (1998). Triple helices formed at oligopyrimidine.oligopurine sequences with base pair inversions: effect of a triplex-specific ligand on stability and selectivity. *Nucl. Acids Res.* **26**, 2179-2184.

44. Chandler, S. P., Strekowski, L., Wilson, W. D. and Fox, K. R. (1995). Footprinting studies on ligands which stabilize DNA triplexes: Effects on stringency within a parallel triple helix. *Biochemistry* **34**, 7234-7242.

45. Thomas, T. J., Faaland, C. A., Gallo, M. A. and Thomas, T. (1995). Suppression of c-myc oncogene expression by a polyamine-complexed triplex forming oligonucleotide in MCF-7 breast cancer cells. *Nucl. Acids Res.* **23**, 3594-3599.

46. Duval-Valentin, G., Debizemont, T., Takasugi, M., Mergny, J. L., Bisagni, E. and Hélène, C. (1995). Triple-helix specific ligands stabilize H-DNA conformation. *J. Mol. Biol.* **247**, 847-858.

20 NEW TARGETS FOR TRIPLE HELIX FORMING OLIGONUCLEOTIDES

J. S. Sun

SUMMARY

The recognition of Watson-Crick base paired double-helical DNA by oligonucleotides *via* triple helix formation was restricted to oligopyrimidine•oligopurine sequences. Progress has been achieved to extend the repertory of DNA sequences which can be recognized by oligonucleotides. This chapter reviews different approaches to solve this molecular recognition problem.

INTRODUCTION

Oligonucleotide-directed triple helix formation is based on the sequence-specific recognition of oligopyrimidine•oligopurine sequences of double-stranded DNA (dsDNA). The third-strand oligonucleotides (often referred to as triple helix forming oligonucleotides: TFOs) bind in the major groove of DNA double helices and recognize the oligopurine strand of the target by establishing a pair of hydrogen bonds with the hydrogen bond donor and acceptor groups available on the major groove edge of the purine bases. Different base triplets can be formed through either Hoogsteen or reverse Hoogsteen hydrogen bond formation (Figure 1). Some bases in the third strand can adopt either Hoogsteen or reverse Hoogsteen configuration, for instance, thymine/uracil, protonated cytosine and guanine. The reverse Hoogsteen base triplets imply an upside-down rotation and a translation (if necessary) of the bases in the third strand. The third strand orientation with respect to the oligopurine strand of the target dsDNA is parallel if the base triplets involved are in Hoogsteen configuration, whereas it is antiparallel for the base triplets in reverse Hoogsteen configuration. This opposite strand orientation is due to an *anti* glycosidic conformation for all base triplets.

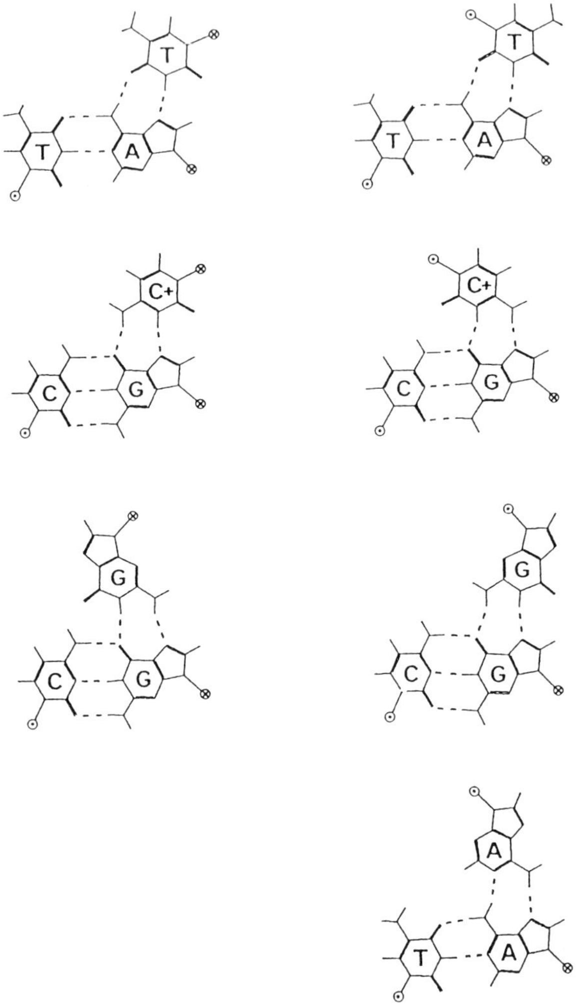

Figure 1. Base triplets formed by natural bases through either Hoogsteen (left column) or reverse Hoogsteen hydrogen bonds (right column).

Triple helix motifs

In theory, a triple helix can be formed by any combination of two base triplets provided they are both in Hoogsteen or in reverse Hoogsteen configuration, since the recognition of oligopyrimidine•oligopurine dsDNA sequences is a two-letter system. In practical, the stability of triple helix depends on the intrinsic stability of base triplets and on whether the triplets are isomorphous or not (*1*), as well as experimental conditions. Therefore, there are mainly three triple helix motifs involving natural bases in the third strand (Table 1). In the literature, the (T, C)-motif is often referred to as pyrimidine-motif, whereas the (G, A)- and (G, T)-motifs are referred to as purine-motifs. It is worth noticing that the (T, C)- and (T, G)-motifs in Hoogsteen configuration can be combined to form a mixed (T, C, G)-motif which is especially useful to replace strongly pH-dependent contiguous C•G x C+ triplets by pH-independent C•G x G triplets (*2*).

Motif	*Base composition*	*Hydrogen bond configuration*	*Strand orientation*	*Isomorphism*
(T, C)	T/U, C	Hoogsteen	parallel	Yes
(T, G)	T, G	Hoogsteen	parallel	No
(G, T)	T, G	reverse Hoogsteen	antiparallel	No
(G, A)	A, G	reverse Hoogsteen	antiparallel	No

Table 1. Classification of triple helix motifs

Target sequence considerations

Several types of non-perfect oligopyrimidine•oligopurine sequences can be potential targets of TFOs. Based on the approaches developed to cope with these sequence imperfections, three categories of sequences are considered: 1) oligopyrimidine•oligopurine sequences interrupted by a single or double purine•pyrimidine base pair(s); 2) sequences composed of alternating oligopyrimidine•oligopurine sequences; and 3) proximal oligopyrimidine•oligopurine sequences.

Due to page limitations, it is not possible to make an exhaustive review of this topic here. Only the principal results and approaches will be briefly reviewed and discussed, with special emphasis on structural aspects

TRIPLE HELIX FORMATION WITH MISMATCHES

The first obvious deviation from the canonical target sequences for triple helix forming oligonucleotides is the interruptions of oligopyrimidine•oligopurine sequences by purine•pyrimidine base pair inversion(s) which cause mismatches in triple helices.

Effect of base triplet mismatches

The effect of base triplet mismatches on triple helix stability has been extensively investigated, mostly in (T, C) motifs (*3-7*) but also in (G, T) motifs (*8*). In general, the triplex destabilization by triplet mismatches depends on: 1) the nature of the base triplet; 2) the location of the mismatch; and 3) its neighborhood. A central mismatch is more destabilizing than an end mismatch. A central single or double mismatch decreases the melting temperature of the triple helix by (at least) about 10° C or more than 20° C, respectively.

Least destabilizing base triplets

It was found that the mismatched A•T x G and G•C x Y (Y = T or C) triplets in Hoogsteen (T, C)-motif, and the A•T x T and G•C x T triplets in reverse Hoogsteen (G, T)-motif are the least destabilizing mismatched triplets. In addition, the A•T x G triplet in (T, C)-motif and the G•C x T in (G, T)-motif still retain some extents of selectivity as judged by their free energy, which is about 0.8-1.5 kcal mol^{-1} lower than that of other mismatched triplets involving a similar purine•pyrimidine base pair inversion in the target sequence (*7, 8*).

Stabilization of triplex mismatches by tethered intercalators or triplex-specific ligand

Mismatched triplets can be stabilized by incorporating an intercalator into the TFO. It was shown that the incorporation of an acridine, a well known double-helical DNA intercalator, in general at the 3'-side of a mismatch site in a TFO (which corresponds to a 5'-pyrimidine-purine-3' step in the target dsDNA, a favorable site for intercalators) reduced mismatch destabilization to about 5° C and 10° C for a single and double central mismatch, respectively (*9,10*).

A recent study shows that the presence of a triple-helix-specific ligand such as benzo[e]pyridoindole (BePI) can, not only markedly stabilize triplexes containing a central mismatch, but also retain some sequence discrimination, especially when certain modified nucleobases were used (*11*). In the presence of BePI, 3-nitropyrrol discriminates between G•C and C•G, A•T and T•A base pairs, whereas an N6-methoxy-2,6-diaminopurine recognizes an A•T base pair better than T•A, G•C or C•G base pairs. The nature of sequence selectivity is unclear at present. It remains to be shown whether the incorporation of BePI in TFOs in conjunction with these modified bases could provide a solution to overcome the triplex destabilization caused by purine•pyrimidine base pair inversion(s) in the target dsDNA sequences.

ALTERNATE-STRAND TRIPLE HELIX FORMATION

DNA sequences composed of alternating oligopyrimidine•oligopurine tracts can also be targeted by TFOs. As several short triple helices form on each oligopyrimidine•oligopurine tract, these short TFOs can be linked together to achieve cooperative binding. Therefore, a TFO containing several triple helix-forming domains can bind to the target dsDNA composed of adjacent and alternating oligopyrimidine•oligopurine tracts, and zigzag along the major groove, switching from one oligopurine strand to another at the $^{5'}$purine-pyrimidine$^{3'}$ or $^{5'}$pyrimidine-purine$^{3'}$ step (hereafter designated to as $^{5'}$RpY$^{3'}$ or $^{5'}$YpR$^{3'}$ junction, respectively).

Considerations on strand orientation and triple helix motifs

Two approaches have been described to form such an alternate-strand triple helix (also called a "switch" triple helix): 1) short triple helices are formed by motifs involving the same type of hydrogen bonding interactions (Hoogsteen or reverse Hoogsteen configurations), mostly the (T, C)-motif; 2) short triple helices are formed by different motifs involving alternately Hoogsteen and reverse Hoogsteen configurations.

In the first approach, when Hoogsteen motifs are used, the orientation of the TFOs is always parallel with respect to the target oligopurine strands, whereas that of different oligopurine strands is antiparallel. Therefore, the 3' or 5' ends of two neighboring short TFOs meet at the $^{5'}$RpY$^{3'}$ or at the $^{5'}$YpR$^{3'}$ junction, respectively (Figure 2, top part). Consequently, a linker is necessary to attach these two TFOs at the junction. Chemical modifications of TFOs are required.

Hoogsteen motifs

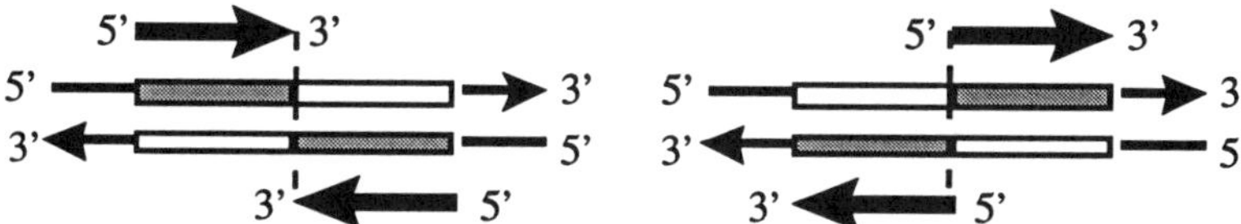

Hoogsteen & reverse Hoogsteen motifs

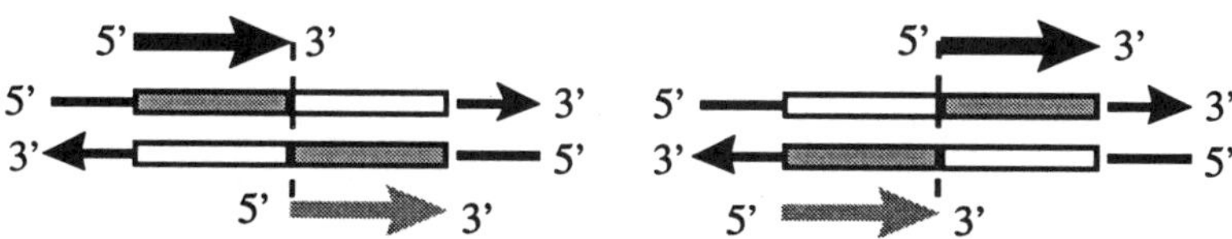

Figure 2. Recognition of alternating oligopurine (gray box)/oligopyrimidine (white box) sequences by combinations of triple helix motifs. The TFOs involving Hoogsteen or reverse Hoogsteen hydrogen bonding are shown by big black or gray arrows (which point from 5' to 3' ends), respectively. The dashed lines at the junctions represent the linkages between TFOs.

In contrast, the second approach consists of combining alternately different motifs in Hoogsteen and in reverse Hoogsteen configurations. Since the orientation of each of the TFOs is parallel with respect to its target oligopurine strand in the motifs involving Hoogsteen hydrogen bonding, and antiparallel for the motifs in reverse Hoogsteen configuration, the 3' (or 5') end of one short TFO encounters the 5' (or 3') end of its neighbor TFO at the junctions (Figure 2, bottom part). Thus a natural phosphodiester linkage could be used to attach the TFOs involving alternating Hoogsteen and reverse Hoogsteen motifs, provided the conformational constraints were properly removed. Therefore, a standard TFO can be used without any chemical modification.

Combination of triple helix motifs with same-strand orientation

Structural considerations. The structures of alternate-strand triple helices using two (T, C)-motifs at the ^{5}RpY^3 and at the ^{5}YpR^3 junctions are distinctly different, as illustrated by the energy-minimized models (Figure 3). The distance between the 3' ends of TFOs at the ^{5}RpY^3 junction is shorter than that between the 5' ends of TFOs at the ^{5}YpR^3 junction. In addition, the shortest distance between two TFOs at the ^{5}RpY^3 junction can be obtained (about 13 Å) when it is measured from the terminal O3' atom of one TFO to the O3' atom belonging to another TFO over two intervening base triplets.

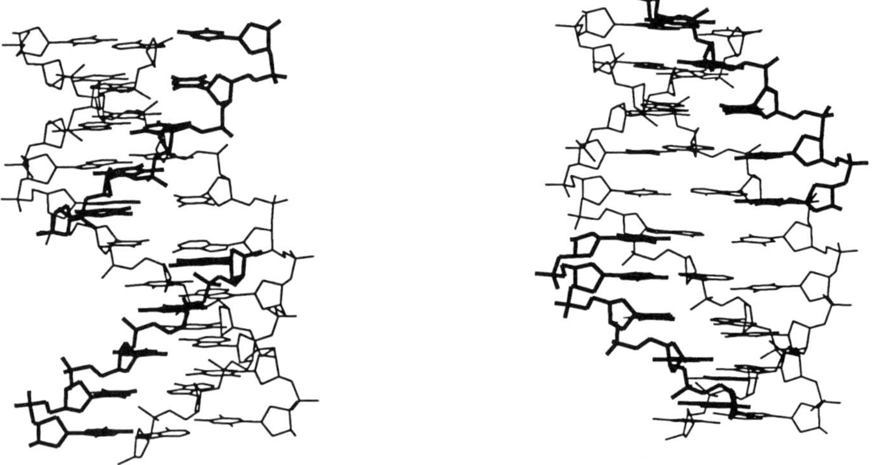

Figure 3. The ^{5}RpY^3 (left) and the ^{5}YpR^3 (right) junction structures made of unlinked alternate-strand (T, C)-motif triple helices. TFOs are drawn in bold lines. Hydrogen atoms have been omitted for clarity.

Chemical modifications. Several linkers have been used, including propylene glycol, 1,2-dideoxyribose, *p*-ribose dimer and *p*-xylose dimer, to attach two TFOs in (T, C)-motif through the 3'-3' or 5'-5' phosphodiester backbone termini (*12-15*). Among them, a 3'-3' linker made of *p*-xylose dimer has been designed to fit into the dinucleotide gap separating two TFOs at the $^5\text{'}\text{RpY}^{3'}$ junction (see above), and has been shown to provide significant cooperative binding of alternate-strand TFO (*15*). On the other hand, there has so far been no satisfactory 5'-5' termini linker which can afford cooperative binding at the $^5\text{'}\text{YpR}^{3'}$ junction.

An alternative approach has been reported which consists in attaching two TFOs through terminal bases by a short linker (6-8 Å) tethered to either the H_2N4 amino group of cytosine or the C5 atom of uracil. The advantage of this base-to-base approach is that it requires a short linker such as a tetra- or penta-methylene, since the terminal bases are localized near the center of major groove (Figure 4). Therefore, the entropic loss due to the use of a long and flexible linker could be greatly reduced. Molecular modeling using conformation energy minimization was used to determine the optimal linker length. Experimental data confirmed expectations both at the $^5\text{'}\text{RpY}^{3'}$ (*16*) and at the $^5\text{'}\text{YpR}^{3'}$ (*17*) junctions.

Combination of triple helix motifs with opposed strand orientation

Structural considerations. The main structural features at the junctions when an alternate-strand triple helix formation involves a Hoogsteen motif (*i.e.*, (T, C)- or (T, G)-motif) and a reverse Hoogsteen motif (*i.e.*, (G, T)- or (G, A)-motif) remain similar to those observed when two (T, C)-motifs are used. However, the distance between the termini (3'-3' or 5'-5') in the former is shorter than that in

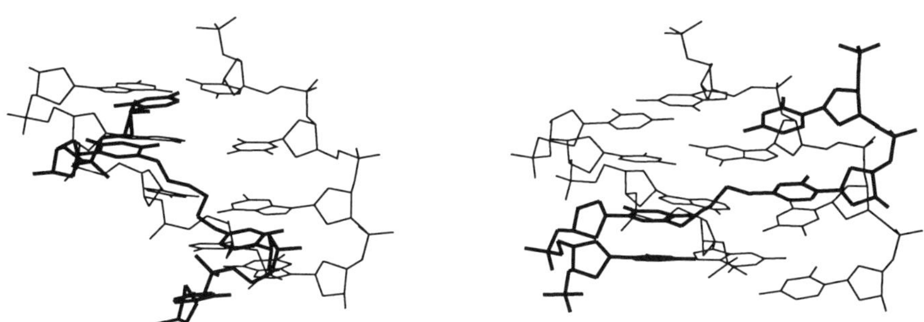

Figure 4. Base-to-base linkage used at the $^5\text{'}\text{GpT}^{3'}/^{3'}\text{CpA}^{5'}$ (left) and at the $^5\text{'}\text{TpG}^{3'}/^{3'}\text{ApC}^{5'}$ (right) junctions. TFOs are drawn in bold lines. Hydrogen atoms have been omitted for clarity. The cytosine and uracil at the junction were linked through a penta-methylene linker.

the latter, due to the near central location of the phosphodiester backbone in reverse Hoogsteen motifs in the major groove (Figure 5).

Sequence effect at the junction. Many studies have shown that alternate-strand triple helixes can be formed by combining motifs in Hoogsteen and reverse Hoogsteen configurations (*18-28*). The additional nucleotides which serve as linkers are generally necessary at the $^5YpR^3$ junction, whereas they are not required at the $^5RpY^3$ junction. Recent works using rational design by molecular modeling indicate that the sequence of an alternate-strand TFO at the junction plays an important role in its binding affinity. This sequence effect depends not only on the nature of the junction but also on the triple helix motifs used (*29-31*). For example, when a combination of a (T, G)-motif and a (G, A)-motif was used, the removal of an A provided the highest binding of the alternate-strand TFO at the $^5GpT^3/^3CpA^5$ junction (Figure 5, bottom left), whereas the addition of two cytosines between the two triple helix-binding domains was the best at the $^5TpG^3/^3ApC^5$ junction (Figure 5, bottom right).

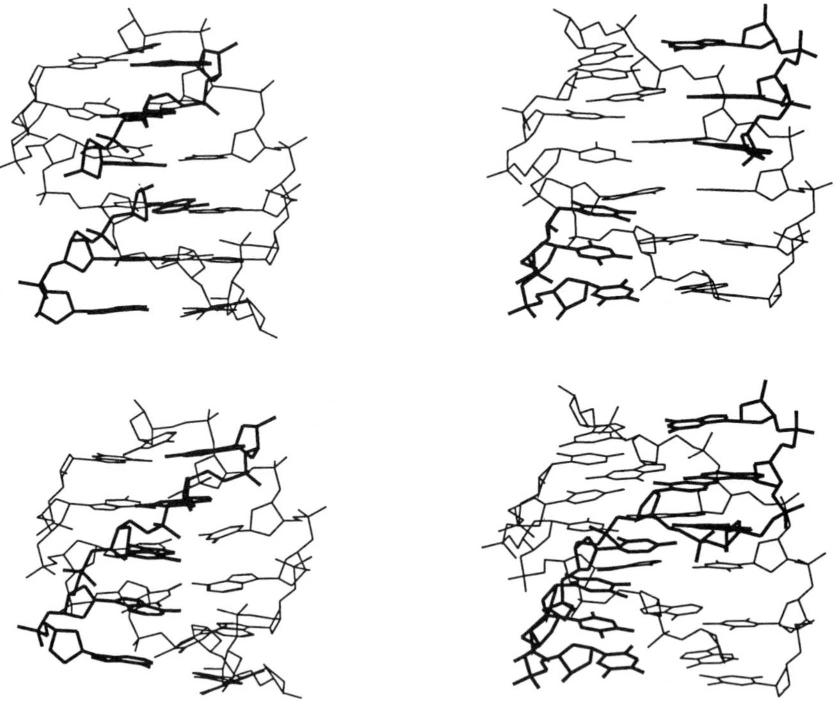

Figure 5. Energy-minimized models of the unlinked alternate-strand triple helices (top) made of a (T, G)-motif and a (G, A)-motif, as well as two optimized alternate-strand triple helices (bottom), at the $^5GpT^3/^3CpA^5$ (left) and at the $^5TpG^3/^3ApC^5$ (right) junctions. TFOs are drawn in bold lines. Hydrogen atoms have been omitted for clarity.

General remarks

It emerged that the binding of alternate-strand TFOs at the $^{5'}RpY^{3'}$ junction is higher than that at the $^{5'}YpR^{3'}$ junction, probably due to a better base-stacking interaction at the $^{5'}RpY^{3'}$ junction. It should be pointed out that self-associated complexes could be formed with alternate-strand TFOs, especially when they involve both (T, C)- and (G, A)-motifs. Therefore, it should be taken into consideration when targeting DNA sequences composed of alternating oligopyrimidine•oligopurine tracts.

There are six junction steps in target sequences: $^{5'}ApT^{3'}$, $^{5'}GpC^{3'}$ and $^{5'}GpT^{3'}$ (which is equivalent to $^{5'}ApC^{3'}$) for the $^{5'}RpY^{3'}$ junction, $^{5'}TpA^{3'}$, $^{5'}CpG^{3'}$ and $^{5'}TpG^{3'}$ (which is equivalent to $^{5'}CpA^{3'}$) for the $^{5'}YpR^{3'}$ junction. A comprehensive knowledge of how to deal with all six junctions would extend the range of DNA sequences which can be recognized by TFOs. Further studies will establish a full "switch" code for all six junction steps.

TRIPLE HELIX FORMATION AT PROXIMAL SITES

Cooperative triple helix formation was observed at the adjacent oligopyrimidine• oligopurine sites. The free energy of the cooperative binding of two adjacent TFOs was estimated to be about 1.5-2.9 kcal mol^{-1} depending on the sequence composition at the junction (*32*). This was rationalized in terms of base stacking at the junction.

Two oligopyrimidine•oligopurine sites separated by about one helical turn (10 bp) can be targeted by a TFO made of two short TFOs joined by an appropriate linker. Although the absence of base stacking between two triplexes is not expected to achieve cooperative binding of the TFOs in terms of enthalpy, the reduced entropy should contribute, to some extent, to cooperative triple helix formation. Such an approach was used to study the flexibility of the DNA sequence between two triplex sites (*33-35*). It was shown by varying the length of the linker and by performing circular permutation phasing analysis and ring closure carried out in non-denaturing electrophoresis that: 1) 50°-70° DNA bending toward the minor groove costs only about 1 kcal mol^{-1}; 2) DNA minor groove bendability is inherently high and in-dependent of temperature or under a wide range of Mg^{2+} concentrations (0.3-10 mM); and 3) the extent of DNA bending toward the minor groove is sequence dependent.

CONCLUSION AND PERSPECTIVES

The joint efforts to overcome purine•pyrimidine base pair inversion(s) in oligopyrimidine•oligopurine sequences, and to target the sequences composed of alternating or closely separated oligopyrimidine•oligopurine tracts have already led to a significant extension of dsDNA sequences which can be recognized by TFOs. It still remains a great challenge to design and synthesize new nucleotide analogues and/or new backbones that would allow modified TFOs to target any sequence of double-helical DNA from the major groove side by unambiguously recognizing all four base pairs.

ACKNOWLEDGMENTS

The author would like thank to Drs. U. Asseline, T. de Bizemont, S. Kukreti, C. Marchand, J. L. Mergny and B. W. Zhou for their contributions to extending the range of DNA sequences for triple helix formation, and Profs. T. Garestier and C. Hélène for valuable discussions and encouragement.

REFERENCES

1. Sun, J. S. and Hélène, C. (1993). Oligonucleotide-directed triple helix formation. *Curr. Opin. Struct. Biol.* **3**, 345-356.
2. Giovannangeli, C., Rougée, M., Montenay-Garestier, T., Thuong, N. T. and Hélène, C. (1992). Triple helix formation by oligonucleotides containing three bases T, C and G. *Proc. Natl. Acad. Sci. USA* **89**, 8631-8635.
3. Griffin, L. C. and Dervan, P. B. (1989). Recognition of thymine•adenine base pairs by guanine in a pyrimidine triple helix motif. *Science* **245**, 967-971.
4. Mergny, J. L., Sun, J. S., Rougée, M., Montenay-Garestier, T., Barcelo, F., Chomilier, J. and Hélène, C. (1991). Sequence-specificity in triple helix formation: experimental and theoretical studies of the effect of mismatches on triple helix stability. *Biochemistry* **30**, 9791-9798.
5. Yoon, K., Hobbs, C. A., Koch, J., Sardaro, M., Kutny, R. and Weis, A. L. (1992). Elucidation of the sequence-specific third strand recognition of four Watson-Crick base pairs in a pyrimidine triple helix motif. *Proc. Natl. Acad. Sci. USA* **89**, 3840-3844.
6. Kiessling, L. L., Griffin, L. C. and Dervan, P. B. (1992). Flanking sequence effects within the pyrimidine triple helix motif characterized by affinity cleaving. *Biochemistry* **31**, 2829-2834.
7. Best, G. C. and Dervan, P. B. (1995). Energetics of formation of sixteen triple helical complexes which vary at a single position within a pyrimidine motif. *J. Am. Chem. Soc.* **117**, 1187-1193.
8. Greenberg, W. A. and Dervan, P. B. (1995). Energetics of formation of sixteen triple helical complexes which vary at a single position within a purine motif. *J. Am. Chem. Soc.* **117**, 5016-5022.
9. Zhou, B. W., Puga, E., Sun, J. S., Garestier, T. and Hélène, C. (1995). Stable triple helices formed by acridine-containing oligonucleotides with oligopurine tracts of DNA interrupted by one or two pyrimidines. *J. Am. Chem. Soc.* **117**, 10425-10428.
10. Kukreti, S., Sun, J. S., Garestier, T. and Hélène, C. (1997). Extension of the range of DNA sequences available for triple helix formation: stabilization of mismatched triplexes by acridine-containing oligonucleotides. *Nucleic Acid Res.* **25**, 4264-4270.
11. Kukreti, S., Sun, J. S., Loakes, D., Brown, D. M., Nguyen, C. H., Bisagni, E., Garestier, T. and Hélène, C. (1998). Triple helices formed at oligopyrimidine•oligopurine sequences with base pair inversions: effect of a triplex-specific ligand on stability and selectivity. *Nucleic Acid Res.* **26**, 2179-2183.
12. Horne, D. A. and Dervan, P. B. (1990). Recognition of mixed sequence duplex DNA by alternate-strand triple helix formation. *J. Am. Chem. Soc.* **112**, 2435-2437.

13. Ono, A., Chen, C. N. and Kan, L. S. (1991). DNA triplex formation of oligonucleotide analogues consisting of linker groups and octamer segments that have opposite sugar-phosphate backbone polarities. *Biochemistry* **30**, 9914-9921.

14. McCurdy, S., Moulds, C. and Froehler, B. (1991). Deoxyoligonucleotides with inverted polarity: synthesis and use in triple helix formation. *Nucleosides Nucleotides* **10**, 287-290.

15. Froehler, B., Terhorst, T., Shaw, J. P. and McCurdy, S. N. (1992). Triple helix formation and cooperative binding by oligodeoxynucleotides with a 3'-3' internucleotide junction. *Biochemistry* **31**, 1603-1609.

16. Zhou, B. W., Marchand, C., Asseline, U., Thuong, N. T., Sun, J. S., Garestier, T. and Hélène, C. (1995). Recognition of alternating oligopurine/oligopyrimidine tracts of DNA by oligonucleotides with base-to-base linkages. *Bioconjugate Chem.* **6**, 516-523.

17. Marchand, C., Cristol, N., Asseline, U., Sun, J. S., Garestier, T. and Hélène, C. (1998). *Manuscript in preparation*.

18. Sun, J. S., de Bizemont, T., Duval-Valentin, G., Montenay-Garestier, T. and Hélène, C. (1991). Extension of the range of recognition sequence for triple helix formation by oligonucleotides containing guanines and thymines. *C. R. Acad. Sci. Paris Sér. III* **313**, 585-590.

19. Beal, P. A. and Dervan, P. B. (1992). Recognition of double helical DNA by alternate-strand triple helix formation. *J. Am. Chem. Soc.* **114**, 4976-4982.

20. Jayasena, S. D. and Johnston, B. H. (1992). Oligonucleotide-directed triple helix formation at adjacent oligopurine and oligopyrimidine DNA tracts by alternate-strand recognition. *Nucleic Acids Res.* **20**, 5279-5288.

21. Jayasena, S. D. and Johnston, B. H. (1993). Sequence limitations of triple helix formation by alternate-strand recognition. *Biochemistry* **32**, 2800-2807.

22. Washbrook, E. and Fox, K. R. (1994). Comparison of antiparallel A•AT and T•AT triplets within an alternate-strand DNA triple helix. *Nucleic Acids Res.* **22**, 3977-3982.

23. Washbrook, E. and Fox, K. R. (1994). Alternate-strand DNA triple helix formation using acridine-linked oligonucleotides. *Biochem. J.* **301**, 569-575.

24. Olivas, W. M. and Maher, L. J., III. (1994). DNA recognition by alternate-strand triple helix formation: affinities of oligonucleotides for a site in the human p53 gene. *Biochemistry* **33**, 983-991.

25. de Bizemont, T., Duval-Valentin, G., Sun, J. S., Bisagni, E., Garestier, T. and Hélène, C. (1996). Alternate-strand recognition of double-helical DNA by (T, G)-containing oligonucleotides in the presence of a triple helix-specific ligand. *Nucleic Acids Res.* **24**, 1136-1143.

26. Olivas, W. M. and Maher, L. J. (1996). Binding of DNA oligonucleotides to sequences in the promoter of the human bcl-2 gene. *Nucleic Acids Res.* **24**, 1758-1764.

27. Balatskaya, S. V., Belotserkovskii, B. P. and Johnston, B. H. (1996). Alternate-strand triplex formation: modulation of binding to matched and mismatched duplexes by sequence choice in the Pu•PuxPy block. *Biochemistry* **35**, 13328-13337.

28. Bouziane, M., Cherny, D. I., Mouscadet, J. F. and Auclair, C. (1996). Alternate-strand DNA triple helix-mediated inhibition of HIV-1 U5 long terminal repeat integration *in vitro*. *J. Biol. Chem.* **271**, 10359-10364.

29. Sun, J. S. (1995). "Rational design of switched triple helix-forming oligonucleotides: extension of sequences for triple helix formation." In *Modeling of Biomolecular Structures and Mechanisms*, Pullman, A., Pullman, B. and Jortner, J. eds. Kluwer Academic Publishers, Amsterdam, 267-288.

30. Marchand, C., Sun, J. S., Bailly, C., Waring, M. J., Garestier, T. and Hélène, C. (1998). Optimization of alternate-strand triple helix formation at the 5'CpG3' and 5'GpC3' junction steps. *Biochemistry* **37**, in press.

31. de Bizemont, T., Sun, J. S., Garestier, T. and Hélène, C. (1998). New junction models for alternate-strand triple helix formation. *Chem. Biol.*, submitted.

32. Colocci, N. and Dervan, P. B. (1995). Cooperative triple helix formation at adjacent DNA sites: sequence composition dependence at the junction. *J. Am. Chem. Soc.* **117**, 4781-4787.

33. Akiyama, T. and Hogan, M. E. (1996). The design of an agent to bend DNA. *Proc. Natl. Acad. Sci. USA* **93**, 12122-12127.

34. Akiyama, T. and Hogan, M. E. (1996). Microscopic DNA flexibility analysis. Probing the base composition and ion dependence of minor groove compression with an artificial DNA bending agent. *J. Biol. Chem.* **271**, 29126-29135.

35. Akiyama, T. and Hogan, M. E. (1997). Structural analysis of DNA bending induced by tethered triple helix forming oligonucleotides. *Biochemistry* **36**, 2307-2315.

21 TRIPLEXES AND BIOTECHNOLOGY

Maxim D. Frank-Kamenetskii

SUMMARY

Triplexes are very attractive for biotechnology applications because they open up ways to manipulate in a sequence-specific manner with double-stranded DNA (dsDNA). This became evident after discovery of the effect of co-migration of a triplex-forming oligonucleotide (TFO) with dsDNA and subsequent development of the triplex affinity capture (TAC) technique for plasmid and cosmid dsDNA. TAC has recently proved to be a very useful approach to purify recombinant plasmids for purposes of gene therapy because it avoids toxic chemicals. Triplexes formed by peptide nucleic acids (PNAs) with dsDNA have opened up new opportunities. Their major advantage over TFO/dsDNA complexes consists in much higher stability without loss of specificity. This makes triplex-forming PNAs (TFPs) a more promising tool than TFOs for some biotechnology applications. Two TFP-based techniques recently developed in the author's laboratory are considered: PNA-assisted rare genome cleavage (PARC) and oligonucleotide/PNA affinity capture (OPAC).

INTRODUCTION

Most biotechnology applications of nucleic acid interactions are grounded on the Watson-Crick pairing. As a result, the corresponding techniques require single-stranded nucleic acids and are not applicable to double-stranded DNA (dsDNA). The triple-helix mode of interaction between dsDNA and an oligonucleotide (its analog or mimic) provides a very attractive opportunity to use DNA in its intact, duplex form. For some applications it allows one to skip time/labor consuming steps of the protocols and significantly improve the techniques; for other applications it creates totally new opportunities.

Lyamichev *et al.* (*1*) were the first to demonstrate that dsDNA carrying a homopurine-homopyrimidine insert is able to capture the homologous homopyrimidine oligonucleotide in an acidic medium. As a result of this capturing, the oligonucleotide co-migrates with dsDNA during gel electrophoresis. This finding opened the way for subsequent development of a procedure for affinity capture of dsDNA by triplex-forming oligonucleotides (TFOs), triplex affinity capture (TAC) (*2, 3*). One of the potential biotechnology applications of TAC stems from the possibility of purifying plasmids for gene therapy, avoiding toxic chemicals and hazardous biochemicals (*4, 5*).

The ability of TFOs to form sequence-specific complexes with dsDNA opened up the possibility of using TFOs for creation of artificial "restriction enzymes" and for increasing the selectivity of the usual restriction enzymes. In the case of the artificial "restriction enzymes", TFOs are tagged by chemicals or non-sequence-specific enzymes capable of cleaving DNA (see chapter 7, this volume). To increase the selectivity of commonly used restriction enzymes, the Achilles' heel approach is being applied. This general approach, which was first proposed by Szybalski (*6*), consists in using any molecule that binds to dsDNA in a sequence-specific manner. The molecule must protect the restriction sites it covers (completely or partially) against methylation with the methyl transferase (methylase). After subjecting the entire genome to methylation with a specific enzyme, the protection molecule is removed from the DNA, and the DNA is digested by a restriction enzyme which recognizes the same sites as the methylase did. Only the restriction sites protected against methylation by the protection molecule will be digested (see Figure 7 below). This approach makes it possible to dramatically increase the selectivity of the common restriction enzymes.

Implementation of peptide nucleic acid (PNA) based techniques promises to open a new chapter in the biotechnology applications of triplexes. Like TFO, a homopyrimidine PNA (we will label such PNA as triplex-forming PNA, or TFP) is able to form a very stable and sequence-specific complex with dsDNA, although the structure of the complex in the TFP case is quite different from that in the TFO case. Whereas a TFO lies in the major groove of dsDNA forming Hoogsteen pairs with the DNA purine strand, the TFP invades the DNA duplex (Figure 1) so that two TFP molecules form a $(PNA)_2$/DNA triplex with the DNA purine strand, displacing the other strand (see chapter 18 and references therein).

To increase the efficiency of TFP binding to dsDNA, bis-PNAs are used, which consist of two TFP molecules connected by a flexible linker. Usually, in bis-PNA all cytosines in one of the two TFPs are replaced by pseudoisocytosines (J bases). In contrast to cytosines, these bases carry hydrogen in the N3 position, and their Hoogsteen pairing with guanines in the triplex structure does not require protonation. As a result, bis-PNAs carrying cytosines in one of the two PNA oligomers and J bases in the other oligomer form triplexes with the purine DNA strand in a virtually pH-independent manner. Binding of bis-PNA to dsDNA is additionally enhanced by tagging positively charged lysine residues to bis-PNA. Such cationic J-containing bis-PNAs have proved to be extremely efficient tools for targeting dsDNA, ssDNA and RNA *via* triplex formation (see below, and chapter 18).

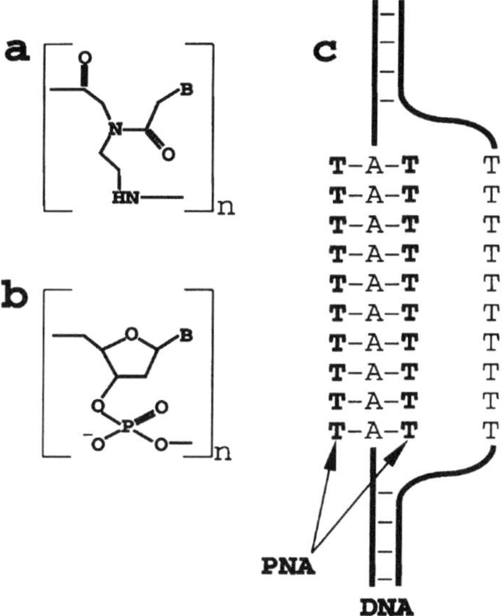

Figure 1. Chemical structure of (**a**) PNA and (**b**) DNA; and (**c**) schematics of complex formation between duplex DNA and a TFP (decathymine-PNA as an example). Such a complex was dubbed the P-loop.

Potential biotechnology applications of triplexes formed by such bis-PNAs with dsDNA include a very efficient technique of PNA-assisted rare cleavage of genomic DNA (PARC), hybridization of oligonucleotide probes with dsDNA (PD-loops) and oligonucleotide/PNA affinity capture (OPAC). The PARC method is grounded on the formation of P-loops (Figure 1), in which the (PNA)$_2$/DNA triplex is a major element. The hybridization and OPAC techniques are grounded on the recently demonstrated ability of two bis-PNA molecules (PNA "openers") to open a DNA region, thus making it possible for an oligonucleotide to bind to the single-stranded part of an extended P-loop (*7*), and so forming a so-called PD-loop (see Figure 2).

AFFINITY CAPTURE

Triplex affinity capture

Using the gel co-migration assay, Lyamichev at al. (*1*) demonstrated that dsDNA carrying a homopurine-homopyrimidine insert may capture a TFO in solution (reviewed in refs. *8, 9*). Ito *et al.* (*2, 3*) demonstrated that the effect can be used for affinity capture of dsDNA using the TFO.

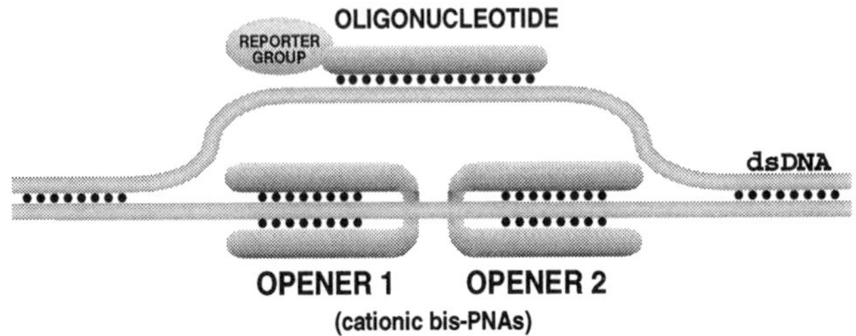

Figure 2. The PD-loop consists of duplex DNA, two PNA openers and an oligonucleotide. Here the oligonucleotide is shown to carry a reporter group (for instance, biotin).

The triplex affinity capture (TAC) method is expected to become useful in biotechnology for enrichment of genome libraries (2, 3), and for purification of plasmids and cosmids carrying genes for nonviral gene therapy (4, 5). To be used in gene therapy, the plasmids, which carry special inserts to be transferred to human cells, cannot be extracted by the usual methods, which include toxic chemicals and hazardous biochemicals. Therefore, a purification technique was developed, called triple-helix affinity chromatography (THAC). In this technique, chromatography columns are used carrying TFOs chemically attached to a support *via* a polymer linker (Figure 3). The THAC technique will make it possible to produce pure DNA samples on a commercial scale (5).

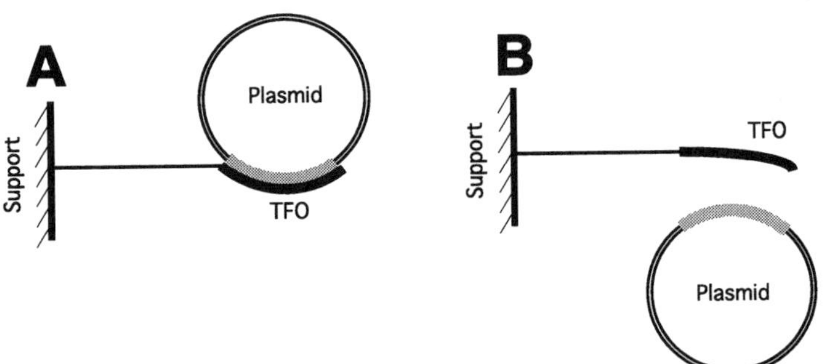

Figure 3. Purification of plasmid DNA by the triple-helix affinity chromatography (THAC) technique. **A**, At acidic pH, only the "triplex" plasmid carrying a specially inserted homopurine-homopyrimidine region is retained in the column because it is captured by the TFO. All contaminants, including DNA molecules without the insert, are washed out from the column. **B**, At alkaline pH, the purified plasmid is eluted from the column.

Thus, the triplex affinity capture is a promising method for biotechnology applications when specially constructed "triplex" plasmids carrying long homopurine-homopyrimidine inserts can be used. At pH close to neutral, short DNA triplexes are known to be unstable (9). As a result, long homopurine-homopyrimidine inserts are required in order to avoid very low pH, which can damage DNA. However, long homopurine-homopyrimidine tracts are not normally encountered in genomes. Special "triplex" vectors are therefore absolutely necessary in order to use the TAC/THAC technique of DNA purification.

Affinity capture via PD-loop

PD-loops (Figure 2) open the way to develop an affinity capture technique applicable to natural genome sequences. This is possible because, in contrast to TFOs, TFPs form very stable complexes with rather short binding sites on dsDNA

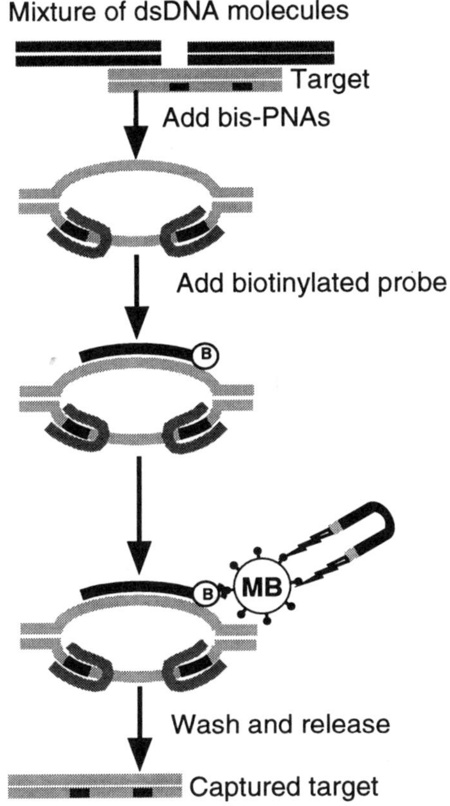

Figure 4. The oligonucleotide/PNA affinity capture (OPAC) technique. B is for a biotin molecule, MB is for a streptavidin-covered magnetic bead.

at neutral pH. As a result, sequences able to form PD-loops are encountered, statistically, quite frequently, after each 750 bp or even more often (7).

The PD-loop affinity capture protocol, or OPAC (oligonucleotide/PNA affinity capture), is schematically shown in Figure 4.

To demonstrate the OPAC technique, Bukanov *et al.* (7) fished out a specific dsDNA fragment from a restriction digest of the entire yeast genome. The *Mse* I restriction enzyme yielded about 16,000 fragments of yeast genome. One of these fragments carried a sequence specially chosen for PD-loop formation (Figure 5).

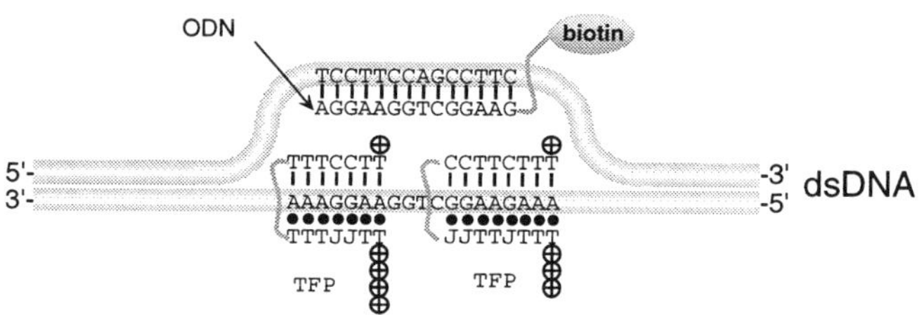

Figure 5. PD-loop used for affinity capture of a fragment of yeast genome. Pluses indicate the positive charges on TFPs due to the lysine residues. Watson-Crick and Hoogsteen pairs are shown by strips and dots, respectively. J is for pseudoisocytosine (şee INTRODUCTION section).

Five rounds of the affinity capture procedure shown in Figure 4 were performed (see Figure 6). Aliquots of captured DNA collected after each round of enrichment were analyzed by non-specific PCR amplification using a primer complementary to a special adapter ligated to both ends of all fragments of the original digest (7).

This assay amplified all DNA fragments captured on magnetic beads after the PD-loop formation. If the desired fragment of yeast DNA were captured by the technique, it would be detected as a fragment with the size of 903 bp (the fragment plus two

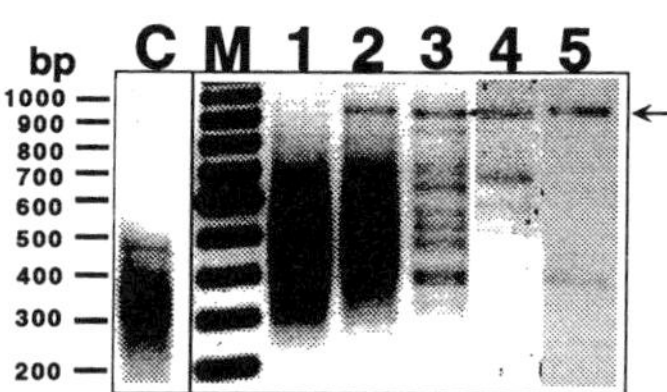

Figure 6. Affinity capture of a yeast DNA fragment from a digest of the entire yeast genome by *Mse* I restriction enzyme. Lanes **1** through **5** correspond to one, two, three, four and five rounds of PD-loop formation/separation procedure, respectively; **M** corresponds to the 100-bp DNA ladder; **C** represents the *Mse* I digest of the yeast genome before capturing. A targeted 903-bp-long fragment is shown by an arrow. The data are from Bukanov *et al.* (7).

adapters). Note that PCR was used only as a detection method at the end of one, two, three, four and five rounds of enrichment. No intervening amplification of captured material was performed between the rounds. The result was the isolation of a virtually pure fragment of the expected length (Figure 6). The restriction and sequence analyses confirmed that the fragment isolated was the expected fragment retrieved from the database.

Thus, the PD-loop affinity capture procedure makes it possible to isolate specific dsDNA fragments from a very complex mixture of DNA fragments. The procedure has significant advantages over the TAC technique because for PD-loops the sequence limitations are much milder. The technique has a potential of purification of specific genes in an intact form with retention of epigenetic modifications. Such modifications are normally lost during PCR amplification, although they play an important role (10-12).

GENOME RARE CUTTING

The Achilles' heel approach

Figure 7 illustrates the general idea of application of TFOs and TFPs in the Achilles' heel approach (for the particular case of a TFP). The same general idea is

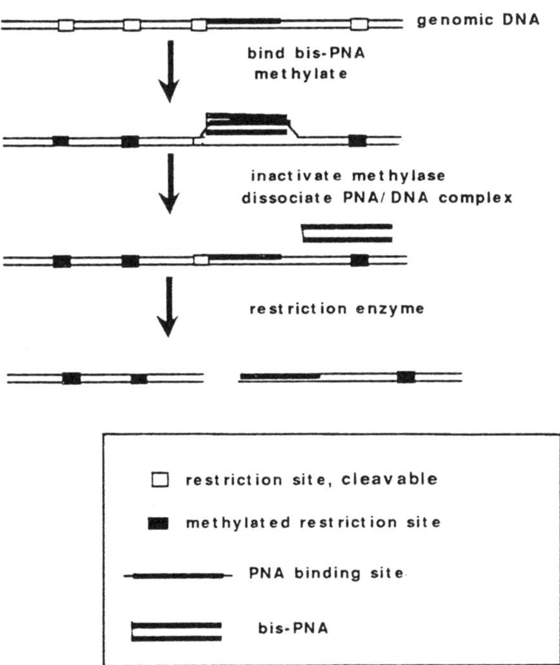

Figure 7. The Achilles' heel principle in the example of TFP-assisted rare cleavage of genomic DNA.

also used in the RARE strategy, in which oligodeoxynucleotide (ODN) bound to RecA protein targets DNA sites, thus protecting them against methylation (*13-15*). The end result of all these approaches consists in a significant increase in the selectivity of action of common restriction enzymes. Each approach has its strong and weak sides.

A great advantage of RARE consists in its possibility to target any sequence. Its disadvantage is the necessity of using a complex mixture at the methylation stage, consisting of RecA protein, magnesium ions, ODN and methylase. Contaminating nucleases in RecA and methylase preparations may be activated by magnesium and lead to non-specific dsDNA cleavage (15).

A disadvantage of using TFOs and TFPs in rare cleavage consists in sequence limitations inherent in their binding to dsDNA. These limitations are especially severe in the case of TFOs, making their use totally impractical. Indeed, to prevent methylation, a TFO must be attached to DNA strongly enough, which requires the TFO to be at least 25 nt long. Such long homopurine-homopyrimidine sequences are virtually not met in genomes. As a result, applicability of TFOs in the Achilles' heel strategy was demonstrated only when a long homopurine-homopyrimidine sequence was specially inserted into yeast chromosomal DNA (16).

Genome rare cutting using TFP

Sequence limitations for TFPs are much milder than for TFOs because, although both require a homopurine-homopyrimidine target, for TFPs the target does not have to be long. Homopurine-homopyrimidine sequences as short as 7-8 bp long form stable P-loops when targeted by TFPs (*17-19*). Because of this, the PARC method has proved to be a very efficient and convenient approach for rare genome cutting.

Up to now, the method has been used mostly in application to yeast genomes in our laboratory (17-19). We have demonstrated that by choosing various combinations of TFPs and methylation/restriction enzymes, different chromosomes can be selectively and quantitatively cleaved. Figure 8 shows examples of such cutting. In these experiments, the same combination of methylation/restriction enzymes was used. In lane 2, PNA I was applied, which resulted in quantitative cutting of chromosomes II and III (Chr II and Chr III), each of which received a single cut. Two fragments of Chr II are indicated by arrowheads. When PNA II was used, chromosomes XIV and VI were cut (lane 4). When both PNA I and PNA II were used, four chromosomes were cut (lane 6). Because the yeast genome is sequenced, we could check that in each case the cleavage pattern corresponded to the pattern expected on the basis of the recognition sequence for the methylation/ restriction enzymes used and the target sequence for the PNA.

We have recently demonstrated (*19*) that the PARC approach may be very useful for segregation of yeast artificial chromosomes (YACs) from endogenous yeast chromosomes in case of their overlapping in pulsed-field gel electrophoresis (PFGE). Specifically, when a strain carrying a YAC overlapping with Chr II was used, we cut Chr II using TFP I and, as a result, we got only YAC DNA at this position. We failed to detect any cleavage of the YAC.

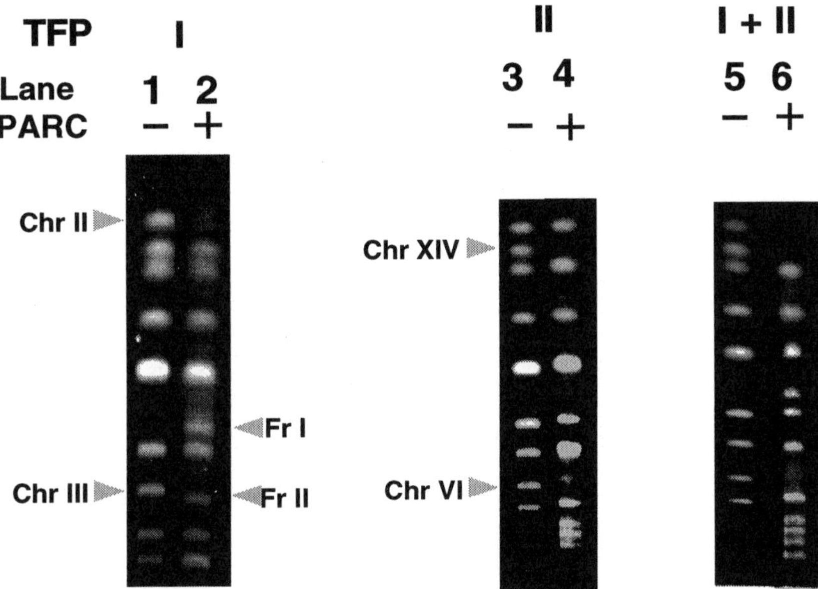

Figure 8. TFP-assisted cleavage of yeast genome. Pulsed-field gel electrophoresis (PFGE) of yeast chromosomal DNA before (odd lanes) and after (even lanes) application of the PARC strategy. Two different TFPs were used in combination with the *Hha* I/*Hae* II pair of methylation/ restriction enzymes. Fr I and Fr II indicate the bands corresponding to the two fragments of cut Chr II.

CONCLUSION

Biotechnology applications of TFOs and TFPs are grounded on their ability to target duplex DNA in a sequence-specific manner. The usefulness of TFOs is hampered by their weak binding to target sites available in genomes. Still, in combination with special 'triplex' vectors, a promising purification technique (the THAC) has been elaborated. TFPs are very attractive for biotechnology applications because of their ability to bind strongly and sequence-specifically to short target sites on dsDNA, which are abundant in genomes. Two promising approaches, the PNA-assisted rare cutting (PARC) and PD-loop hybridization/affinity capture, are proposed.

ACKNOWLEDGMENTS

I thank Vadim Demidov for numerous comments and suggestions. This work was supported by grant GM54434 from the National Institutes of Health.

REFERENCES

1. Lyamichev, V. I., Mirkin, S. M., Frank-Kamenetskii, M. D. and Cantor, C. R. (1988). A stable complex between homopyrimidine oligomers and the homologous regions of duplex DNA. *Nucl. Acids Res.* **16**, 2165-2178.
2. Ito, T., Smith, C. L. and Cantor, C. R. (1992). Sequence-specific DNA purification by triplex affinity capture. *Proc. Natl. Acad. Sci. USA* **89**, 495-498.
3. Ito, T., Smith, C. L., and Cantor, C. R. (1992). Affinity capture electrophoresis for sequence-specific DNA purification. *Genet. Anal. Tech. Appl.* **9**, 96-99.
4. Ledley, F. D. (1995). Nonviral gene therapy: the promise of genes as pharmaceutical products. *Human Gene Therapy* **6**, 1129-1144.
5. Wils, P., Escriou, V., Warnery, A., Lacroix, F., Lagneaux, D., Ollivier, M., Crouzet, J., Mayaux, J. F. and Scherman, D. (1997). Efficient purification of plasmid DNA for gene transfer using triple-helix affinity chromatography. *Gene Therapy* **4**, 323-330.
6. Koob, M. and Szybalski, W. (1990). Cleaving yeast and *E. coli* genomes at a single site. *Science* **250**, 271-273.
7. Bukanov, N. O., Demidov, V. V., Nielsen, P. E. and Frank-Kamenetskii, M. D. (1998). PD-Loop: A complex of duplex DNA with an oligonucleotide. *Proc. Natl. Acad. Sci. USA* **95**, 5516-5520
8. Frank-Kamenetskii, M. D. (1992). Protonated DNA structures. *Methods Enzymol.* **211**, 180-191.
9. Frank-Kamenetskii, M. D. and Mirkin, S. M. (1995). Triplex DNA structures. *Annu. Rev. Biochem.* **64**, 65-95.
10. Reik, W. and Maher, E. R. (1997). Imprinting in clusters: lessons from Beckwith-Wiedemann syndrome. *Trends Genet.* **13**, 330-334.
11. Sun, F.-L., Dean, W. L., Kelsey, G., Allen, N. D. and Reik, W. (1997). Transactivation of *Igf2* in a mouse model of Beckwith-Wiedemann syndrome. *Nature* **389**, 809-815.
12. Hastie, N. (1997). Disomy and disease resolved? *Nature* **389**, 785-787.
13. Ferrin, L. J. and Camerini-Otero, R. D. (1991). Selective cleavage of human DNA: RecA-assisted restriction endonuclease (RARE) cleavage. *Science* **254**, 1494-1497.
14. Ferrin, L. J. and Camerini-Otero, R. D. (1994). Long-range mapping of gaps and telomers with RecA-assisted restriction endonuclease (RARE) cleavage. *Nature Genetics* **6**, 379-383.
15. Gnirke, A., Huxley, C., Peterson, K. and Olson, M. V. (1993). Microinjection of intact 200- to 500-kb fragments of YAC DNA into mammalian cells. *Genomics* **15**, 659-667.
16. Strobel, S. A. and Dervan, P. B. (1991). Single-site enzymatic cleavage of yeast genomic DNA mediated by triple helix formation. *Nature* **350**, 172-174.
17. Veselkov, A. G., Demidov, V. V., Frank-Kamenetskii, M. D. and Nielsen, P. E. (1996). PNA as a rare genome-cutter. *Nature* **379**, 214.
18. Veselkov, A. G., Demidov, V. V., Nielsen, P. E. and Frank-Kamenetskii, M. D. (1996). A new class of genome rare cutters. *Nucleic Acids Res.* **24**, 2483-2488.
19. Izvolsky, K. I. , Demidov, V. V., Bukanov, N. O. and Frank-Kamenetskii, M. D. (1998). Yeast artificial chromosome segregation from host chromosomes with similar lengths. *Nucleic Acids Res.* (*in press*).

Index